Biological and Quantum Computing for Human Vision:
Holonomic Models and Applications

Mitja Peruš, University of Ljubljana, Slovenia

Chu Kiong Loo, Multimedia University, Malaysia

Medical Information Science REFERENCE

MEDICAL INFORMATION SCIENCE REFERENCE

Hershey · New York

Director of Editorial Content: Kristin Klinger
Director of Book Publications: Julia Mosemann
Acquisitions Editor: Lindsay Johnston
Development Editor: Julia Mosemann
Publishing Assistant: Deanna Jo Zombro
Typesetter: Natalie Pronio
Production Editor: Jamie Snavely
Cover Design: Lisa Tosheff

Published in the United States of America by
Medical Information Science Reference (an imprint of IGI Global)
701 E. Chocolate Avenue
Hershey PA 17033
Tel: 717-533-8845
Fax: 717-533-8661
E-mail: cust@igi-global.com
Web site: http://www.igi-global.com

Library of Congress Cataloging-in-Publication Data

Biological and quantum computing for human vision : holonomic models and applications / Mitja Peruš and Chu Kiong Loo, editors.
p. ; cm.
Includes bibliographical references and index.
Summary: "This book presents an integrated model of human image processing and conscious visual experience, based mainly on the Holonomic Brain Theory by Karl Pribram. This work researches possibilities for complementing neural models of early vision with the new preliminary quantum models of consciousness in order to construct a model of human image processing"--Provided by publisher.
ISBN 978-1-61520-785-5 (h/c) -- ISBN 978-1-61520-786-2 (e-ISBN) 1. Visual cortex--Mathematical models. 2. Computational neuroscience. 3. Quantum computers. 4. Neural networks (Neurobiology) I. Peruš, Mitja, 1969- II. Loo, Chu Kiong, 1971-
[DNLM: 1. Visual Perception--physiology. 2. Models, Psychological. 3. Quantum Theory. WW 105]
QP383.15.B56 2011
612.8'255--dc22
2010036659

British Cataloguing in Publication Data
A Cataloguing in Publication record for this book is available from the British Library.

To Mila, Manca, and Alexei

Table of Contents

Section 2

Foreword

Mitja Peruš and Chu Kiong Loo have written a most interesting monograph regarding a model that they have worked out using currently available massively parallel "neural net" architecture. They introduce the model by reviewing their understanding of the views of established neuroscience. They then go on to indicate where these views need to be modified on the basis of neurophysiological data that are currently largely ignored. These data are more supportive of the models used in developing neural nets and fit especially well Perus' and Loo's modifications.

Peruš' and Loo's book, as is their work, is a valiant foray into much needed transdisciplinary research, much needed to help us to understand the relationship between our experience, our behavior and the place that our brain has in organizing both. I fully share these aims. The following quotation eloquently states the issue.

We have inherited from our forefathers the keen longing for unified, all embracing knowledge —. But the multifarious branches of Knowledge during the past hundred odd years has confronted us with a – dilemma. We feel clearly that we are now beginning to acquire reliable material for welding together — all that is known into a whole: but on the other hand, it has become next to impossible for a single mind fully to command more than a small specialized portion of it.

I see no escape from this dilemma (lest our true aim be lost forever) than that some of us should venture to embark on a synthesis of facts and theories, albeit with second-hand and incomplete knowledge of some of them — and at the risk of making fools of ourselves. - Erwin Schrödinger, 1944, What is Life

Karl H. Pribram
Georgetown University, USA
George Mason University, USA
Stanford University, USA
Eminent Scholar, Commonwealth of Virginia, USA

***Karl H. Pribram**, MD, is one of the legendary pioneers of several fields of experimental brain science. After his studies with the early founder of neuroscience Prof. Karl Lashley, whom he succeeded as director at the Yerkes Laboratory for Primate Biology, Pribram researched and lectured for many years at the US-universities of Yale, Stanford, Radford and Georgetown, among many others, published several influential books and hundreds of papers. He lead large neuroexperimental groups which provided results of historic importance. For example, he discovered the functions of the limbic system, the frontal lobes and the inferotemporal cortex. He triggered the cognitive revolution in psychology. Later he also developed several very important theories, the most far-reaching being the Holonomic Brain Theory, presented most comprehensively in the book "Brain and Perception". He is one of the most cited authors of science and is world-famous. He educated some of the most important neuroscientists of the present time, received many most prestigious awards and two honorary PhDs. After his "retirement", Prof. Pribram publishes much also on his hypothesis of the role of quantum holography in brain-mind. The authors of this book are indebted to him, and to Dr. John Sutherland, for the "holographic" (holography-like computation) interpretation of certain information processing in brain (memory-storage and image recognition, especially), and in its computer imitations.*

Summary

The book presents an integrated model of human image processing up to conscious visual experience, based mainly on the Holonomic Brain Theory by Karl H. Pribram. Founded on extensive studies of experimental and computational literature, and on author's experience with neural-net simulations, the model is partly descriptive and partly computer-implementable (a set of "algorithms", exhibiting the core of information processing at various stages). Image processing is presented as largely spectral, holography-like, parallel-distributed processing (PDP), mainly along the retino-geniculo-striate pathway and in the striate cortex (V1), which provides representations for subsequent object perception in extrastiate cortices.

Hypotheses on how we become conscious of the results of image processing are discussed. In this context, quantum and subcellular models are presented, but also essentials of processing in the extrastriate and inferotemporal cortices. In the context of conscious image processing, data on the following processes of visual perception are reviewed: sensation, figure/ground segmentation, object perception / recognition, visual memory (and retinotopic mapping onto cortex), perceptual binding, a sketch of color processing, visual attention and conscious experience.

Gestalt formation is implemented, in the model, in the attractor dynamics of brain networks like the neural, dendritic and quantum nets. Attractors can migrate between different areas and scale-levels. Special attention is devoted to the holonomic dynamics of slow wave potentials of the electric polarization fields inside the webs of cris-crossed dendrites and their membranes, because they are considered as the neural correlates of conscious experience by the holonomic theory. Since the holonomic theory proposes perceptual convolutions using the Gabor-wavelet-like receptive-field profiles, which maximally preserve information ("infomax") along the visual pathway, the "infomax" phase-processing of higher-order statistics is discussed as essential. It is also shown to be necessary for detection of edges.

The following relevant PDP models are considered, compared and roughly integrated (where possible): the Holonomic Brain Theory, "infomax"-models like Independent Component Analysis (ICA) and the sparseness-maximization network (by Bell, Sejnowski, Olshausen, Field, and others), MacLennan's dendritic field computing, quantum neurodynamics by Jibu & Yasue, Holographic Neural Technology (HNeT) by J. Sutherland, and quantum associative network by Peruš. Biological plausibility of the models is considered especially.

Three sorts of image representation are proposed to be used in the cortex: 1. the Gabor wavelets, rooted in dendritic webs, seem to be used for associative processes underlying visual cognition; 2. their Gabor coefficients represent neural-net's sparse codes which serve for automatic processing; 3. the spatial image, reconstructed in the extrastriate area, is those which is then consciously perceived "with shapes and colors". Because the third image representation is even perceptually projected back into external space, so that it coincides precisely with the original object, I propose that quantum holographic process is necessary, since neural nets cannot realize that alone.

As the core of image processing in V1, essentially modulated by various (sub) neuronal processes, holography-like implementation of pattern recognition and content-addressable associative memory in an original model called Quantum Associative Network is presented. The dynamics of the Hopfield net is directly translated into isomorphous quantum wave dynamics. The quantum interfering waves could be the Gabor wavelets. The so-called Hebb rule is generalized into a form named the phase-Hebb rule, since it mediates oscillatory (phase) coupling or coherence phenomena. The Green-function propagator, which has a phase-Hebb structure incorporating phase-differences, is harnessed for the quantum memory-storage and associations. The quantum wave-function "collapse" is proposed as the dynamic correlate of image recognition, based on selective recall of the most similar memory "traces", which results in pictorial conscious experience.

Applications for recognition of images and objects are presented.

Keywords: *vision, image processing, image/object recognition, conscious, holonomic, Pribram, V1, gestalt, attractor, (wave) phase, dendritic processing, ICA, infomax, holography, Gabor wavelet, quantum associative net, neural/brain modeling*

Research fields: *computational cognitive neuroscience; neuropsychology; neuro-computing; quantum information science*

IMPORTANT NOTE

In the first section (Main Chapters), the topics are presented following the Holonomic Brain Theory of Karl H. Pribram. In the second section (Auxiliary Chapters), some additional and complementing topics are added which are not necessarily related to the Holonomic Brain Theory.

The main chapters are explicitly presenting the central issues of the book in direct relation with the Holonomic Brain Theory of Karl Pribram, whereas the Auxiliary chapters provide additional or detailed discussion of some issues which are either needed to understand the main chapters better or to go into depths which might not be interesting to all readers. In both cases, the Auxiliary chapters do NOT constitute the Holonomic Brain Theory. They enable a comparison of the holonomic theory and the broad or mainstream neuroscientific framework of vision research and some quantum analogies which might in the future turn out to be significant for it.

Preface

There was a kingdom where people wanted to make a map which would depict all the things in the world. When the map, which they made, turned out to be as large as the whole world and covered the whole earth surface, trees and animals started to die beneath, and people had to dig holes into it to take breath... (from a tale)

In science, especially in science of complex systems and phenomena, we have to make approximate models which are never perfect. It is often very hard or even impossible to falsify them, but they are anyway indispensable. We believe that a model is good if there is no better model available in present circumstances. However, we can always only guess how much it fits reality.

We tried to make as *much a biologically-plausible model as possible* — at least on a global scale, i.e. in spite of unavoidably having to neglect numerous known and yet unknown details. Because of brain's complexity, experimental evidence is still not rich at all, but we tried to select best of it and take it into account as much as possible. Reports with lack of parallel evidence have been treated carefully and with reservation. Special attention has been paid to (dis)agreements in various reports. If there was some mismatch in different reports, all were considered "neutrally" at best – if they were not ignored entirely as in cases of large disagreements.

It has been a delight to uncover, month by month, the exquisite and subtle structure of the Holonomic Brain Theory and the brain systems and processing in general, as we can imagine them, to build a panoramic view – in spite of limitations, open questions and inability to provide a sufficient link with the topic of main concern of psychology – mental and conscious phenomena.

Acknowledgment

Heart-felt thanks to Professor Karl H. Pribram for numerous very valuable comments and suggestions after reading various versions of the manuscript more carefully than one might expect from a pioneering and internationally-leading scientist of intense and world-wide activity. The first author benefited also very much during days of intensive discussions while he was enjoying warm hospitality of Prof. Pribram and wife Mrs. Katherine Neville in Georgetown, DC, but also during other meetings in Ljubljana and at conferences.

Thanks for discussions and help to: Professors Andrej O. Župančič, Janek Musek, Laurence I. Gould, Andrej Ule, John Bickle, Dobilas Kirvelis, Dušan Hadži, Paul Kainen, Andrew Vogt, James Glazebrook, Giuseppe Vitiello, Anton P. Železnikar, Wieslaw Sienko, John Cramer, Alexandr A. Ezhov, Bruno Olshausen, Bruce MacLennan, Aleksandr D. Linkevich, Dejan Raković, Masanori Ohya, Ron Chrisley, Dan Ventura, Elaine Funnell, H. John Caulfield, Vladimir Bužek, Alexandr V. Pavlov, Germano Resconi, Alex T. Granik, Marion Bartlett, to Doctors Peter Marcer, Ron Spencer, Metod Škarja, Carlo Trugenberger, Rafal Bogacz, Aleksei Ju. Ippa, Te-Won Lee, Takashi Matsuoka, Alexander Vlasov, and to many members of the *Slovene Society for Cognitive Sciences* and the *Society for Neuropsychology and Cognitive Neurology of Slovenia*. I appreciate cooperation and help of Professors Manca Košir, Igor Kononenko, Daniel Dubois, Elisabeth Behrman, Subhash Kak, Walter Schempp, Basil Hiley, Suhrit Dey, Selma Supek, and of Doctors Mila Božič, Bono Lučić and Tarik Hadžibeganović.

Finally, special thanks to Professors Horst Bischof and Franz Leberl, Drs. Friedrich Fraundorfer, Roland Perko, and to other researchers of the *Institute for Computer Vision and Graphics* of the *Graz University of Technology*, Graz, Austria, for collaboration and support during M.P.'s Marie Curie post-doc fellowship, contract HPMF-CT-2002-01808, funded by the European Commission. Applications were tested in this framework.

Support of Slovenian institutions – *the BION Institute, the Ministry of Education, Science and Sports*, and the *National Institute of Chemistry* (especially its former director Prof. S. Pejovnik) – is gratefully acknowledged. Thanks for support from Multimedia University, eScienceFund funded by Ministry of Science, Technology & Innovation, Malaysia and Mr. Nuo Wi Tay for the computer simulations in this book.

Introduction

Many-body interactions have been successfully described through models based on classical or quantum physics. More recently, some of the models have been related to cognitive science by researchers who are interested in describing brain activity through the use of artificial neural networks (ANNs). But most of those networks are described through classical physics.

A popular one, the Hopfield network, is usually modeled on the physics of a system called a "spin glass" — a bunch of particles with spin that settles down to a final set of states derived from a minimization of the system's total magnetic-interaction energy. However, in contrast to a relatively unconstrained system, such as a bunch of magnetic compasses laying on a table in close proximity to each other, it is possible for the spin-glass system to have *many* energy minima. (As a rough analogy to a spin glass: imagine a collection of people at a party who seek to talk to others in the same profession. Gradually, the different professional groups are built up. In order to see the state of affairs as time passes, a person in another room surveys all groups, using a ceiling camera, to see what percent of the people in each group is in a given profession — lawyers, journalists, biologists, physicists, etc. When all persons have found their group the number of people outside a give group has been minimized. Hence the many minima; each associated with a different group — like a mountain range with many valleys!)

Although the Hopfield model has been extensively applied to ANNs using classical physics, this book goes forth in new directions by extending the model for ANNs into the quantum domain. One of their main contributions here (based on extensive experience with classical synergetic associative neural nets) is showing how the Hopfield prescription can be cast into a quantum-mechanical form, which the authors refer to as the Quantum Associative Network. The algorithm, which describes it, has four main parts: Step 1 is the encoding of external information (e.g., light information from an object) into quantum state functions. Step 2 is the construction of a quantum-memory-storage function (analogous to a Hebb memory-storage matrix for the Hopfield network) by using the state functions from the first step. Step 3 then employs the matrix from the preceding step to create a formula, which shows how a new quantum state function is associated with an arbitrary input function. Step 4, the last part of the algorithm, performs the function of quantum-memory recall by using the memory-storage function of step 2, along with the formula of step 3 in order to take new external information, transform it into a new quantum state function, and then determine how similar it is to the original information — only new information sufficiently similar to the original will survive the process.

The Holonomic Brain Theory of Karl Pribram is included as an important part of understanding how the brain may process visual information at a macroscopic (biologically plausible, computational) level as well as at a microscopic (synapto-dendritic) level. The phase information contained in waves associated with dendritic processing is thought to be rooted in quantum mechanisms; pointing up the relevance of Pribram's Theory to Peruš's Quantum Associative Network. There are, however, issues of whether quantum mechanics can be used to describe mental processes. But perhaps the Holonomic theory is an intermediate way; for here there is the quantum-mechanical–like, yet *classical*, neural wave equation.

This book by Dr. Peruš and Dr. Loo is, technically, not as formidable as one might think. This is because the authors have taken great pains to explain the specialist's terms. Hence the book should be accessible to the diligent intelligent layperson.

Biological and Quantum Computing for Human Vision is a novel approach to the field of neural networks which attempts to model biological systems. It could be of benefit for the increasingly prominent area of computer-simulated biophysical studies of information processing in neuronal systems; and thus yield a possible way of understanding the biological "brain net" through a neural-net model that has both classical and quantum features. It may also be useful for the application of neural nets in technology as well as in modeling aspects of cognition.

Laurence I. Gould
University of Hartford, USA

Laurence I. Gould, *after studies at the University of Temple, USA, became Professor of Physics at the University of Hartford, Connecticut, USA. He has very broad research interests covering also complex (bio)systems, foundations of quantum mechanics and non-locality, conservation laws, and symmetry in the broadest sense (writing a book on it).Since 1994/5, in pure coincidence with Peruš, he presented mathematicalanalogies of quantum physics, using the Bohm and Hamilton-Jacobi formulations, and description of dendritic neurodynamics by Jibu, Yasue and Pribram.*

Section 1

This section serves as the comprehensive introduction for the Holonomic Brain Model by Karl H. Pribram and other computation models such as the Independent Component Analysis (ICA), the sparseness-maximization net of Olshausen and Field and the quantum-implementable associative network. The essential issues related to the holonomic brain model are discussed. Within this section, the reader can learn integrated model of human image processing up to conscious visual experience, based mainly on the Holonomic Brain Theory by Karl H. Pribram.

Chapter 1
Introduction to Holonomic-Compatible Models for Vision

1.1 STARTING REMARKS

The purpose of this book is to research possibilities for complementing neural models of early vision with the new preliminary quantum models of consciousness in order to construct a model of human image processing.

Contents. The contents of this book are the following. In the first chapter, a comprehensive introduction to the problem and the basic "tools" for tackling it will be presented and extended. These are the holonomic brain model and some computational models like the Independent Component Analysis (ICA) and quantum-implementable associative networks of coupled oscillators with phase-encoding. In the second chapter, the *Holonomic Brain Theory* of Karl Pribram, especially its part on vision, will be presented. In the third chapter, computational information-

DOI: 10.4018/978-1-61520-785-5.ch001

maximization models like the version of Independent Component Analysis (ICA) of Bell and Sejnowski and the sparseness-maximization net of Olshausen and Field will be presented and discussed in the context of Holonomic Brain Theory and related models. In the fourth chapter, the models of image processing are summarized and integrated into a complete, partially hypothetic, global picture. Some speculations about higher associative, attractor-based and quantum-rooted, visual processing, which gives results (images, episodes) that we are conscious of, are presented. For this purpose, models of dendritic image-processing in V1 and hypotheses on subcellular (e.g., microtubular) and quantum brain processing and binding are reviewed. Consciousness, in particular visual consciousness, and its neural correlates are discussed at the end.

The *second part* of the book provides additional or detailed information – auxiliary to the chapters of the first, main part. In the first auxiliary chapter, no. 6, information processing along visual pathways from retina to cortical areas will be presented as described by the modern main-stream experimental neuroscience (including neurophysiology, neuropsychology, cognitive neuroscience). The neuropsychology of image processing and related processes (like early vision, figure/ground segmentation, visual memory, object perception, visual attention and binding) will be presented. Recognition of edges and other features using contrast- and color-processing will be especially discussed. A comparison with the holonomic alternative will be made.

The majority of so-called auxiliary chapters mainly review neural-net-like quantum associative processing, starting from my list of neuro-quantum analogies which is then used for translation of a neural information-processing "algorithm" to quantum formalism. Some essential improvements where quantum associative processing transcends the neural-net basis are highlighted. These improvements are mainly based on holography-like phase-relationship processing.

Auxil. Chapter 6 can be skipped by neuroscience experts. The mathematical auxiliary chapters and appendices are complements to the main text and to the Auxil. Chapter 6 which, however, alone incorporate *all* the relevant neuropsychology. Mathematical Auxiliary Chapters provide theoretical and computational details mathematically and physically more rigorously, but are *not essential* for understanding of the main text (or of Auxil. Ch. 6).

Computer-implementable core. The central part of this work is to provide a global (rough) mathematically-supported, computer-implementable model, i.e. a "skeleton" of an information-processing "algorithm", for image processing giving a conscious result – a consciously perceived image. We will start with a "skeleton" based on neural-net models, enrich it with holography-like phase-dynamics, and embed it into quantum dynamics.

Here I will use the term *neural networks* for those *real (biological) networks or circuits of neural cells which are optimally modeled by computer simulations*, i.e. *we take into account as much experimental biological data as possible* in the model.

That is to say, we do not know very well how the real visual brain looks like. But, anatomical research shows that there are a lot of densely-connected cells. Our mathematical models transformed into computer simulations try to imitate as much physiological findings as possible. Such models become our metaphor or our "internal image" of the brain. Everybody of us has an optimal "image" of it which she/he uses for her/his research purposes. When I will talk about neural networks here, I will have in mind this optimal "map" of the brain interior at the level of more or less large neural circuits (many cells connected in a web). I do not know how much we share such a "virtual image" in its details, but I will try to increase agreement by occasionally providing a more detailed specification of my "image".

Beyond neural nets. Neural networks[1] realize the basis for automatic behavior and are good in processes like pattern recognition and associative memory. On the other hand, they do not explain awareness. Therefore the question remains how much are neural networks responsible for real brain functioning which is (also) essentially related to the phenomenon of (self)consciousness. I consider awareness as for all practical purposes synonymous (identical in our model) to self-consciousness (i.e., phenomenal auto-reflective consciousness). And I also suppose, following arguments by Gennaro (1995), that consciousness entails self-consciousness.

Neural networks surely are very important, but our hypothesis is that they also essentially exploit relatively more microscopic (i.e., sub-cellular) processes. I will take into account that such sub-neuronal processes might be of different, but essentially interdependent, types or levels: dendritic (Savtchenko & Korogod, 1997), microtubular, quantum-biological, sub-quantum (Globus, Hiley in Pylkkänen & Pylkkö, 1995). This means that I will consider different, but hopefully approximately compatible, models which subjectively put the center of scientific attention to a particular level, not entirely neglecting the others. I believe that the "objective truth" is a sort of "combination" of these specialized views. However, in particular I will follow the holonomic[2] brain model of Pribram (1986, 1991-1998). Namely, it seems presently to have more experimental support, regarding the question of consciousness, than "neighboring" models like *neural network models* (for example, by Churchland & Sejnowski (1992), Crick, J.G. Taylor), the *microtubular model* (by Hameroff with quantum extensions by Penrose – see Hameroff *et al.* (1996, 1998; in Pribram, 1993)), and *quantum-biological models* – i.e., models with some relatively-purely quantum ingredients: by Vitiello (2001, 1996), Alfinito & Vitiello (2000) and Pessa & Vitiello (1999, and refs. within), Jibu & Yasue (1995, 1997; in Pribram, 1991, 1993) and other quantum field physicists[3]. Of course, almost everybody of the above-mentioned agrees that beside their preferred level also (all) "neighboring" levels are important, therefore the "names" of the models refer just to the center of attention of the model. My hypothesis is that too specialized models (devoted to a particular level only) are irrelevant for conscious experience which seems to be an essentially global, multi-level and holistic phenomenon.

Pribram's holonomic model, using the *holographic metaphor* (Pribram, 1971), gives the center of attention to *synapto-dendritic networks* (therefore I added it among dendritic models). Webs of dendrites with their synaptic connections are embedded in dense interactive polarization fields. Dendritic junctional microprocesses[4] have background in quantum coherent states (Fröhlich, 1968; cf., Matsuno, 2002), according to Pribram's hypothesis (later shared and extended to quantum domain with Jibu and Yasue) (Pribram, 1991, 1993). Hypothetically, peri-membranous[5] ordered water dynamics and Bose condensates[6] (Marshall; *et al.* in Hameroff *et al.*, 1996) realize ultimate perceptual binding.

Note that Karl Pribram's original views will be presented under keyword "holonomic theory". My extensions of the holonomic theory will be presented usually in separated sections or paragraphs with first-person statements.

1.2 THE HOLONOMIC MODEL OF VISUAL PERCEPTION

Aims. Pribram expresses a hypothesis, with serious theoretical reasons and with experimental support, that holonomic principles govern several parts of the cortex, especially visual, auditory and somatosensory. Discussion will from now on be confined to the visual cortex which is central in the holonomic theory. According to my knowledge, Pribram's Holonomic Brain Theory provides, among all theories, the most global physiological experimental data and holistic theoretical interpretation for modeling visual perception – therefore such a choice. I combine this theory with own observation (orig. Peruš in Wang *et al.*, 1998) how holography-like information-processing principles are implementable in quantum systems also. Holonomic Brain Theory seems to provide a firm framework for such quantum–neural complementary and comparative studies.

V1 and surround. The *striate cortex* (i.e., V1 area of the visual cortex)[7] is the most important one for image processing, although for higher-order visual and synthetic or synesthetic perceptions the extrastriate cortices (V2 and higher) seem to be very relevant also. It is debatable which of the above-mentioned visual cortices is the central one for *conscious* visual perception. V1 is *necessary* for conscious image processing, because if V1 is damaged, eye-inputs are processed unconsciously (Baars, 1997). However, Koch and Crick claim that primates might *be aware of* processing in extrastriate cortices, but *not directly* of those in the striate cortex (Koch in Hameroff *et al.*, 1996).

Figure 1. Each sub-figure depicts a Gabor wavelet. Wavelets have different orientations and spatial frequencies (i.e., "densities of waves"). (From Lee, 1996, fig. 1)

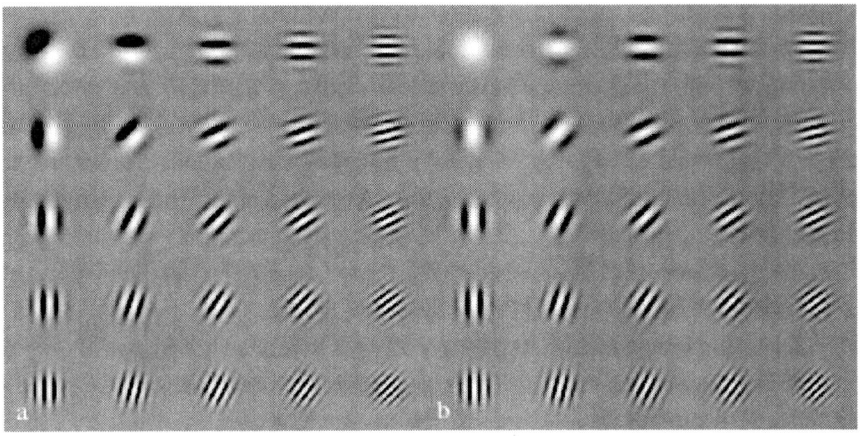

The aim of this introductive presentation is to provide a biocybernetic framework for a computer-implementable model which attempts to take into account as much significant neuropsychological and psychophysical evidence as possible, as reviewed in the book "Brain and Perception" (Pribram, 1991). Following this reference, the form of receptive fields[8] of the striate-cortex neurons are *Gabor wavelets* (Figure 1). They minimize the uncertainty in the so-called phase-space of space-time and spatial and temporal frequencies, and thus maximize information. Gabor wavelets serve as "patterns" which interfere in order to realize a Hebb-like memory-storage (presented in sec. 6.8). We propose a hypothesis that *holography-like image-processing and related Gabor transforms are implementable in synapto-dendritic and/or quantum substrates*. Because holonomic principles seem to be to a significant extend level-invariant, somewhat similarly to fractals[9], their synapto-dendritic and quantum-biological implementations could coexist and cooperate.

Convolutions. The Holonomic Brain Theory (henceforth: holonomic theory) provides a macroscopic as well as microscopic view on brain processes. At the *macroscopic* level, the theory suggests biologically-plausible computational mechanisms for a triple convolution-process[10] along the visual pathway. The visual pathway is: *retina — lateral geniculate nucleus (LGN) — striate cortex*. At each of these three stages, inputs from the previous stage are convoluted with the receptive fields of neurons receiving those input at the present stage. For example, signals from many retinal neurons (retina – first stage) arrive into a LGN neuron (LGN – second stage). These retinal neurons constitute the *receptive field* of the LGN neuron.

The first convolution process is executed by the optics of the eye: Pupillary input (optical image) is convoluted with activities of retinal receptors. The second convolution is the following: Retinal signals are convoluted with the receptive fields of the signal-receiving LGN-neurons, and this "overall weighted sum" (such a sum is convolution!) of signals now determines the states of the LGN neurons. Similar signal-summation process, involving Gabor transforms, is realized by a striate-cortex neuron (striate cortex – third stage) while receiving signals from many LGN neurons (These LGN neurons constitute the receptive field of the striate neuron.). (Pribram, 1991)

Roughly the same or at least similar processes continue inside the visual cortex, at its various hierarchical levels (Perret & Oram, 1998).

Dendritic implementation. At the *microscopic* level, the holonomic theory considers synapto-dendritic networks (where synapses and dendrites belong to neurons at the above-mentioned three stages of the visual pathway, especially V1) as the most relevant information-processing executive level.[11] So, image processing in the striate cortex using Gabor wavelets may be implemented by interacting dendritic polarization-fields[12] arising from spine-produced[13] oriented electrical dipoles.[14] In retina and LGN a similar image *pre*-processing is done. The difference is that such pre-processing is realized using "Mexican-hat"-shaped "filters" corresponding to receptive fields of retinal neurons and later, at the second stage, of LGN neurons. On the other hand, cortical neurons have receptive fields described by Gabor wavelets. This means that along the visual pathway between retina (or, better to put, LGN) and striate cortex, a "transformation" to Hilbert space[15] is realized (Pribram & Carlton, 1986). In the striate cortex, the image is then processed "to a full extent" in a (quantum)-holography-like way.

Why does the holonomic theory consider processes at the synapto-dendritic-network[16] level more relevant than processes at the usual neural-network level (or neural-circuit level[17], respectively)? The reason is that dendritic processes are more flexible, faster, realize a higher rate of connectivity, or interaction, and greater parallelism than conventional nerve-signal processing. Dendritic processes are also "more directly connected" with quantum holism and unity – over the citoskeleton, especially microtubules (sec. 4.7). That is to say, as everybody agrees, nerve impulses between somas propagate through dendrites and axons, and over synaptic clefts between them, but the usual neural-network theories consider just the out-of-soma axonal signal propagation, without adding the sub-cellular microstructure of interactional processes in dendrites and their webs[18] into account. These dendritic microprocesses can modulate the macroscopical nerve-signal propagation and are thus relevant.

It has been observed that the duration of awareness is proportional to the delay in dendritic-network processing (Pribram, 1991). Especially relevant seem dendritic networks to be because they could be essentially influenced by underlying quantum processes (AuxLit 21). Quantum holism is needed, we believe, for explaining subjective unity and "envelopal" stability[19] of bound-together multi-modal conscious experience. Classical binding models using synchronous neuronal oscillations (e.g., Gray *et al.*, 1989; Roelfsema, 1998) are probably not sufficient.

Thus, the holonomic theory postulates quantum-rooted dendritic processes realizing *conscious* perception, not merely information dynamics. The "involvement" of consciousness starts (or to say alternatively: conscious experience emerges) probably not earlier than in the visual cortex.[20]

The holonomic theory is optimally experiment-fitted (Pribram, 1995) (as will be shown later) and allows at least approximative mathematical and computational modeling of visual perception up to a semi-phenomenal (pre-qualitative) level. Conventional neural-net models (LeCun & Bengio, Fukushima in Arbib, 1995; ASSC, 1998; Gupta & Knopf, 1994) do not come so far, I would like to argue. We will pay our interest to quantum-neural holographic processes, like Fourier decomposition, optimal correlation-matrix memory-storage and associative recall, which explicitly include phase-information.

Gabor wavelet transforms. According to the holonomic theory, the task of primary visual cortex (or its simulation) is to determine the Gabor coefficients s corresponding to elementary Gabor functions (also called Gabor wavelets) g which together constitute a superposition, the Gabor field γ, describing the cortical receptive fields: $\gamma = \sum_k s_k g_k$.

Following Pribram & Carlton (1986), let us make an approximation of the visual-pathway processing and concentrate on the main part of its multi-stage convolutional mechanism, that is on the retina-cortex-convolution process. It is described by the following integral over the 2-dimensional area of retinal cells:

$$Gf(p,u) = \iint_{retina} g(p,u;x) f(x) dx_1 dx_2 \ . \qquad (1.1)$$

Gf is the Gabor transform of the instantaneous retinal image *f*. Coordinates x and p correspond to points on retina (p falls into the center of the Gabor wavelet $g(p,u)$) measured in distance units from the origin at retinal foveola[21]; u corresponds to spatial frequency.[22] They are all two-dimensional coordinates, as the retina is. The response of the striate-cortex neuron to stimulation with signals from retina, described by $f(x)$, is determined by such a Gabor transform. The above convolution integral sums all contributions from retinal neurons weighted by the elementary Gabor function or wavelet *g(p,u)* having the following form:

$$g(p, u; x) = \exp\left(-\frac{(x-p)^2}{\sigma^2}\right)\exp\left(-iu(x+p)\right). \qquad (1.2)$$

Such a *Gabor wavelet* or wave-packet (details in Lee, 1996) describes the form of the receptive field of a striate cell. It has a gaussian bell-shaped envelope, centered at *p,* (first exponential factor) and a sinusoid "interior" (second exponential term). The wavelet has sinusoidal excitatory and inhibitory regions (seen from above: "stripes"; with "density of stripes per wavelet" *u*) which are a result of lateral inhibition.[23] It ensures maximal selectivity to spatial frequency *u* (R. & K. DeValois, 1990) and consequently to stimulus-orientation angle given by $\arctan\frac{u_y}{u_x}$.

V1 columns. Each component *Gf(p,u)* of the Gabor transform *Gf*, made by the network of striate neurons as anatomically connected with groups of retinal neurons while receiving retinal input *f,* represents the activity in a *cortical column* tuned to spatial frequency *u* and responding maximally to an input from the retinal position *p.* So, columns of the striate cortex are specialized, local neural-net structures which respond maximally to a stimulus which has a particular orientation and arrives from a particular origin as detected by retina.

Outputs from striate (V1) columns project to peri-striate (V2) neurons. It is probably there that retinal image *f* is recovered in the inverted form (turned upside-down): *F* = *f(-p)*, where *F* corresponds to the V2 image-reconstruction. This happens after a new, inverse, Gabor transform has been made. (Pribram & Carlton, 1986)

Figure 2. Examples of two different groups of "spatial plain-waves" (above) with the same spatial frequency, but different amplitudes (or also intensity-contrasts). By comparing the upper with the bottom parts of both sub-figures, one can understand the origin of the term "spatial frequency". (From Stillings et al., 1995, fig. 12.6)

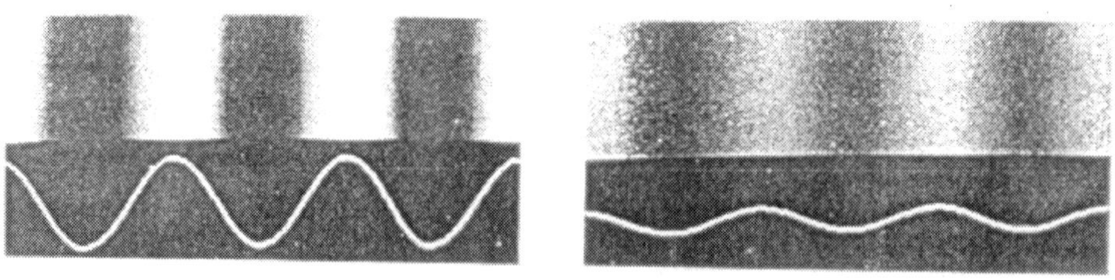

1.3 SEARCH FOR QUANTUM NEURAL SUBSTRATES OF VISION AND THEIR MODELS

The question of biological implementation of holonomic principles in various sub-cellular environments remains the central problem. Due to enormous complexity and interdependence of brain structures, experimental evidence is often poor in many respects, but there was some relatively reliable and precious experimental work done. Constraints, which a mathematical or/and computational model should satisfy as much as possible while realizing image processing, are being derived from experimental reports.

The multi-stage convolution of advancing input-signals with the Gabor "filters" is, in principle, equivalent to a special[24] *multi-layer perceptron*-like artificial neural network (ANN) with an important exception. ANN (Haykin, 1994; AuxLit 10) usually incorporate only "amplitude"-information which is quantified using real-valued numbers.[25] Such ANN do not incorporate phase-information, because they do not contain wave-dynamics. There are some new ANN-models which incorporate wave-dynamics and wavelet processing using complex-valued numbers and phase-information (in the Euler exponents).[26] If the wavelets are Gabor wavelets, then such ANN come quite close to the holonomic model. (This is no wonder if that imitation followed results of the same psychophysical research which lead Marčelja[27] to describe receptive fields by the Gabor elementary functions.)

Phase models. To proceed with modeling of image processing, it is thus useful to incorporate phase-information into amplitude-correlation (second-order statistics) models like the model of Hopfield (1982) or *Principal Component Analysis* (PCA) (see Appendix A). Such associative-memory models use correlation-matrices (Hebb memory-storage matrices).[28] There are some modern models which successfully exploit *phase-information* in addition to amplitude-information. Some of them are very interesting in our holonomic context, because they share many features with the holonomic theory and may complement it. These models are:

- **Independent Component Analysis** (ICA): a perspective method (e.g., Comon, 1994) brought to vision research by Anthony Bell and Terrence Sejnowski (1995, 1996, 1997);
- **field computation** model by Bruce MacLennan (1990, 1991, 1994; in Pribram, 1993; in Wang *et al.*, 1998; Wolpert & MacLennan, 1993) which also uses Gabor wavelets and many physics-inspired techniques;
- **Holographic Neural Technology** (HNeT)[29] by John Sutherland (1990): an application-effective neurobiology- and holography-inspired artificial model[30] (in other words, simulated holography-based neurocomputation) which was recently renamed to **Holographic / Quantum Neural Technology**, because

analogies between mathematical models of holography, associative neural networks and quantum fields were recently systematically listed and discussed (Gould, Peruš in Pylkkänen & Pylkkö, 1995; Peruš, 1996, 1998a; Schempp in Pribram, 1993; Marcer & Schempp, 1998);

- **Quantum neurodynamics** by Mari Jibu and Kunio Yasue (in Pribram, 1991, 1993);
- **Quantum associative network** model by Peruš (2000b,c; in Wang, 1998): a quantum information-processing "algorithm", inspired by the Hopfield associative neural-net model (Hopfield, 1982), and partly by the Haken synergetic-computer model (Haken, 1991), rewritten to quantum-mathematical formalism.

1.4 QUANTUM ASSOCIATIVE NETWORK

Origins. My model, the Quantum Associative Network, is deliberately "shockingly"-simple and fundamental (Peruš, 2000b,c; Peruš in Wang, 1998). It bases on observations that the mathematical formalism of associative-neural-network theory, as usually simulated, and the mathematical formalism of quantum theory are very similar (Peruš, 1997a, 1998). To construct the quantum associative net[31] from Hopfield-like simulated ANN, the real-valued "neural" variables have been translated to analogous quantum complex-valued (phase-information-carrying) variables (Peruš, 1997a). The translation to quantum formalism was deliberately and purposely done in the most "natural" way, i.e. so that the neural-net-like associative information-processing can be *implemented in a usual quantum-physical system, not necessarily in some artificial quantum-based device*. This characteristic distinguishes the quantum associative net from all similar quantum-computer models (Ezhov *et al.*, 2000, 2001; Berman *et al.*, 1998; Ventura, 1999; Bonnell & Papini, 1997; Behrman *et al.*, Ventura in Wang *et al.*, 1998; Chrisley in Pylkkänen & Pylkkö, 1995) which are, at least in part, hypothetically-implementable only using some artificial quantum devices or procedures.

Significance. The quantum associative net is the *simplest* or the *basic* quantum model which realizes content-addressable associative memory and pattern recognition. It uses fundamental and completely-natural quantum characteristics as described in the first chapters of any basic textbook on quantum mechanics (e.g., Messiah, 1965; AuxLit 16). I believe that many more advanced, more complicated, but also more effective quantum information-processing implementations will be found in the near future – natural, artificial, or hybrid. But the purpose of the quantum associative net is to serve as a *primitive prototype* for all future quantum associative PDP-models[32] which will necessarily be advanced extensions of the present model. They will probably exploit "deeper" (e.g., quantum-field-based) or/and more artificial, technological ingredients, e.g. NMR.[33] (For purposes like the present one, i.e. modeling human vision, artificial models are, of course, not directly acceptable.)

Furthermore, I argue that the quantum associative net processing suggests that *any quantum-physical system is a potential associative-information-processing system if properly manipulated.*[34] With proper manipulation I mean that something or somebody, usually an intelligent being, interacts with the system, triggers its transformations, and gives interpretations to its initial and final states. *By giving such an interpretation or meaning to system's states, physical dynamics "transforms" into information-processing dynamics.* So, the (quantum) system has to be open and flexible enough that we can force it to manifest specific states, and we must be able to attribute specific meaning to these states (Peruš & Dey, 2000). Brain is such a system, but not only brain. Brain is a system which even interprets its own states and even more (exhibits conscious experience etc.). Usual quantum systems could also be harnessed for associative information processing after we give interpretations to their states.

"Algorithm". I will now present briefly the core of processing of the *Quantum Associative Network* as a Hopfield-like "algorithm":

STEP 1: **encoding images into a quantum system**

Quantum eigen-waves are given the role of encoding our patterns or images. Their number shall be P. Eigen-waves will be denoted by Ψ with index k (in contrast to Ψ which shall denote the current quantum state):

$$\psi_k(\vec{r},t) = A_k(\vec{r},t)\exp(i\phi_k(\vec{r},t)) \ . \tag{1.3}$$

STEP 2: **quantum memory construction**

Hebb-like quantum interference pattern is produced, e.g. by performing a holography-like interference of (quantum) image-modulating waves. **G** is the matrix of all Green functions G which describe the response of the system to perturbations (in our case, "inputs of data"). Global result of this interaction is the emergence of interference-pattern. **G** describes the interference-pattern which encodes the *quantum associative memory.*[35]**G** is thus the direct analogue to the Hebb memory-storage matrix from ANN-models. **G**, like the Hebb memory matrix, has a form of a correlation (which is similar to convolution!) matrix, but additionally implicitly encodes local phase-differences (in the exponent of the bottom part of equation below). The elements of **G** are given by:

$$G(\vec{r}_1, t_1, \vec{r}_2, t_2) = \sum_{k=1}^{P} \psi_k(\vec{r}_1, t_1)^* \psi_k(\vec{r}_2, t_2) = \\ = \sum_{k=1}^{P} A_k(\vec{r}_1, t_1) A_k(\vec{r}_2, t_2) \exp\left(i(\phi_k(\vec{r}_2, t_2) - \phi_k(\vec{r}_1, t_1)) \right) . \tag{1.4}$$

STEP 3: **quantum associative computing**

G also serves as so-called quantum propagator, therefore the usual quantum propagation can be harnessed as a quantum associator (but only if **G** has the form given in STEP 2; in fact it has!). Such quantum computation is best described by the Feynman interpretation of the Schrödinger equation:

$$\Psi(\vec{r}_2, t_2) = \iint G(\vec{r}_1, t_1, \vec{r}_2 . t_2) \Psi(\vec{r}_1, t_1) d\vec{r}_1 dt_1 \ . \tag{1.5}$$

STEP 4: **quantum memory recall or image recognition**

G has one more capability: it is a projection-operator: It projects to a subspace which corresponds to those eigen-image which is the most similar to the newly-presented "input" Ψ'. For such a quantum image-recall process, the celebrated *wave-function "collapse"* is used:

$$\Psi(\vec{r}_2, t_2 = t_1 + \delta t) = \int G(\vec{r}_1, \vec{r}_2) \Psi'(\vec{r}_1, t_1) d\vec{r}_1 = \int \left(\sum_{k=1}^{P} \psi_k(\vec{r}_1) \psi_k(\vec{r}_2) \right) \Psi'(\vec{r}_1, t_1) d\vec{r}_1 = \\ = \left(\int \psi_1(\vec{r}_1)^* \Psi'(\vec{r}_1, t_1)\, d\vec{r}_1 \right) \psi_1(\vec{r}_2) + \ldots + \left(\int \psi_P(\vec{r}_1)^* \Psi'(\vec{r}_1, t_1)\, d\vec{r}_1 \right) \psi_P(\vec{r}_2) = A \psi_1(\vec{r}_2) + B \tag{1.6}$$

where $A \approx 1$ ("extracted image") and $B \approx 0$ ("noise").

Thus, those eigen-image is recalled from memory (i.e., from the interference-pattern of all images) which has the highest correlation with the current input Ψ'. This is because one of the terms in the series goes to 1 (constructive interference), all the others decay to 0 (destructive interference). For details see Auxil. Chapter 10 and Peruš (2000b,c) or Peruš & Dey (2000).

The quantum associative memory-storage and pattern recognition are essentially nothing else but the most fundamental quantum feature — in the simplest way described (in Dirac notation, in three equivalent versions, the third one being the most suited for our purposes) by

$$\left| \Psi \right\rangle = \left| \Psi \right\rangle \left\langle \Psi \right| \left| \Psi \right\rangle = \sum_k \left| \psi_k \right\rangle \left\langle \psi_k \right| \left| \Psi \right\rangle = \left(\sum_k \left| \psi_k \right\rangle \left\langle \psi_k \right| \right) \left| \Psi \right\rangle . \tag{1.7}$$

1.5 TRYING TO INTEGRATE THE MODELS

I have already listed some models of visual perception which have, in my opinion, potentials for tackling the problem of visual conscious experience, not merely the early vision analysis (like "feature detecting"[36], orientation selectivity) and automatic image processing (Stillings *et al.*, 1995). Because I share the belief that the future models of conscious vision have to be compatible with quantum-biological models, I concentrated on the holonomic theory by Pribram and its compatible models (listed in sec. 1.3). A sketch of the main specific purpose of this book is given here, i.e. search for a potential unified framework of these models, which all seem to have some merit for modeling conscious vision, and the potential role of the quantum associative net as a "primitive prototype" model (as sketched in sec. 1.4) in this context.

ICA. The model of Independent Component Analysis (ICA) is a relatively new (Comon, 1994), but perspective one. Pribram believes[37] that it is very relevant for the holonomic theory of vision, because it may provide an experimentally-supported solution to the problem of *edges* of perceived objects (which has not been well solved by classical neural models). Namely, the ICA-filters resemble the receptive fields of so-called simple cells in the primary visual cortex (Bell & Sejnowski, 1997; van Hateren & Ruderman, 1998). The "independent components", selected by ICA-filters, are localized and oriented, as was also experimentally found for those simple cells which were denoted as edge-filters. Furthermore, the unsupervised ICA learning algorithm, based on information-maximization (*"infomax"*), provides maximal statistical independence of resulting neural states (outputs). This happens when their probability-distribution gets factorized. This is equivalent to getting mutual information between the output-states zero (details in: Bell & Sejnowski, 1995, 1996, 1997; McKeown *et al.*, 1998).

Maximization of information (and minimization of mutual information) is a characteristic of Gabor wavelets. Gabor wavelets are a possible connection between ICA and the holonomic theory. Another connection is the fact that ICA as well as the holonomic theory encode and exploit the phase-information in addition to the amplitude-information. Actually, as ICA-simulations show, the amplitude-information is less necessary for good image-recall than the phase-information. This is similar to HNeT-processing and to findings by MacLennan (in Wang, 1998) for some Gabor-based field-computation models. In HNeT phases are considered as the main output; amplitudes merely measure the rate of importance, confidence (or reliability, accuracy, respectively) or urgency of the phase-output.

Phase-information is necessary for edge detection, because edges are situations where many sine-waves of different spatial frequencies "get in phase", i.e. they are all aligned in phase and sum together precisely where the edge lies. Such a coherent constructive interference is not realizable without wavelets, so the usual Hebbian amplitude-correlation methods (without phase-differences incorporated) cannot be successful for edge detection. Without detection of edges the image-processing capability of the system can be severely impaired.

Comparison. ICA, HNeT and field computation are not necessarily quantum-implemented, but are classical (partially holography-like) computational models. However, quantum ideas might improve them all, and might also make them relevant for modeling conscious vision. The basic core of HNeT is mathematically almost identical to the quantum associative net, as shown by Peruš (in Wang, 1998), although they were developed independently – the HNeT model from classical (holographic) foundations, but the quantum associative net from quantum ones (although ANN-inspired). The relations between ICA, HNeT and the quantum associative net will be explored and put into the context of vision research in my following chapters. However, although effective computationally and having biologically-plausible *results* (outputs), the ICA algorithm does not have any biologically-plausible *mechanism* to support it as a brain procedure. That is to say, up to now it has not been found where and how the nonlinear ICA "infomax"-algorithm could be implemented in brain tissue. But this is a possibility which will be explored.

Table 1 compares characteristics of chosen models in the context of vision problem.

My hypothesis. Since holography is an universal process, the microstructure of junctional slow-wave potentials, attributed to polarization-fields in dendritic nets (Pribram, 1971, 1991), may have various implementations, but the basic one would

Table 1.

	HNeT	**Quan.Assoc.Net**	**ICA**	**field computing**
effectiveness	very effective	effective, potentially very effective	very effective	a general model with potentially very effective "sub-branches"
biological plausibility	biol. plausible at the fundamental level only	biol. plausible at the fundamental level only	biol. implausible mechanism, but plausible output	biol. plausible at the fundamental level only
possible quantum implementation	indirect, similar core as Q.A.N.	direct	not yet known	indirect, partially direct
main strength	very applicative and developed	fundamentally quantum, "natural"	fits experiments on receptive field profiles	a general model with a concrete vision sub-model
main weakness	a mixture of natural and artificial features	limited to assoc. memory and pattern recognit.	algorithm's biol. implementation unknown	(consciousness still missing)

Note: With "fundamental" I mean "fundamental (bio)physical implementation" – without specific macroscopic biological characteristics (structural and functional), although compatible with them.

be quantum. The quantum associative net could realize it *if Gabor wavelets would be used as quantum associative-net's eigenstates* Ψ to be processed as in section 1.4.[38] If this is the case, then *the neural brain serves as an encoder and decoder into and out of the Gabor representation*, and such Gabor wave-packets are processed in quantum interference dynamics as in section 1.4. Detailed biological structures and processes (ICA-like?) which maximize *(de)coding* performance (as ICA does) remain to be researched. The *core* of information processing would be of quantum nature (and therefore ultimately conscious?).

ENDNOTES

1 Artificial neural net (ANN) models are overviewed in Haykin (1994) and Peretto (1992). Details see in AuxLit 10.

2 In simplified terms, holonomic refers to holography-like, spectral (additionally to space-time), but law-full (constrained with biological laws), and bio-dynamic (in contrast to invariant, unconstrained optical holograms). Spectral here refers to description in terms of distribution (spectrum) of wave characteristics, like frequency.

3 Here we must not forget that they do not have intention to present a self-standing brain model, but collaborate in this respect with neuroscientists.

4 With the term "junctions", all those interaction-sites are meant which can implement the associative (i.e., correlational / convolutional) processing as will be described later. These sites, collective carriers of the interference-pattern ("hologram"), could be various synapses (especially between dendrites), or any more microscopical or less materialized *mediators of interaction.*

5 Peri-membranous means near to or along the membrane.

6 Bose-(Einstein) condensates are quantum structures where bosons (quantum particles with integer spin) behave collectively as a coherent whole – they enter into a state where they all have the same quantum numbers (i.e., characteristic parameters describing a quantum state).

7 Striate cortex, primary visual cortex, V1 and area 17 (which must not be confused with the citoarchitectonic Brodmann area 17) are synonymous expressions for the earliest cortical vision center.

8 Let's begin with the most general definition; its variations will be precised later. A receptive field of a neuron is everything that influences its output *after* inputs from its environment have entered it, i.e. all transformations of all inputs along its own arborization (dendritic tree). With the term "receptive field", one might mean the whole surrounding space or network where stimuli effect the neuron from, or a mathematical function describing the effect of transforma-

tions upon neuron's inputs (the "weights" of inputs) before the axonal output is "calculated".

9 Fractals are self-similar structures emerging from recursive, collective dynamics, usually in complex systems.

10 Before giving a more precise mathematical definition later, i.e. in Equation (1.1), let us describe convolution $g*f$ as a sum of all inputs f, where each input f gets its own "weight" g (i.e., the "amplifying factor" for each particular input signal) (details in MacLennan, 1990).

11 When we will talk about activity of a *neuron*, we will mainly mean the activity of its body (*soma*) and its nerve impulses sent along its *axon*. (Processing in its dendrites will usually be considered separately.)

12 (Electric) polarization is the strength of the electric field between a group of positive charges and a group of negative charges. It may constrain the electromagnetic waves to oscillate in specific directions only.

13 Spines are small excrescences on dendrites.

14 Electric dipoles are structures, usually molecules, with two opposite endings carrying positive and negative electric charge. Magnetic dipoles (related term: spins) are small "magnetic needles" (which have two poles).

15 Hilbert space is the characteristic mathematical framework of quantum mechanics (e.g., Messiah, 1965). States in Hilbert space (here: images) can be described as a superposition, i.e. as a weighted sum of elementary so-called eigen-states (eigen-images).

16 In dendrites, a signal-summation process is performed (a sub-cellular one). It is similar to summation in neural bodies (i.e., somas) (an inter-cellular one) – "a fractal-like replica" of the summation process in somas.

17 Here we mean the level of neural bodies (somas) and signal-summation within somas (or, to be more precise, in axon hillocks).

18 Meant are dendritic signals before they reach the soma of the neuron.

19 By "envelopal" stability I mean that although there is a lot of hidden dynamics inside the process, e.g. as consciousness, also a stable overall aspect (i.e., the "envelope" – a term from wave analysis) of the process emerges.

20 As mentioned before, it is still open whether consciousness starts at the level of striate cortex (V1) or at some later extrastriate cortex level (V2 and above), if such localization attempts are relevant at all.

21 Central fovea coincides with the yellow spot in the (most sensitive) center of retina.

22 Spatial frequency is a measure of frequency or density of periodic events (e.g., lines, waves) across space (receptive field) in a given direction (Figure 2).

23 Lateral inhibition is a suppressing influence of neighbouring neurons onto a particular neuron.

24 Here "special" means that the Gabor elementary functions replace Mexican-hat functions or radial basis functions (RBF) which are more usual in perceptrons (a sort of hierarchical artificial neural net). Mexican-hat "filters" are used as models for receptive fields of retinal and LGN neurons, but Gabor "filters" are appropriate as models for receptive fields of cortical neurons. For some other possibilities, like Hermite, Fresnel, Fourier or Walsh transforms, see Aleksenko & Kirvelis *et al.* (1987).

25 Amplitude is the highest value or intensity (height) that a wave can reach. The word "amplitude" is in quotes, because in non-wave ANN there is computation using just "intensities of signals", or just some numerical values (integers or reals), respectively.

26 Wave phenomena are conveniently described by complex numbers: Real components are cosines and imaginary components are sines (Both function-types, sines and cosines, look like "waves".). After the Euler formula $exp\ (ix) = sin\ x + i\ cos\ x$ (where i is the square-root of –1), the exponent (here ix) incorporates the *phase* x. Phases are delays between peaks of waves. They are measured as angles (in degrees), due to a mathematical observation that the waving process can be geometrically projected to an analogous pendulum-like oscillating process. Waves are thus mathematically described by expressions like $A\ exp\ (ix)$, where A is the amplitude and x is the phase.

27 S. Marčelja (1980): *J. Opt. Soc. Amer.* **70**, 1297-1300.

28 I have presented these well-known second-order-statistics models, supported by own computer simulations, in several previous works (Peruš, 1995a,b, 1997d, 2002; Peruš & Ečimovič, 1998), therefore they will not be discussed here. Hebbian learning is briefly presented in Auxil. Section 6.8.

29 HNeT has been developed by AND Corporation, Toronto, Canada. (Now renamed to HQNeT.)

30 In this context, see also Prideaux (1996) and Psaltis *et al.* (1990).

31 This name will henceforth be used for *my model* which is here presented in detail in Auxil. Ch. 8 (and also 10).

32 PDP=parallel-distributed processing (like in networks where many interacting units process in parallel).

33 Some attempts in this direction, beside those mentioned in the previous paragraph, see in: Cory et al. (1997); Gershenfeld & Chuang (1997); Pessa & Vitiello (1999); Fatmi & Resconi (1988). NMR is nuclear magnetic resonance.

34 Similar views see in Chapline (1999), Cahill (2001) and Landauer (1991 – especially p. 29). Here are relevant citations from Chapline (1999). "Quantum mechanics can be regarded as a fundamental theory of distributed parallel information processing and pattern recognition." (p. 97) "It appears that our [Chapline's] holographic formulation of quantum mechanics may indeed be

pointing us in the direction of a fundamental structure of theoretical physics." (p. 103) "We are led to suggest that the fundamental connecting link between mathematics and theoretical physics is the pattern recognition capabilities of the human brain." (p. 104)

[35] Memory and associations could be realized also using the *quantum density-matrix* description (Auxil. Ch. 7, paragraph no. 7; Peruš, 1997a, and references within; Robertson, 1966; basic theory in Ballentine, 1970).

[36] The term (a not-up-to-date name, but still used in practice) is in quotes, because neurons do not really behave as feature detectors. There is no one-to-one correspondence between a stimulus feature and the firing of a specific neuron, or its spike train, respectively (Pribram, 1991).

[37] Personal communication.

[38] For quantum associative nets, as a general associative-information-processing model, it is not necessary that eigenstates are Gabor wavelets. But without Gabor wavelets, quantum associative nets would be biologically implausible; they would only be physical.

Chapter 2
Holonomic Brain Processes

2.1 GENERAL SCOPE OF THE HOLONOMIC BRAIN THEORY

Comparative introduction. Pribram's Holonomic Brain Theory[1] emphasizes that brain processes are parallel-distributed (as many contemporary computational models like Hopfield's, synergetic and other so-called associative neural nets mimic them[2]) as much as localized (as the main-stream neuroscientific experiments show). According to the holonomic theory, brain processes are multi-level and operate top-down (like Haken's synergetic computer model) as well as bottom-up (as the reductionist majority states). In contrast to conventional cognitive neuroscience, which advocated neurons and their networks, and in contrast to PDP neural-net models which are model-networks of "units" or "nodes" or "(formal) neurons" (and

DOI: 10.4018/978-1-61520-785-5.ch002

the true biological nature is usually not specified), the holonomic theory advocates parallel-distributed processing (PDP) in *dendritic networks*.

Polarization patterns. Thus, the connectionist models[3] are in general compatible with the holonomic theory under two essential re-interpretations. First, the "focus" of the multi-level brain model is transferred from neural cells to their dendritic trees and the "matrices of interaction-junctions" of these trees. *The activity of a "unit" or "formal neuron" does not correspond to a (biological) neuron any more*[4]*, but to the fluctuating electric polarization at a location within the dendritic crisscrossings.* Instead of a neuron firing or staying quiescent, we have depolarization (+) or hyperpolarization (–)[5] in parts of the dendrites, including their spines, especially in membranes and the surrounding (peri-membranous) space.

Phase-Hebbian memory-storage. Secondly, in addition to Hebbian computation based on a matrix of amplitude-correlations, the holonomic theory suggests that *phase*-encoding is essential. Thus, a combination or an essential extension of the memory-storage process, which I will call *phase-Hebbian*, is postulated to be much more effective.[6] This postulate of the holonomic theory has origin in the holographic analogy[7], because it is essentially connected with collective *oscillatory* phenomena.

Recently, many neuroscientists report relatively autonomous changes of complex *intrinsic* temporal dynamics, mainly involving membrane currents, inside a neuron with oscillatory activity (Marder *et al.*, 1996).[8] Intrinsic changes with memory effects may occur independently of, or concurrently with, changes in the synaptic transmission-rate, thus complementing them. Different synaptic strengths can cause opposite changes in an oscillator's response to a synaptic input at an oscillator's particular phase. Thus, modulation and control of neuronal oscillations might be very subtle. Marder *et al.* (1996, p. 13481) continue:

"Some neurons display intrinsic oscillatory properties involving periodic bursts of action potentials. Modification in the number of each channel type present in the membrane can change a variety of neuronal properties, including firing frequency and threshold, rate of spike repolarization, degree of postinhibitory rebound, burst amplitude and burst period."

Dendritic nets. What are dendritic networks in biological detail? Each neuron has tree-like dendrites. These branches of two or more neurons cross each other many times on very many sites, called junctions. They do not cross literally, of course, but they have contacts – synapses. Dendrites have membranes on their cylindric surfaces, and the charge-transfer processes on/through the membrane are naturally accompanied by electric fields.[9] So, dense bunches of dendrites (Hellwig, 2000) and their synaptic junctions are "swamped into" their own electric field. Relative differences of the electric field, called electric polarizations, constitute a pattern all-over the densely-connected webs of dendrites and other neuronal outgrows (dendritic spines, axons).

There are various dendro-dendritic and axo-dendritic (on dendrites or their small outgrows – spines)[10], axo-somatic, axo-axonic, etc., chemical synapses and electrical gap junctions (Damask & Swenberg, 1984). So, in contrast to relatively regular structure on the level of whole neurons, the more microscopic structure on the level of dendritic arborization (i.e., tree-like structure) is less fixed. This means that it is more "free", flexible, adaptable (Pribram, 1991).[11]

Thus, dendritic network is a name for a relatively *irregular, unconstrained and heterogeneous, multi-level connection-array* (modeled usually as a "matrix") of extreme density. It is multi-level in the sense that many levels are involved in a (roughly self-similar) fractal-like manner: neuronal parts, dendrites, dendritic parts like spines, synapses, and electric fields of all mentioned structures. These electric fields, producing a "bio-plasma", accompany mainly the membranes at all neuron's boundaries (their largest amount is around dendrites). Transfer of ionic charge through the membrane is of essential importance.[12]

Maletić-Savatić *et al.* (1999) found by a new fluorescence microscopy method that patterns of electrical activity may shape the developing neuronal morphology by promoting new dendritic extensions, called filopodia, using cytoskeletal mechanisms. These dendritic sub-branches, which might transform into spines when getting mature (Shepherd *et al.*, 1985), may form new synapses with axons they would meet. Altered dendritic length and growth of new dendritic branches participate in long-term memory-storage.[13] An associative or "Hebbian" (we might say "phase-Hebbian") memory-storage[14] could take place there, the authors write (see also Smith, 1999).

Presynaptic arrival-patterns of sinusoidally-fluctuating hyper- and de-polarizations, triggered by incoming nerve impulses, provide bases for dendritic microprocessing which can be approximated as linear. Neurons act as thresholding gates that spatially and temporally segment outputs of the dendritic microprocessing into discrete axonal impulses which serve for communication and control of neighboring processes, not for information procesing. Thus, the dendritic microprocesses seem to be central for "computing" cross-correlations that is essential for associative processing. Moreover, these dendritic processing-patterns are supposed, by the holonomic theory, to be correlates of awareness (sec. 4.8). The rate of successive change in awareness could be correlated with the duration of correlation-process (Pribram, 1971, 1991). Processing delays increase the richness and complexity of computation.

Slow potentials versus spikes. Pribram (1971) lists reasons[15] why many observed processes and behavior cannot be attributed to processing in soma and to axonal spike conduction, including transitions and balancing of ions, but to junctions and slow oscillating potentials in dendritic webs. Slow potential changes persist also if there are no nerve impulses sent. Anyway, intra-neuronal recordings and gross recordings from whole neuronal assemblies, performed simultaneously at the same location, show coinciding fluctuation patterns. This indicates that slow potential oscillations, which might sometimes exhibit wave-fronts, are (co)responsible for global electric fluctuations as measured by EEG (Pribram, 1971). So, EEG may show relatively large-scale brain processes also in the case of absence of axonal signals.

Initial layers of retina (and olfactory bulb networks) do not incorporate nerve impulses (Pribram, 1971). Many neurons do not have axons at all. Damask & Swenberg (1984) state that *probably 50% of neurons of the central nervous system are without axonal processes*. Furthermore, in those neurons that possess axons, the action potential propagates undiminished in amplitude. This indicates that no (significant) information processing could be involved in axonal process, and that it serves only for communication of result or output of processing in earlier stages (dendritic net, soma).

Damask & Swenberg (1984, vol. III, p. 88) write: "The *information content* of some types of neurons are represented by the state of *polarization* of their somas. By the state of polarization we mean both the spatial and temporal characteristics of the soma's transmembrane potential. If the soma's membrane is depolarized sufficiently, then this depolarization is converted at the axon hillock into an action potential, which then travels undiminished to the axon terminals.[16] […] The past history of the synapse, the state of depolarization of the terminal, and the intracellular and extracellular environment of the synapse influence the operation of synaptic transmission. In contrast to nerve impulses, which are subject to little variability, the transmission of impulses at synapses is a labile process."

Pribram concluded that digital nerve impulses and absolute stimulus properties are relatively *un*important for behavior. What is important (as Lashley already pointed out) are *ratios of excitations*, and here not only excitations of neuronal assemblies are meant, but more microscopic ones, i.e. those of dendritic electric fields.

Pribram (1971, p. 18) summarized: "Slow potential pattern 'computes' both the spatial neighborhood interactions among neural elements and, to some extent, the temporal interactions over a range of sites by a continuous (analogue) rather than discrete, all-or-none (digital) mechanism."

Graded slow potentials, performing analogue processing, seem to be more fundamental, more general, more permanently modifiable, more flexible and adaptable, and thus more suitable for information storage and processing than all-or-none axonal spike propagation, in spite of (the latter) being marvelously described by the Hodgkin-Huxley equations (Damask & Swenberg, 1984).

Isopotential abstract "gestalts". Making a cross-section of a dendritic net, we can draw isopotential contours on it: All locations on the net cross-section, which have the same electric polarization, when connected with lines, form a contour. Seen from above, it is loop-like. Such a family of isopotential contours forms a holoscape where the roughly-concentric contours concentrate around the potential minimum (Pribram, 1991; cf., Freeman, 1991). This is actually an alternative, "ground-plan" manifestation of the notion of *attractor* (Pribram, 1991; Hopfield, 1984). The (family of) "concentric" isopotential holoscape-contours (or attractor with its *basin of attraction*, respectively) is the secondary, algebraic isomorphism[17] of the perceptual process – in contrast to the gestalt-theorists' geometric isomorphism. Namely, the gestaltists thought that the electric isopotential contours correspond to the geometric shapes of object-images (the contour shape was thought to be determined by, and to be equal to, object's shape). Pribram's (1991, 1995) experiments, among others, rejected that, and supported another view that isopotential contours are the representations which *are* isomorphic to percepts, but in an "abstract"[18], non-geometric manner. (From now on, I will use the term "gestalt" in the holonomic sense: as an abstract, topologically-*in*correct *form*ation of a perceptual whole from the sensed elements.)

2.2 MODELER'S DISCUSSION

Re-interpretation of ANN. A dendritic net easily appears as a more random, flexible and microscopic *replica of a prototypic neural network of connectionists* — because such ANN are models focused on connectivity, the nature of "units" is unspecified. I believe, dendritic net can also realize almost all the computational characteristics which brought big applicative success to ANN models. These include, for example, the Hopfield model, based on the Hebb rule, c.f. equation (6.1),[19] and its attractor dynamics, content-addressable associative memory, pattern recognition and so on (e.g., Amari & Maginu, 1998). Thus, *we can keep all early successes of Hebbian associative ANN models* (as described in detail in: AuxLit 10; Peruš, 1995a; Peruš & Ečimovič, 1998) *and apply them, with plausible extensions (like phase-processing) allowed, to dendritic nets.* The result, to be presented later, is the phase-Hebbian memory-storage.

Deeper connections are less constrained. On the level of neurons, computational models[20] have problems with *uni-directionality of signal flow* (from soma along the axon and not back) in chemical synapses. Exceptions are bi-directional electrical synapses (which are more rare than chemical ones) (Damask & Swenberg, 1984).[21] On the subcellular level, *interactions go in both directions*. At any site of the dendritic network, inputs from "all" other sites can be received, and outputs to "all" other sites can be sent (details in Pribram, 1998b). In many computer simulations (Hopfield-based and many others), it turned out that relevant and controlable information processing is easier to get in symmetrically (i.e., bi-directionally) connected networks of "(neural) units". Processing symmetry might become greater, thus potentially increasing computational efficiency, as much as we go deeper into subcellular structure, especially into (quantum) electrodynamic background processing as considered by quantum field theory.

The "holographic" "one-to-all, and all-to-one" connectivity of the dendritic field is to be taken *literally* at the quantum level, at the latest. The multi-level brain system is becoming more and more interconnected as deeper we go into the microscopic levels. At the macroscopic (biological) level, the above-mentioned sites, which process, send and/or receive signals, are located on the synapto-dendritic *skeleton* of the dendritic field (therefore, "all" was in quotes). At the microscopic (quantum) level, however, those sites (connectionist units) can be at *any* point.[22] If they do not coincide with the synapto-dendritic skeleton, they can effect it indirectly – through polarization dynamics.

Multi-level view. Important facts about subcellular processing are now increasingly researched by main-stream neuroscientists also. Dendrites are not passive cables, but they contain voltage-dependent membrane conductances. Such calcium and potassium membrane-conductances can selectively amplify the input on the way from synapse toward soma. So, it seems now that dendrites are far more complex than just linear cables are. They perform linear and also complex non-linear operations (e.g., multiplication and amplification of signals, saturation, thresholding) used in information processing.[23] This was written in Koch (1997) where it is concluded:

"And what about memory? It is everywhere (but can't be randomly accessed). It resides in the concentration of free calcium in dendrites and the cell body; in the presynaptic terminal; in the density and exact voltage-dependency of the various ionic conductances; and in the density and configuration of specific proteins in the postsynaptic terminals." (Koch, 1997, p. 210)

This list of memory-bearing sites, states and processes is surely not complete, I believe. Very probably, memorized patterns–attractors are not only distributed across the net, but are also distributed over (rooted in) many "levels" or "scales" – from neuronal activites, over dendritic fields, down to quantum levels. Furthermore, short-term synaptic changes involve alterations in existing proteins (without macromolecular synthesis of various sorts), but synthesis of RNA and new proteins and expressions of genes are necessary for long-term changes (Goldman-Rakić, 1996, after Bailey *et al.*, same proc.). *What is important for a memory-trace is not so much its material implementation, a (sub)neuronal correlate, but the fact that a percept isomorphically triggers a homeostatic reorganization of the PDP system in such a way that it newly stabilizes in a specific, characteristic state, called eigenstate, which acts as an attractor of system's dynamics.* Peruš (1995a) preferred to say that attractors are thus "beyond matter" (cf., Haken, 1996) and manifest "virtually".[24]

Nevertheless, attractors have material roots, and the holonomic theory emphasizes that the *"bulk"* of those "roots" are in dendritic polarization-field patterns, not in neuronal signals triggered in neuronal somas or, more precisely, in axon hillocks. Namely, the dendritic field dynamics, as will be shown later, enables more efficient and flexible processing, although with significant interventions of "big brothers" – the (somas of) neurons (Each of them "acts like a bee-queen among numerous bees".). Some neurons are without axons and do not send nerve impulses, but merely effect the polarizations.

The field neurodynamics is, because of collective oscillatory phenomena like phase-coupling[25], significantly more similar to holography than to conventional neural nets. However, these fields, mainly electric, are rooted in even more microscopic quantum fields where we can directly speak about quantum holography. Holography (AuxLit 12), based on interference, can be directly derived from quantum interference and superposition phenomena which are the very core of quantum nature (Bohm, 1954; Hiley & Peat, 1987; AuxLit 16; details in Auxil. Chapters 7 and 8).

2.3 GLOBAL VIEW: FOURIER AND GABOR TRANSFORMS; "INFOMAX ATTRACTORS"

Fourier decomposition. The *Fourier analysis* is of essential importance in physics in general, and in wave dynamics and holography in particular (AuxLit 12). The Fourier theorem says that any mathematical function / physical state / informational pattern can be considered as a so-called linear combination or superposition of basic *periodic oscillatory* functions / eigen-states / eigen-patterns.[26] The functions describing these basic wave phenomena are *sines* and *cosines*. If they are combined having all possible multiples of the basic frequency ω, and various amplitudes (i.e., maximal oscillation rates) a and b, or d, then such a superposition can approximate (i.e., effectively replace) any function / state / pattern f:

$$f(t) = \sum_{n=0}^{\infty} (a_n \cos(n\omega t) + b_n \sin(n\omega t)) = \sum_{n=-\infty}^{\infty} d_n e^{in\omega t} . \quad (2.1)$$

Here, t refers to time. However, this Fourier analysis, which can be considered as a decomposition of a wave-conglomerate into fundamental waves, is valid also if we replace t (time) by x (position), and ω (frequency, which is in quantum theory proportional to energy E) by k (wave-vector, which is in quantum theory proportional to momentum p).[27] We thus have "yin-yang-like" Fourier dualities or dynamic complementarities: space-time (x-t) versus energy-momentum (E-p).[28] In the holonomic theory, this is considered as the *configural* versus *spectral* duality.

Fourier pairs (x,p) and (E,t) coincide with the Heisenberg uncertainty: x and p cannot be simultaneously quantum-measured. We have similar situation with the

(E,t)-pair and the (angular momentum, (phase)-angle)-pair, and their derivations. Fourier pairs like (x,ω), which are derived from the previously-mentioned pairs,[29] are essential actors in Fourier transforms. Function of the (pair's) first variable, e.g. $f(x)$, can be *Fourier transformed* to a function of the (pair's) second variable, e.g. $F(\omega)$:

$$f(x) = \frac{1}{2\pi}\int_{\omega=-\infty}^{\infty} F(\omega)e^{i\omega x}d\omega\,, \tag{2.2}$$

and *inverse Fourier transformed*:

$$F(\omega) = \int_{x=-\infty}^{\infty} f(x)e^{-i\omega x}dx\,. \tag{2.3}$$

These were spectrum-to-space Fourier transform and its inverse.[30] The spectrum-to-spacetime Fourier transform and inverse Fourier transform, look as follows:

$$f(x,t) = \frac{1}{(2\pi)^2}\int_{k=-\infty}^{\infty}\int_{\omega=-\infty}^{\infty} F(k,\omega)e^{i(kx+\omega t)}dkd\omega\,, \tag{2.4}$$

$$F(k,\omega) = \int_{x=-\infty}^{\infty}\int_{t=-\infty}^{\infty} f(x,t)e^{-i(kx+\omega t)}dxdt\,. \tag{2.5}$$

In the case of image analysis, $f(x,t)$ can represent a static image if $t=0$, else the image is changing, and x is considered to be a 2-dimensional vector-variable. $F(k,\omega)$ can then represent the Fourier-decomposition of the image (static if the temporal frequency $\omega=0$, else dynamic) if k is considered to be a 2-dimensional vector-variable denoting spatial frequency with components (k_x, k_y).

Holonomic theory: beyond Fourier. In holography, information is encoded *either* in space-time (configural aspect) *or* in Fourier coefficients, hidden in $F(k,\omega)$ (spectral aspect). In the holonomic theory, a "synthesis of the Fourier pairs" is incorporated by using the Gabor wavelets. This synthesis enables embedding of *both* space-time *and* spectrum into a joint frame as its two *orthogonal* subspaces.[31] This joint *spacetime–spectrum* "phase-space" unites both dimensions — in contrast to Fourier dual-spaces which "alternate" from one to another aspect, but never get conjoined. In vision, space-time dynamics constrain mainly spectral computation (Pribram, 1991). Comparing with ANN, the duality of neurons and connections is roughly similar: Neuronal activities, when first getting input, encode space-time aspects of patterns; on the other hand, connections or weights encode spectral aspects, like holograms do.[32]

As with Heisenberg's amount of uncertainty in measurements of quantum systems, in brain's visual apparatus with Gabor wavelets, a similar minimal uncertainty, called Gabor uncertainty, determines maximum information-content of the "synthetic space" of spacetime–spectrum. Here, a generalization of Shannon's definition of information as equivalent to (neg)entropy or to reduction of uncertainty is made.[33] The holonomic theory considers entropy as potential information. The basic information unit (a sort of *bit*) called *logon*, or "quantum of information", is the minimum element of the spacetime–spectrum which can still encode information. It is based on a "*sine* trade-off" in the spacetime–spectrum chart (Pribram, 1991, Figure 2) ensuring minimal uncertainty.

Shannon's definition is more appropriate for cognitive aspects, but Gabor's or Pribram's definition fits better the configural aspects of information.[34] Spectral as well as spatial aspects, e.g. colors and tones as well as shapes and forms, can be processed by Fourier decomposition and Fourier transforms with Gaborian optimisation constraints. Roughly, colors are encoded as oscillations, and forms as modulations of oscillations.

Information maximization. Logons are Gabor wavelets or Gabor elementary functions, as in equation (1.2).[35] They are composed by the processing of dendritic trees. Each logon or Gabor wavelet is composed of several receptive fields.

In the Hopfield associative neural net, the so-called energy-function of the system is minimized (as described in: Peruš, 1995a; Peruš & Ečimovič, 1998; orig. Hopfield, 1982; Amit, 1989; Geszti, 1990). In contrast, in a Gabor-filtered net, *entropy of the system is minimized, or information is maximized.* So, instead of attractors as minima of the energy-function (Hamiltonian) as in Hopfield nets, Gabor-filtered nets exhibit *attractors as minima of entropy* which are equivalent to maxima of information (Pribram, 1991). This choice is supported by evidence that processing efficiency depends mainly on band-width, not on energy, and that the resolving power of the visual system is dependent on spatial-frequency resolution, not on the amplitude modulation of the system. Image-processing efficiency is thus determined by the spectral resolution of overlapping Gabor wavelets based on band-width, not on the average amplitudes.[36]

The information maximization principle is incorporated in many so-called *infomax* models, in the sense of maximum information preservation or maximum information content (e.g., Linsker, 1988), as well as in ICA (ch. 3). In many models, the goal is to maximize the mutual information between network's input and output, or equivalently, minimizing information loss during network processing (Intrator & Cooper in Arbib, 1995).

Energy versus entropy minima/attractors. Hopfield's energy-function minimization and the holonomic entropy minimization, or information maximization, are both special cases of the least-action principle, a fundamental law of physics.

It generally leads to formation of relatively stable states, called eigenstates (i.e., "natural states" of the system), which act as attractors (the system stabilizes into them). The holonomic option ensures the so-called *dissipative*[37]*processing* far from thermodynamic equilibrium (which is characteristic for many complex processes, especially in living matter) more explicitly than the Hopfield model. In the Hopfield model, the energy-function is a theoretical construct which is not entirely equivalent to the physical energy. Energy is related to force-based influences (where intensity matters) rather than subtle, informational ones (where the content or "form", a sort of order, of a system's state matters).[38] The holonomic theory finds the "subtle" option more appropriate, because relationships, coded by phases, are more important than intensities of each state alone (the latter only are taken into account in the Hopfield model).

My remarks. In biophysical systems, another quantity, the *free energy* F, is minimized. It is a "compromise" between energy E and entropy S in trade-off with temperature (stochastic dynamics) T, as given by formula: $F = E - T\,S$. Energy minimization ensures more permanent stability. But for efficient information processing, a *dynamic* informational trade-off is needed — therefore, entropy, as *potential* information, is minimized to make some selected information (i.e., the *actual*, currently relevant information) more explicit. The stochastic (sub)neuronal dynamics, hidden in T, which stimulates such trade-off dynamics and constantly refreshes extraction of information, have been researched and found useful in simulations, but not enough exhaustively *in vivo*. Therefore, the free-energy minimization has not (yet) taken place in brain-processing models, due to unknown effects of the TS-term.

I must unfortunately express an opinion that transfers of physical quantities (in particular: entropy, the most chameleonic one, also in pure physics) to theories of living matter, and especially to information processing theories, are all quite equivocal, paradoxical and without general agreement. This situation is not by chance, since it arises from the fundamental Fourier dualities or complementarities in process. It potentially hides interesting, uncovered processing capabilities. Therefore anyway, several attempts, including the holonomic theory and ICA, were productive (Is this just the beginning?).

Significant processes accompanying perception. As long as the response-triggering stimulus remains unaltered, the neural response decreases due to *inhibitory mechanisms* and the organism *habituates*. However, encoding of the stimulus has been sharpened, and allows its optimal reconstruction in enhanced stimulus–background contrast. Also, any change in the stimulus alerts the organism again to be sensitive to novelty (Pribram, 1971; Luria, 1973).

Current input is compared with the memory representation of previous inputs, and an updated response is made after the representation is corrected proportionally to the stimulus–memory difference (i.e., *novelty* of the new input). This has been well reproduced in numerous computer simulations (e.g., Kohonen, 1984; Haykin, 1994; cf., Appendix B), including author's.[39] Pribram's (1971) idea that an array of interneurons, which connect input and output neurons, is responsible for habituation is similar to these simulations made later. He suggested that habituation is *not* a consequence of the junctional slow-potential changes. Sokolov suggested that habituation arises in the nerve-impulse generator-mechanism.

In vision, perceptual *adaptation* (not to be confused with habituation) parallels the image stabilization. The perceptual apparatus exactly corrects the retinal signals in order to, in spite of eye and head moves, finally keep the image still. For this, feedbacks and top-down control are needed, involving not only higher cortical regions, but also motor centers in an essential way. When adaptation is achieved, the just-stabilized image becomes unobservable.

A *test-operate-test-exit* (TOTE) *servomechanism* (Pribram, 1971)[40], incorporating feedforward-feedback-loop with a comparator, is proposed as a model of the perceptuo-motor complex. The *Test*-comparison should be performed by the interefference pattern of junctional potentials; and on the other hand, *Operate* should be realized by nerve impulses generated by receptors or by the central nervous system.

Low-level processing is perception; high-level processing is cognition. In general, top-down influences, based on prior experience, shape the processing by structural constraining (Pribram, 1991). Top-down action is mediated by midbrain and thalamic structures, and operates on primary sensory cortices. If we focus on vision now: In the primary visual (or striate) cortex (V1) there are neurons which project to the extrastriate cortex and at the same time receive feedback from it, and are strongly connected to each other. Such a net is modeled in Baird (1990); similar nets are those by Olshausen & Field (1996a,b; 1997), which is an ICA alternative, and by Harpur & Prager (1996) — see Auxil. Chapter 11.

The latest experiment. Pribram assumes that the perceptual encoding into Gabor wavelets is quite ubiquitous in cortex. His latest experiment in a series of experiments trying to prove the Gabor encoding is done on the whisker-field, which is relatively similar to retina (Pribram, 1998b), to make a comparative study of vision and somasensation. Beside substantial evidence from Pribram's group and other authors on Gabor receptive fields, especially in vision[41], the following experiment revealed the spectral encoding, primacy of slow wave potentials and, again, agreement of Gabor functions with measurements. Rat whiskers were stimulated and responses in somatosensory cortex (which was shown to be similar enough to V1 for present purposes) were displayed in diagrams with the following axes: x – spatial frequency, y – temporal frequency of stimulation[42]; z – number of resulting bursts (spiking intensity – proportional to the field amplitude). These receptive-fields fit well with those obtained from simulations using the Gabor wavelets (Pribram, 1995, figs. 4, 6; Pribram, 1998b, figs. 1-3). Thus, Gabor-based holonomic theory was again in agreement with measurements which, in this experiment, showed also that slow wave potentials precede the corresponding spikes (Pribram, 1995, figs. 1-3). This indicates that the dendritic slow wave potentials (from the "deep structure", which manifests in PDP) are more fundamental than spikes from the soma along the axon (the "surface structure", which is locally manifested). The *slow electrical potentials seem even to generate spikes*, not vice versa. (Cf.: King *et al.*, 2000)

Actually, the relevance of Gabor wavelets (especially in vision – the next section) has already become widely accepted, but the primacy of slow wave potentials is still not a generally-accepted view, although the importance (but not primacy) of dendritic PDP is acknowledged.

2.4 HOLONOMIC THEORY OF VISION

Holonomic theory on optical and retinal processing. Retinal receptors are sensitive to individual quanta of light (i.e., minimal wave-packets called photons), therefore they are also sensitive to their phase, not only magnitude.[43] The receptive field of a retinal ganglion cell is obtained by measuring response to a small light-spot flashed (shaping any virtual figure, or moving at any velocity at any orientation) in various parts of the visual field in front of the eye. The shape of the receptive field (Figure 1-left) is then found to be a convolution of the first derivative (i.e., rate of change) of the shape of the optical stimulus-configuration (Pribram, 1971).[44] (Similarly for brain neurons, the receptive field is obtained by measuring input-signals from precursive neurons, not from light.)

Figure 1. Left: The graph of the receptive-field profile, called "Mexican hat". Its height at a point gives the number of ganglion-cell's firings resulting from light-flashes at that point. Right: 2-dimensional Gabor wavelet (cosine component). (Courtesy: MacLennan, 1991, fig. 14)

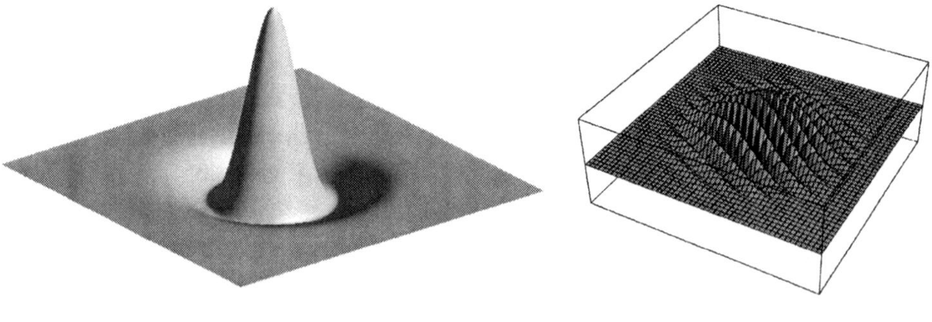

A Fourier transform exactly describes the optical "image" at the back focal plane of the eye (Pribram, 1991, after C.A. Taylor). The optical apparatus, especially the lens, adjusts the phases to get the optical image from many reflected and diffracted waves. Spatial and temporal *relations* among magnitudes, in each infinitesimal point, of the electromagnetic field transmitting radiant energy, or the spatial and temporal derivatives of the resulting luminance, respectively, are then started to be processed in retina. The holonomic theory emphasizes that the perceptual process is relational and transformational in a holistic (holography-like) sense[45]; it is neither a construction from elements nor direct in J. Gibson's sense (although he was right, from Pribram's viewpoint, that visual form perception initially is 3-dimensional rather than a conglomerate of 1- and 2-dimensional components).

Visual processing can be globally modeled as a series of transformations from the spatial (localized) to holographic (spectral, distributed) representations and vice-versa, i.e. as a series of Fourier transforms and its inverses (Figure 2). Concretely: ambiental *light* (distributed, i.e. hologram-like)[46] scatters on an *object* (localized in space-time) towards the eye's *lens* (hologram-like diffraction pattern appears on it) which collects the rays on the *retina* (localized image of the object appears on it). Then, the image is encoded into the spectral (i.e. spatial-frequency-based) repre-

Figure 2. The serial encoding / decoding, or enfolding / unfolding, or Fourier / inverse-Fourier transform process exchanging space-time and quasi-holographic representations. Note that the lines depict light-rays, following the laws of geometrical optics, only until retina. The lines through the retinal layers (P–H–A–G) and further-on depict neural information-flow. Where the lines are converging to points on a layer (that's "decoding"!), the representation is localized. Where the lines are diverging from a point to spread over a layer (vertical "screen") (that's spectral "encoding"!), the representation is hologram-like distributed. Li = light; O = object; SL = scattered light; R = retina, which includes: P = photoreceptor layer, H = horizontal-cell layer, A = amacrine-cell layer, G = ganglion-cell layer; ON = optic nerve; OT = optic tract; OC = optic chiasm; LGN = lateral geniculate nucleus layers; OR = optic radiation. Areas V1–V5 (and beyond) each has several layers.

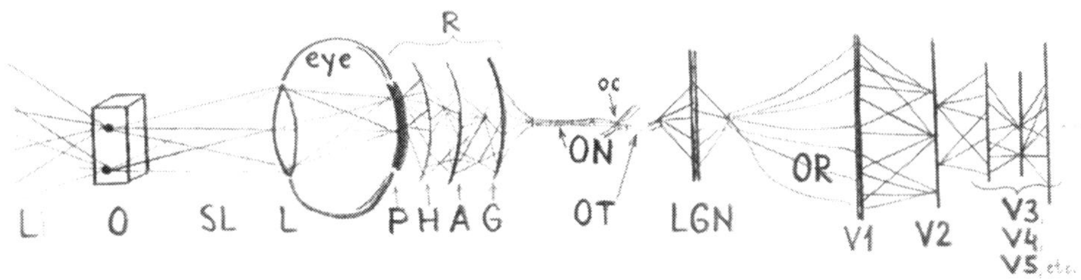

sentation (hologram-like), processed through LGN and V1, [47] and is reconstructed in the space-time representation in V2 (localized). This sequence of "enfoldings" and "unfoldings" is "interrupted" by propagation along the optic nerve (between retina and LGN) and by the so-called optic radiation (between LGN and V1); but then the representation-exchange sequence probably proceeds, at least in a similar (e.g., convolutional) way, also beyond V2 (cf., Tootell *et al.*, 1998).[48]

Enfolding/unfolding. The light (i.e., the visible part of electro-magnetic radiation) reflects from each point on the object's surface to all directions (*scattering*). The information about the object is thus distributed to the entire surrounding space where light-waves propagate. This is *enfolding*, acting also as encoding. The structure, hidden in the light-hologram, i.e. in an "optic array" distributed across the whole space surrounding the eye, is then *unfolded* (decoded) by the pupil and lens. These optic transformers in the eye thus perform an inverse transform or decoding (inverse relative to the scattering, encoding transformation) which extracts or explicates (by collecting the rays) the information about the object into a space(-time)-localized image of it. Such (at least)-3-dimensional optical images are inputs to the retinal neural net. Similar enfolding-unfolding (encoding-decoding) transformations, i.e. transforms from space-time to spectral representation (hologram) and back to space-time, are sequentially repeated along the visual pathways deep into the cortex.

This is analogous to a multi-layer ANN (Peretto, 1992; Haykin, 1994; AuxLit 10).

Optical hologram. *Each* point on a screen receives reflected waves or rays from *each* point of the object. The pattern that emerges on the screen is the hologram (Denisyuk, 1984; Hariharan, 1996; AuxLit 12). It is determined by interference of all the numerous incident light-rays. *Every* part of the hologram contains the *whole* information about *every* part of the source of the reflected light (object).

On the retina, because the lens is located in-between, the pattern is not the hologram, but the object-image *reconstructed* from the hologram. We may put the lens to *any* location in-between the object and retina, the lens will collect the rays, "make a synthesis of their information" and *completely reconstruct* object's image on the retina. The lens will explicate or unfold object's image in space(-time) from the hologram which is a distributed pattern of interfering (summed) scattered light-rays. This is because *any* part of the similarly-illuminated screen (hologram) contains the *complete* information about *all* parts of the object.

Beyond optics. In general, hologram is a record of an interference pattern made of intersections of two (light, sound, quantum) waves. Optical holography has two phases. In phase one, storage, an object is illuminated by coherent light directly and indirectly (from another angle or over a mirror to achieve time-delay), and these two waves interfere at the hologram-plate leaving a trace on it. In phase two, reconstruction, the hologram-array is illuminated again with the same coherent light-wave as used in phase one. The hologram-plate acts as a filter for this light-wave, producing jointly a veridical 3-dimensional image of the object as if it was stored in the hologram (plate). (For many variants see: Denisyuk, 1984; Hariharan, 1996; AuxLit 12.)

My general observations. A hologram is a perfect manifestation of Nature's complete and fundamental holism in the sense of a distributed, non-local, enfolded encoding of all environmental influences which may be of electro-magnetic (optic), acoustic, quantum or other kind. The hologram is usually implemented on a piece of matter which is, internally, always a web of interactions – similar to Hopfield-like Hebbian ANN models (Hopfield, 1982; Amit, 1989; Peruš, 1995a, 2000c, 2001) or, even more, to phase-Hebbian network models. In the latter models exhibiting a sort of interference of wave-patterns, like in equation (1.4), Hopfield-like associative ANN, holography and quantum dynamics are basically the same thing (Peruš, 1997c)!

Any physical system which can correlate or convolve pairs of patterns (i.e., which exhibits *interference* of pattern-bearing waves) *can imitate holography*, and consequently also the associative information processing and content-addressable memory.[49] Fourier transforms (mimicked by tree-like perceptrons in ANN) and holograms (mimicked by Hebbian or, better, by phase-Hebbian memory matrices) are thus *fundamental in physics*.[50] In contrast to discrete neural nets, a hologram is continuous or quasi-discrete: its points are not fixed to a location of its medium. This is just like in the case of a water-wave: it is not fixed to particular water-molecules. The wave is a sort of physical gestalt which uses many different masses of molecules as it travels through the medium (water, or microscopic brain tissue).

Holonomic theory on image processing. The optic enfolding (scattering of rays, later sent through the lens) and the optic unfolding (focusing by the lens) are describable by the Fourier transform and its inverse. In the first paragraph of this section, we have already continued the description of the series of enfoldings and unfoldings (Figure 2). The mathematical details of such a series of convolutions, which is the core of holonomic image processing, have already been presented in sec. 1.2. Some other models[51] also, at least roughly, provide such a description.

The retino-striate convolution-process, as described by eq. (1.1) after eq. (1.2) had been inserted, provides a spectral representation of image(s), available in V1. As mentioned in sec. 1.2, an individual image is reconstructed from the V1 spectral representation Gf by an integration over V1-columns. This integration over a "manifold" M (array) of striate cortical columns realizes an inverse Gabor transform, which is inverse of the Gabor transform Gf of eq. (1.1). It reproduces an upside-down image $f(-p)$ in V2 (Pribram & Carlton, 1986):[52]

$$\int_{M} Gf(p,u)du = \int_{M}\int_{retina} f(x)g(p,u;x)dx = \int_{M}\int_{retina} f(x)e^{-(x-p)^2}e^{iu(x+p)}dxdu =$$
$$= \int_{retina} f(x)e^{-(x-p)^2}\left(\int_{M} e^{iu(x+p)}du\right)dx = \int_{retina} f(x)e^{-(x-p)^2}[2\pi\delta(x+p)]dx = f(-p)e^{-4p^2} \tag{2.6}$$

$\delta(x+p)$ is the Dirac delta-function (the result of $\int_M$) which effectively prescribes: $x = -p$. The image-representation is now ready in Hilbert space to be processed in associative networks, probably (also) quantum ones.

Spectral selectivity. Better than talking about neurons as feature-detectors, is talking about neuronal receptive fields being selective for spectral characteristics, e.g. to collections of parallel stripes with different stripe-densities and widths[53] (not to individual lines). Such a stripe-density is equivalent to spatial frequency of the patterned stimulus. Pribram (1991, pp. 75-78) reviews substantial evidence[54] that *V1 simple cells are tuned to limited band-widths of spatial frequency*, rather than to the intensity of the stimulus. Pribram (1991, p. 81) concludes that the activity of a V1 neuron is determined by "its resonant frequency modulated by the space-time constraints of orientation selectivity". Neuron's spike-sequences, modulated by its dendritic net, encode both the spatial frequencies and orientations of stimulus-patterns (in experiments: gratings, "chess-like boards").

"Holographic" brain processes. Pribram (1971, pp. 119-124 & 159-163) describes many experiments (own and by others)[55] proving that the visual system, especially cortical, processes in a holography-like way (i.e., distributing, then filtering) rather than in a photography-like way (i.e., isomorphic projection – in a microscopic sense here, not conceptual). Large parts of visual mechanisms (80-98%) can be removed[56] without disturbing the visual function (cf., Lewin, 1980)! Only fractal hologram-like structures have such amazing regenerative capabilities.[57]

However, there is no global Fourier transform which would distribute the retinal signals over the V1-sheet (Pribram, 1991, after Caelli & Julesz). Nevertheless, there is much evidence that neurons in V1 change their stimulus-responses depending on the *global* stimulus-structure (Kovacs, 1996).

Dendritic "holograms". The receptive fields of sensory cortex are determined by interacting polarizations in junctions of dendritic networks, as described in sec. 2.1. Pribram (1991, p. 85) describes:

"[Dendritic outgrowths called] spines have bulbs at their endings, knob-like heads that make contact with branches of axons and other dendrites to form synapses. Activity in axons and in other dendrites such as those stemming from reciprocal synapses produce depolarizations and hyperpolarizations in the dendritic spines. The postsynaptic effects are ordinarily invoked by chemical transmitters whose action is modified by other chemicals that act as regulators and modulators.

Shepherd, Rall, Perkel, and their colleagues [...] modeled the process whereby these postsynaptic events occurring in spine heads interact. [...] The stalks of the spines are narrow and therefore impose a high resistance to conduction (active or passive) toward the dendritic branch. Spine head depolarizations (as well as hyperpolarizations) must therefore interact with one another if they are to influence the action potentials generated at the axon hillock of the parent cell of the dendrite. The interactions (dromic and antidromic[58]) among dendritic potentials (by means of which the signal becomes effective at the next stage of processing) thus depend on the simultaneous activation of both pre- and postsynaptic sites."

The last sentence refers to a polarization-based *Hebb-like plasticity in the synapto-dendritic net*, or its field, respectively. It appears that a spine can spread its polarization to neighboring spines (cf., Shepherd *et al.*, 1985) and be used for collective storage and processing of information, whereby timing is important.

Dendritic processing reflected in spike statistics. Berger *et al.* (1990) and Berger & Pribram (1992) reported about their analysis of the probability density function of inter-spike intervals recorded from visual cortical neurons. It was studied how the neuronal output-code in the axonal spike-sequence is effected by specific input-conditions. In simplified terms: After a period of spiking, the neuron is reset and begins a repolarization "drift" (with a "drift rate") toward the threshold-voltage ("barrier height") where the extreme depolarization triggers a spike. Dendritic inputs were found to effect these parameters in the following way. The first parameter, drift-rate (which reflects the voltage-change due to input-currents and membrane-conductance), depends on spatial frequency of the stimulus (moving bars). The second parameter, barrier-height (which depends on the extremes of hyperpolarization and depolarization), is determined by the orientation of the stimulus. A further analysis related the two parameters to the functional characteristics of dendritic fields as described by Gabor filters of dendritic signals to a neuron's soma. Pribram (1991, p. 68) writes in this context:

"Neurophysiologically, the value of polarizations within such an ensemble of receptive fields is represented by these [Fourier] coefficients whose weighting is determined by the spatial frequency of the displayed patterns. These coefficients are cross-multiplied by probabilities which are represented by amplitudes, boundary conditions [are] determined by the orientation of the display. The result is, of course, the probability amplitude modulated set of Fourier coefficients, which describe a class of four-dimensional informational hyperspaces such as Hermite polynomials constrained by Gaussians, of which the Gabor function is an elementary example." (Cf., sec. 1.2).

The above experiment of Berger, Pribram, Wild & Bridges (1990) allows them to "conceive neuron as a processor of stochastic dendritic events which display its computed output as the statistics of the sequence of inter-spike intervals." (p. 133)

Images, objects, categories. Pribram (1991) emphasizes that features like edges *are already* gestalts, not ingredients from which gestalts emerge. Images, objects and categories are gestalts of "the second, third, fourth orders". They have to be distinguished, also because they are processed by separate brain systems. *Images* are forms that change during relative movement between oculocentric[59] and egocentric frames, but *objects* are forms that preserve their identity during movements and changes of view-points (cf., Auxil. Sec. 6.7). *Categories* are a result of top-down-constrained grouping of objects, focusing on prototypical objects. Such definitions of distinctions between the three notions already emphasize the essential *importance of movement* in perception.

Movements essential for perception. Pribram (1991) emphasizes the essential importance of rapid eye movements (REM), like saccades and nystagmus[60], for visual perception. Movements are not only providing images of objects from various points of view, which are necessary for object perception, but more. Oscillatory micrometer-movements are hypothesized to perform an *inverse Gabor-transform of the micro-processes*, which are processing the image-patches or pixels, *into the macro-image*, when the environment is scanned. The inversion is, as Pribram assumes, determined by corticofugal circuits that act from V1 back to LGN and even retina.[61] Their role should also be to enhance relevant and suppress irrelevant features of images, and to complete them using associations and imagination.

For object perception, and corresponding gestalt-formation aided by the motorics of the eyes (as well as head and body), the peristriate area V2, together with the prestriate cortex (V3 and others), are probably responsible.[62] The retinal image is *re*constructed there by the Gabor-transformational inversion. There are thus (at least) three Gabor transforms along the geniculo-striate pathway: first, optical by lens and pupil; second, neural by circuits of retina and LGN and V1; third, neural by V2 with V1 (Pribram, 1991). Such transforms take place also in V4, V5 and ITC.

Pribram on object perception. This book focuses on image processing, therefore merely some essential points of the holonomic theory of object perception and higher vision will be mentioned. Pribram (1991) stresses that *object-constancy is achieved through movement by iteratively impressing on V1 sequences of images which have to match* if their belonging to the same object should be recognized. Thus, in addition to early-vision theories of Auxil. Sec. 6.7., the holonomic theory emphasizes that the object is perceived as the constancy in the superposition of images, sequentially offered by movement. (Here Pribram follows J. Gibson, among others, but stresses also the memory-based, cognitive, top-down influences, as manifested with optical illusions.) Movement also gives the cues for binding of features that belong together, and introduces differences which help in segregation of these features (belonging to an object) from the background. To constitute an object, points must first be grouped together by proximity; and for a moving object, points must also be grouped together by common motion.

Relation to attractors. Object perception as movement-stimulated gestalt-formation is understandable also from the position of attractor theory (Peruš, 1995a, 2000a, 2001): Movement-sampling of different viewpoints of an object, which are always correlated, as needed for correlation-based memory-storage (like Hebbian and phase-Hebbian), "digs-out" the potential-minima (minimal informational disagreement) accompanying the attractors of system's dynamics. Variations of correlated information, as ensured by movement, for example, are ideal for formation of attractors corresponding to gestalts.

Use of groups. Pribram (1991, p. 108) writes: "[...] Object constancy is achieved by calculations largely in the spectral domain rather than in space-time: centers of symmetry [of the Gabor transformed optical image] rather than centers of gravity; Fourier descriptors (tangent vectors) of contours rather than boundaries; the extraction of invariances by superposition to form groups among transformed patterns. [...] By way of the inverse [Gabor] transformation, these calculations result in cross correlation among ratios of shadings of surfaces so that invariances are extracted." (In [...] are additions by M.P.)

Since the Lie-group theory, as described in Auxil. Sec. 6.7, is limited to space-time, the holonomic theory takes it into account in a generalized way involving Gabor wavelets. The theory of mathematical groups brings the theory of perception close to quantum theory: Like particles are described in the Hilbert space, using the Lie groups, as invariant quantum states (eigenstates) composed of quantum (Gabor-like) wave-packets, objects are described in the Hilbert space, using the Lie groups, as invariant states (eigenstates), composed of Gabor wavelets, of the perceptual networks.

Categories. A category is a composition of relationships among object-forms in a similar way an object-form is a composition of images. The "central, average" object becomes the *prototype* of the category. ITC is important for (visual) category-related *comprehension*. ITC executes the first step in prototype-formation – it chooses the appropriate category. Memory has increasingly significant influence here, in particular as (general) context, including novelty[63], contributed by the hippocampus. ITC is also involved in experience-dependent processing of *global* contour-forms (cf.: Gochin, 1996), based on V1-processing of local contour-elements like edges.

I can say that a prototype (of a category) acts as an attractor (category-gestalt) of attractors (object-gestalts). This indicates a hierarchy which is virtual (emerging from system's collective dynamics) as well as physiological or anatomical (ANN model in Peruš, 1995a). Hebbian or phase-Hebbian memory-storage (see equation

(6.1)) is able to model, I believe, the associative comparisons (correlation searches) between objects of the same type that, by all being "averaged out", form the prototype.

Beyond vision; its application. The holonomic theory extends "holographic" ideas to other brain processes, like motor, auditory and somatosensory[64], as well as to higher associative processes, and beyond. To repeat, holography-like processes manifest on the level of slow wave potentials at dendritic junctions exhibiting isopotential-contours that concentrically encircle the potential-minima which act as attractors, as described in sec. 2.1. Let us just "telegraphically" review some essential points from Pribram (1971, 1991):

Motorics. Visual data are applied for movement. In fact, there is essential reciprocal sensory-motor interdependence. The neural regulation of behavior is determined by the regulation of receptor processes, and by "holographic" images-of-achievement or plans, rather than by direct control of muscle contraction. Motoric action is conceived as a rough inverse of perceptual process, and the motor cortex as a "sensory" cortex for action. Movements (where acting forces are involved) rather than muscles or their contractions are represented in the Rolandic motor cortex. Cerebellar cortex, for example, incorporates Hebbian comparison of inputs with central control states.

Emotion. Internal adjustments of the psycho-neural system, following environmental challenges and their internal consequences like feelings, appetites, affects, interests, motivations, are felt as emotions. Familiarization and evaluation, with cooperation of amygdala and related reinforcement processes, also have an attractor-dynamic background which the holonomic theory locates to the dendritic polarization fields. Attractor-dynamics can become sequential, e.g. episodic.

Executive control. The far frontal cortex is the executive control over all underlying brain areas. Its orbital portions determine "what" (propriety), its lateral portions "how" (practicality), and its dorsal portions determine "when" (priority) these centers should do or process (Pribram in Hameroff *et al*., 1996, ch. 10).

ENDNOTES

1. The whole chapter 2 presents essentials of Pribram's holonomic theory, *except* where explicitly denoted.
2. Sources: Hopfield (1982, 1984); Haken (1991); Amit (1989); Kohonen (1984); McClelland *et al*. (1986).
3. I.e., models of cognitive science which base on parallel-distributed processing (PDP) in networks.
4. This does not mean that the conventional models are not valid for the neuronal level any more, or that neurons are not important, but that neurons realize just

the gross level of processing which is accompanied by essential microscopic *co*processing (or *pre*processing, detailed processing) in dendritic nets.

5 *Depolarization* corresponds to excitatory (+) postsynaptic potential (EPSP), and *hyperpolarization* to inhibitory (–) postsynaptic potential (IPSP). PSPs are proportional to the number of neurotransmitters (or better to say, neuromodulators) getting bound to postsynaptic receptors. PSPs are temporally summed into constant depolarization. Its magnitude is proportional to the frequency (number per second) of nerve impulses. In such a way, soma's membrane depolarization amplitude decodes the information encoded in the frequency of nerve impulses (Damask & Swenberg, 1984).

6 To relate my quantum or quantum-like associative net model with the holonomic theory: My equation (1.4) is the mathematical expression of the phase-Hebbian memory-storage (synaptic learning). It improves the well-known Hebb rule, cf. eq. (6.1), by processing differences in phase.

7 The idea that holography (Denisyuk, 1984; Hariharan, 1996), discovered in the fourties by Dennis Gabor (an engineering physicist with deep theoretical insights), can be useful in brain modeling was pioneered by Van Heerden and Pribram in early sixties. So, the holistic intuition of Lashley ("All behavior seems to be determined by masses of excitation, by the form or relations or proportions of excitation within general fields of activity, without regard to particular nerve cells," he wrote in 1942) got theoretical frame. The holographic metaphor later triggered a branch of early developments of ANN, but this has been forgotten, because their authors used another interpretation (i.e., D.J. Willshaw, O.P. Buneman & H.C. Longuet-Higgins: Non-holographic associative memory. *Nature* 222 (1969) 960-962; 225 (1970) 178).

8 Some modelers understand "oscillatory activities" as oscillatory changes in the probability to fire a spike.

9 The (bio)physical field is a characteristic state of the (bio)physical system (e.g., in our case, a dendrite or its surroundings) manifested at and determined by *all* the point-elements of that system, and by a specific force acting at a specific location. The intensity of the field is given for/at each point. At each point also the direction (described by a vector) of the corresponding force is given. The field can be electric / magnetic / etc. if the force is electric / magnetic / etc. Electric / magnetic polarization field arises from the relative differences in the electric / magnetic field at various locations.

10 Statistical details on net's anatomy and connectivity, as found in the rat visual cortex by Hellwig (2000), reveal that probability for direct connection between neurons decreases with distance. Connection-probability of neighboring neurons is up to 0.8; but if they are 500 μm apart, the probability is about 0.1.

There are more intra-layer inter-neuronal connections than inter-layer ones (axon–dendrite synapses counted). Contacts seemed to be mainly non-specific.

11 Evidence on activity-dependent plasticity of dendrites on the visual pathway is reported, for instance, in: S.R. Bodnarenko & L.M. Chalupa (1993): *Nature* 364, 144-146; A.K. McAllister, D.C. Lo & L.C. Katz (1995): *Neuron* 15, 791-803.

12 Details are, e.g., in: Kandel *et al.* (1991), Damask & Swenberg (1984), AuxLit 2.

13 This process was found in cells with acetylcholine neurotransmitters and is enabled by the degradation of MAP-2 (microtubule-associated protein), as reported in: N.J. Woolf (1998): *Progress Neurobiol.* 55, 59-77. The author recognizes high-energy interactions between dendrites by quantum-optical coherence. Also, a quantum-mechanical "exchange interaction is found in the neurotransmitter-gated ion channels located on the plasma membrane of the muscle cell in the vicinity of secretory vesicles containing acetylcholine near the nerve terminal," as reported in: K. Matsuno & R. Paton (1999): *BioSys.* 49, 3229-3237.

14 *Synaptic memory-storage* (or synaptic "learning") is defined as stimulus-triggered changing of the rate of transmission of neurotransmitters (or neuro-modulators, to use a more precise word) across the synaptic cleft – so in each synapse of a large group.

15 Among them is a comprehensive list of early electrophysiological results, cited in Pribram (1971, p. 7), by C. Sherrington (1947): *The Integrative Action of the Nervous System*; New Haven: Yale Univ. Press; pp. 13-14.

16 Terminals are at synapses. However, Pribram (pers. commun.) corrects that a spike propagates undiminished in speed and amplitude only to the point where the axon branches into its teledendrons. The speed and amplitude of a teledendron spike are then smaller proportionally to decrease-rate of its fiber-radius.

17 Isomorphism is a one-to-one correspondence between two sets. So, each member of the starting set (e.g., external object) corresponds to, or maps to, precisely one member of the final set (e.g., image or attractor-state).

18 In Peruš (1995a) and Peruš & Ečimovič (1998), this is in detail (and metaphorically) described using the term "virtual"... So, instead of "pictures" in the brain, we have polarization patterns (specific states of the subcellular matter) acting as attractors (virtual structures with top-down influence), and they encode the percepts.

19 The Hopfield model is applicable to general systems with a "matrix interaction scheme" which has become called Hebbian, although it also has counterparts in (bio)physical systems like systems of spins (AuxLit 14) and goes far be-

yond Hebb's original idea based only on (local-circuit) neural mechanisms in synaptic connections.

20 In many ANN models, e.g. Hopfield's, "interactions / connections" are bi-directional (signals are exchanged) – because this is efficient and easier to study.

21 Another remarkable fact, which is also an exception, is that a LGN neuron (as depicted on p. 58 of Pribram, 1971) both sends and receives signals "symmetrically". Namely, bi-directional exchange of signals is similar in its axon and dendrites, because dendrites and axon are all similar in this case.

22 Quantum phenomena seem contra-intuitive, because they are so different from classical ones. — However, see later sections (e.g., secs. 4.4, 4.7) and Auxil. Chapters 7—10.

23 Last four statements by Koch can be compared with the following view in accord with the holonomic theory (Pribram, 1991, p. 19, after W. Freeman, except last statement Pribram's):
"...The operations of filtering, integration, and transmission can be described with linear differential equations. [...] Pulse to wave conversion at synapses is commonly thought to be non-linear, but in fact, in the normal range of cortical operation is linear. It is only at the axon hillock where nerve impulses are generated that wave to pulse conversion is non-linear."
Details of neuronal and dendritic processing (e.g., Koch & Segev, 2000) still seem to be quite open. There are indications that a neuron effectively performs multiplication rather than summation of signals. That is, its output firing rate is a product of inputs rather than their weighted sum. This was shown in computer simulations, as reported in: B.W. Mel (1994): *Neural Computat.* 6, 1031-1085.

24 Using metaphors, like a multi-national company or Church which extend beyond borders. Or (using Dennett's metaphor), like fame of a person, which is a collective, virtual, informational state which transcends the famous person. That person is like an "aerosol" triggering a "cyclon of dynamics" in the whole society.

25 Coupling refers to connected dynamics, usually oscillation — like with two pendulums bound together with a thread. Such interactive or collective dynamics manifests often also in synchronization of phase (phase coupling) or of frequency (frequency coupling) of many oscillators.

26 Superposition or linear combination is a sum of state-describing variables (S) where each of them is pre-multiplied by a coefficient called "weight" (c) — e.g.: $c_1S_1+c_2S_2+c_2S_3$. In simple words: this is a sort of "mixture" where each weight c describes how much of each corresponding ingredient S is there in the mixture.

27 Basically, $p=mv$; m = mass, v = velocity; here generalized.
Fourier decomposition (2.1) is generally valid for any variable. Physically, we can have waves changing in time (t as the argument) and we can have spatial waves (x as the argument).

28 x-t and E-p are refered to as mathematical spaces. They are abstract in contrast to the usual geometric space x. E-p space is thus a derivation from the frequency-wavevector space, since $E=h\omega$ and $p=hk$, where the constant of proportionality h is the Plack constant (divided by 2π) which is characteristic for quantum theory. All these variables are connected, because it is also $\omega=kc$ (generally valid), where c is the wave-velocity (light-velocity in electromagnetism and in quantum theory where we have photons), and consequently $E=pc$ (in quantum theory).

29 Because p, k, E and ω are proportional in quantum theory: $p = hk = E/c = (h/c)\omega$. (Remarkable symmetry!)

30 Note that a Fourier-transform and its inverse are always named relatively to each-other and not absolutely, so the names "Fourier transform" and "inverse Fourier transform" could be "exchanged" in another occasion when talking about one or both of the Fourier-processes.

31 Orthogonal (sub)spaces (in other cases: vectors) are (sub)spaces / vectors which "do not share common features". The term comes from the mathematical fact that their scalar product is zero.

32 This is valid for models like Hopfield's and mainly also Sutherland's HNeT, but not so directly for brain.

33 Definition of entropy, because of connections with information (and even consciousness), is one of the most puzzling problems of theoretical physics. In very simplified terms, it can be refered to as a *measure of disorder or inhomogeneity* (1) of the system which is, however, not necessarily random – it can incorporate implicit structure which organizes energy and matter of the system (2). Informationally, entropy can be considered as a *measure of uncertainty or lack of knowledge* (3) about the system, or directly as *potential* (or implicit) *information* (4). Physicists mainly prefer (1); information theorists (Shannon) and quantum information theorists prefer (3); (2) is based on often neglected observations faced by both groups (e.g., Bohm, 1980; Haken, 1991; Detela in *Proc. IS'98: Cog. Sci. conf.*, Ljubljana); (4) is suggested by the holonomic theory and is promising for pattern processing (Pribram, 1991). Definition (4) will be used in this work.

34 This is how Pribram (1991, p. 41-42) shows the holonomic alternative to the conventional Shannon view: "For the holonomic modification of the thermodynamic process [...] not only noise but structure, as inherent in redundancy, can be added to the system in order to maximize efficiency [...]. And redun-

dancy can be structured by experience [...]. This indicates once again that for information processing the measure of efficiency, that is, of entropy, denotes not only randomness but tacit structure."

35 According to Pribram (1991), this is based on: (1) experimental work by D.J. Field & D.J. Tolhurst (1986): *Proc. Roy. Soc. London B* 228, 379-400; and (2) finding that adjacent neurons in the visual cortex respond best to gratings 90° out of phase (like *sine*-waves relative to *cosine*-waves), so that they make a "*sine-cosine* quadriture", giving a module of receptive fields which encodes the *sine* and *cosine* components that establish Fourier and Gabor coefficients – as reported in: D.A. Pollen & S.E. Ronner (1980): *Exp. Brain Res. A* 41, 14-15.

36 These conclusions by Pribram (1991) are based on the following experimental report: T. Caelli & M. Hubner (1983): *Vision Res.* 23, 1053-1055.

37 After Prigogine, dissipative systems are those *self-organizing* systems, having much exchange of energy (and information) with environment, which find stability away from thermodynamic equilibrium, and for which an area of the phase-space (i.e., a mathematical space of physical variables which specify system's dynamical state) does not remain conserved. This also means that parts of the phase-space get shrinked to a point called attractor.

38 "Form" is evident also in the deBroglie-Bohm-Hiley interpretation of quantum theory (Bohm & Hiley, 1993).

39 Peruš (1995a, sec. 13.5.4); details in: M. Peruš (1994): *Programer* 21, 40 & 22, 32 (in Slovene)

40 Introduced in: G.A. Miller, E.H. Galanter & K.H. Pribram (1960): *Plans and the Structure of Behavior*. New York: Holt, Rinehart & Winston.

41 Lists of references to be found in works by Pribram, Daugman and ICA modelers – see ch. 3.

42 They were proportional to some parameters describing the stimulation process with a special rotating device.

43 Storage of phases (i.e., relations of interacting information-encoding oscillations) in addition to amplitudes of waves makes holograms different from photographs which merely encode intensity of waves, but not their phase. Hologram is a multi-view correlogram, giving a 3-dimensional impression when reconstructed. A photograph is just a one-to-one 2-dimensional projection/cross-section of waves onto/at a 2-dimensional film. (Denisyuk, 1984; Hariharan, 1996; AuxLit 12).

44 "The interaction of many such convolutional integrals may produce hologram-like interference patterns within the visual system and elsewhere in the brain." (Pribram, 1971, p. 143) "...This choice is reinforced by evidence that in the retina, at least, neighborhood interactions in the horizontal cells are due exclusively to inhibitory interactions, by slow potential hyperpolarizations and

not [due] to depolarizations which lead to nerve impulses." (Pribram, 1971, p. 142).

45 Such "enfolding-unfolding" PDP (which are compatible with the holonomic theory), based on complex-system physics and ANN simulations, were in detail described also in Bohm (1980) and Peruš (2000a).

46 Without the lens of the eye or video-projector, no image is produced on the screen. This demonstrates that the light itself is "holographic" (i.e., electromagnetic waves are distributed). (Further details in Prideaux, 1996.)

47 Retinal filtering is best described by a difference-of-gaussians (very similar to Mexican hat). The representation in LGN is similar to a *hologram based on near-field Fresnel* optical image procesing. The representation in V1 is similar to a constrained *full Fourier patch hologram*. (Pribram, 1991).

48 In sec. 4.1, see the difference between the convolutional process (where the scalar or inner product is used for code-formation) and the matrix-process (where the tensor or outer product is used for distributed encoding). For the optic nerve (where economic transmission is necessary) and for V1, a convolution-made code, which is "*compact*" and "*focused*" (like Haken's "order parameters") or *sparse*, is appropriate. In V2 and beyond, Hebbian matrix-codes, which are distributed (hologram-like), share their role with (relatively more localized) image-vector representations.

49 As described in detail in Peruš (1995a,b) and Peruš & Ečimovič (1998), and references within.

50 This is well-known (from Gabor *et al.*), but is not taken into account enough in spite of big importance. We believe, the hologram (where "one is in all, and all is in one") is the best global *metaphor* available, not only for the brain, but for the Universe, especially quantum Universe, as a whole (cf., e.g., Chapline, 1999; Marcer in AuxLit 27; Mitchell, Marcer in Dubois, 2000a; Bohm, 1980; Hiley & Peat, 1987; Bohm & Hiley, 1993; Laszlo).

51 E.g., LeCun & Bengio, Fukushima in Arbib (1995). Fukushima's "neocognitron" is a convolution-cascade.

52 A non-expert can skip the following proof-calculation. All variables are as in sec. 1.2. An example of 2-dimensional Gabor elementary function $g(p,u;x) = exp(-(x-p)^2/\sigma^2)exp(iu(x+p))$ is depicted in Figure 1-right.

53 All stripes of the same pattern have the same width, but different stripe-patterns have different stripe-widths.

54 Experiments by Campbell, the DeValois pair, Pollen and their collaborators, and many others.

55 E.g., M. Kraft, W. Obrist & K. Pribram (1960): *J. Compar. Physiol. Psych.* 53, 17-22; J. Stamm & K. Pribram (1961): *J. Compar. Physiol. Psych.* 54, 614-618; E. John *et al.* (1967): *Science* 155, 1439-1442.

56 80-90% of V1 (Lashley); 98% of optic tract (Galambos); 75% of optic tract and 75% of visual cortex (Chow).

57 However, *after learning*, this distribution of information does not involve all parts of the "holographic" brain, as evident from experiments by: K. Pribram, D. Spinelli & M. Kamback (1967): *Science* 157, 94-96.

58 A single action potential can propagate not only forward (from dendritic tree over soma to and along axons), but also backwards (from axon through soma into dendrites) (Koch & Segev, 2000; Suarez Araujo in Dubois, 2000b). This is called antidromic spike invasion (G.J. Stuart & B. Sakmann (1994): *Nature* 367, 69-72.). It is still unknown whether dendrites can trigger action potentials themselves (Koch, 1997).

59 I.e., view-dependent, or view-centered, respectively.

60 Nystagmus is a rapid, persistent, rhythmical, jerky or wobbling movement of the eyes, usually together.

61 About 8% of the nerve fibers in the optic nerve transmit signals from the visual cortex back to the retina. These reciprocal signals modulate the input in 80% of the optic-nerve (retina-to-cortex) fibers (Pribram, 1991).

62 If V2 was resected, monkeys responded to optical image size of the object, ignoring the context, and thus being deceived by illusions — after L. Ungerleider, L. Ganz & K. Pribram (1977): *Exp. Brain Res.* 27, 251-269. Namely, they considered the nearest (small) object as the largest, because their V1 processed images without V2.

63 I.e., information which is found *new* when current input is compared to the stored contents (the context).

64 E.g., distributions of field-potentials recorded from the somatosensory cortex (in the whisker experiment of Pribram's group, described earlier) were similar to those recorded from V1 (Pribram, 1995, 1998b).

Chapter 3
Computational Information-Maximization Models

Some information-maximization models, especially examples of the Independent Component Analysis (ICA), and their importance for the holonomic theory will be presented. Their biological plausibility will be discussed.

3.1 MAXIMAL PRESERVATION OF INFORMATION ("INFOMAX")

Linsker's infomax-net. The early "infomax" perception-model by Linsker (1988) was established with an observation that Hebbian adjustment of connections is able to evolve feature-selective neurons which collectively preserve information "as much as possible" during input-to-output processing. Linsker (1988) used a multi-layer network with local feed-forward connections which are Hebb-like – determined by

DOI: 10.4018/978-1-61520-785-5.ch003

covariance. In this case, local means that each neuron in layer *l+1* receives inputs from neurons in a confined circular region of layer *l*. The "infomax" ideas used in the classical PCA-like[1] self-organizing net by Linsker (1988, and later works) have later been much developed in direction of ICA.

From second- to higher-order statistics. As has been mentioned in section 1.5, ICA is supposed to be important for cortical image processing because of taking into account the statistics of input data which is also of higher order than the second order.[2] *Correlation* and *convolution*, for example, are "learning (i.e., memory storage) rules" of *second order*. Such mathematical expressions, used in unsupervised learning models, have been named *Hebbian* (directly or in a generalized sense, e.g. the *covariance* "learning rule" and *PCA*). ICA goes beyond second-order statistics of PCA, because it also processes higher-order terms in a series of statistical quantities – moments or cumulants. This is related with neuronal interactions of higher order, i.e. more than two neurons may interact at the same time because they may be directly connected (e.g., having synapses close to each-other and affecting one another), or their activity may be coupled in another way (e.g., as in coherent oscillations).

In the case of *oscillatory activities* in neurons and subcellular structures including dendritic fields, the relevant variables become local phases and their relations (King *et al.*, 2000). Oscillations are described by functions of complex-valued exponents (the exponent without imaginary unit *i* is the phase).[3]

Phases needed for edges. Phase-information is needed for successful approximation of V1-filters which were found to be *localized*, *oriented* and *band-pass* (i.e., selective to structure at different spatial scales). Local angle of orientation is described by local phase. Such filters are needed to trace segments of edges which are themselves oriented. Experiments show that individual *simple* cells of V1 with their specific receptive fields act as such filters, i.e. as selectors or edge-segments by having maximal response to specifically oriented stimuli.

An edge (of an object), as a crucial element of an image, manifests specific relationships among many pixels being encoded in neurons or receptors, not only two (neighboring) ones. Second-order statistics (as in PCA) is sensitive *only to pair-wise relationships*, like correlations encoded in the Hebb rule. Higher-order statistics (as in ICA) is sensitive to multi-neuronal relationships, reflecting *multi-pixel gestalt-structures*, and thus goes beyond the two-pixel (or two-neuron, respectively) relations of PCA. In PCA, gestalts are formed from feature-segments by global organization from local *bi*lateral connections (usual simple synapses). In ICA, however, gestalts are formed directly by semi-local *multi*lateral connections (complex synaptic structures[4]). This allows ICA to detect edges by *filters* which are essentially more *localized* (*wavelet*-like) than it would be achievable by PCA performing *global* (*Fourier*-like) spatial and spatial-frequency analysis.

Bell & Sejnowski (1996, pp. 261-262) write: "[Hebbian models] reflect only the amplitude spectrum of the signal and ignore the phase spectrum where most of the suspicious *local* coincidences in natural signals take place. An edge in an image, for example, is a coincidence in the phase spectrum, since if we were to Fourier analyse it, we would see many sine waves of different frequencies, all aligned in phase where the edge occurred. Correlation-based methods cannot learn edge-detectors, though they often may seem to be doing so by local-windowing of the learnt Fourier components, turning them into Gabor-like filters..."

Bell & Sejnowski (1996) furthermore propose a test. If one Fourier-transforms the image, removes the phase-information and transforms this back by inverse Fourier-transform, the starting image will not be reconstructed, but a "noise" will result. If, on the other hand, one makes Fourier-transform of the image, removes the "amplitude"-information, and then performs inverse Fourier-transform, the result will be a distorted, but still recognizable, variation of the starting image. *Phases are thus necessary for image processing; "amplitudes", on the other hand, are useful but not necessary*. Therefore, to mention by the way, "amplitudes" are used in phase-Hebbian models (Sutherland, 1990; MacLennan, 1999) as measures of the level of importance, urgency and accuracy (or, confidence or reliability, respectively) of the information mainly encoded in phases and their relations.

Redundancy reduction. Visual processing also has to be maximally efficient in the sense of *maximally preserving information by using as little structure as possible*. So, information has to be optimally condensed, not by reducing quality and relevance, but only by reducing quantity (where it is unnecessary), or by eliminating redundancy (i.e., double or multiple coding), respectively. This would enable faster processing and decrease the number of errors because of getting trapped into spurious (irrelevant) attractors.

It turns out that information-maximization ("infomax") is realizable if neurons have receptive fields shaped like Gabor functions. Computer simulations were done where model-neurons (as filters) resembled the Gabor-like receptive-field form which enables "infomax" *and* orientation- or edge-selectivity at the same time. 2-dimensional Gabor wavelets are functions which have oriented stripes as seen from above. This means that the response of a *simple* cell with such a receptive field will be the biggest if stimulated by an edge with such an orientation.

Two infomax methods. To realize these characteristics of the primary visual system, several methods turned out to be successful. The most important ones are *Independent Component Analysis (ICA)* and *sparseness-maximization network*.[5] ICA is a new PDP method for searching unknown sources in complex data-structures, i.e. so-called blind separation into independent components. Among many other general-purpose examples, there is also search of signal-sources in EEG- and fMRI-data, as in McKeown *et al.* (1998). Starting from earlier ICA-models like Comon

(1994), ICA has been developed for vision modeling mainly by Bell & Sejnowski (1995, 1996, 1997). Their simulation-results emphasized statistical independence more than sparseness of coding. The sparseness-maximization method, that also resembled Gabor-like infomax orientation-filters, has been developed by Olshausen & Field (1996a,b; 1997). They achieved such results by emphasizing sparse coding over information maximization. *Sparse coding means that information is encoded by a small minority of neurons in a net.* These coding neurons are active ("non-zero") only, while all others are passive ("zero") in the end, although they also participate during the process of memory-storage (i.e., during adaptation of network's parameters). Field (in Arbib, 1995) defines sparse codes as codes which distribute the information across the whole network, but which use its parts relatively rarely.

Works by Olshausen & Field (1996a,b) and Bell & Sejnowski (1995, 1996, 1997) agree that (certain) simple cells in V1 are best modeled as information-maximizing cell-filters with Gabor-like receptive fields for detection of orientation and consequently edges. There is also agreement about necessity of phase-information that hides higher-order statistical structure, i.e. the so-called higher moments of input-data distribution. Bell & Sejnowski as well as Olshausen & Field prove in detail, mathematically and with graphics of resulting simulated receptive fields[6], how their phase-based methods transcend the limitations of PCA. Namely, in PCA neuronal configuration converges toward a pattern-vector (attractor) from a basis of such vectors which are mutually orthogonal. Pairs of PCA output-patterns are therefore decorrelated ("have nothing in common"), but are not statistically independent, because the structure of input-data is usually not mutually orthogonal, especially not in natural scenes (Harpur & Prager, 1996).

In statistical terms, PCA finds orthogonal directions (along the "clowd" of data in data-space) with maximal variance. ICA, furthermore, performs not only the usual decorrelation (i.e., two-point-wise correlation is zero, as in PCA), but also statistical decomposition where higher-order (several-point-wise) correlations are zero (in ideal case) (McKeown *et al.*, 1998).

Statistically-independent coding (ICA). Statistical independence is more than just decorrelation which ensures that stored images do not disturb each other when recalled, and that they can be selectively recognized in novel, similar circumstances. Such decorrelation, as achieved by Hebbian memory-storage and PCA (see Appendix A), is perfect in the case of orthogonal images. However, natural-scene images are rarely orthogonal. As numerous computer simulations (including Peruš, 2002) have shown, Hebbian nets and PCA still have some success in quasi-orthogonal cases, but in non-orthogonal cases they generally fail much earlier than ICA (see Auxil. Ch. 8). Such cases also usually demand non-linear processing, like at the edges. This is the case also because PCA has no explicit lateral inhibition, and consequently not enough enhanced competition between neurons.

ICA solves more difficult problems of vision by decomposing the natural image into its statistically independent component-images. The natural scene usually contains many intermixed features and parts of objects which cannot be stiffly categorized into orthogonal eigen-images, but share the data-structure. ICA achieves this more subtle decomposition on the level of finer data-statistics. Statistical independence means mathematically that the probability distribution of input-data can be factorized, i.e. written as a multiple product of probability distributions of some specific sub-sets of data. These sub-sets are then called statistically independent components. Decomposition of the probability distribution of a natural image into independent factors is equivalent to decreasing their mutual information toward zero. This is, however, *not* completely achievable in practice. Authors recognize that their algorithms do not achieve complete statistical independence, but merely try to get close to it as much as possible. (Such a practical maximum will from now on be meant when speaking about "independent components").

Table 1 summarizes the comparison of the ICA-net by Bell & Sejnowski and the sparseness-maximization net by Olshausen & Field *versus* PCA and Hebbian ANN:

Blind separation of images. Bell & Sejnowski and Olshausen & Field perform a so-called blind separation of natural images X searching for the *amplitude-coefficients* or *codes* ("causes" or "sources"), s, *of the independent image-components*, Y. The image X is in general representable by a linear combination of independent image-components Y which form the *columns* of a fixed matrix $\boldsymbol{A}$. If we represent this relationship with equation $X=\boldsymbol{A}S$, then each vector-component of vector S acts as a *weight* of the corresponding basis-function Y. Each pixel (denoted by index i) of the image X can so be described as follows:

Table 1.

P.C.A., Hebbian ANN	**I.C.A., Sparseness-max.-net**
second-order statistics: bi-linear coupling, e.g. correlation, convolution, covariance	higher-order statistics: multi-linear coupling and/or non-linear interdependence
decorrelation into orthogonal eigen-images	statist. separation of non-orthogonal mixtures
optimal for gaussian input-data	optimal for super-gaussian data (kurtosis > 0)
"static activities of units (neurons)"	"oscillatory activities of units"
"amplitude"-information (activity rates) only	plus phase-information, frequency-spectrum
edges and V1 receptive fields - problematic	edges detected; receptive fields reproduced
filters distributed, not Gabor-like	filters sparsely-distributed, Gabor-like
algorithm up-to-now satisfactorily efficient	algorithm usually more efficient than P.C.A.
implementation biologically more plausible	biological implementation unknown
outputs biologically not plausible (in vision)	outputs biologically plausible (in vision)

$$x_i = \sum_{k=1}^{P} a_{ik} s_k = a_{i1} s_1 + a_{i2} s_2 + a_{i3} s_3 + ... + a_{iP} s_P \,. \qquad (3.1)$$

Each index *i* or *j* corresponds to a Cartesian coordinate-pair, like *(x,y)*, of the 2-dimensional image-space. For all pixels *(x,y)*, $X = AS$ is (in our notation using elements of vectors *X*, *S* (its elements are *s*) and *Y*, hidden as column-vector in matrix ***A***) translated to

$$X(x,y) = \sum_{k=1}^{P} s_k Y_k(x,y)\,, \qquad (3.2)$$

where sum Σ adds all *P* independent-component terms.

If we "look in the opposite direction", matrix ***W*** decomposes an image *X* into independent image-components *Y*, represented by their amplitudes or image-codes *s* (which are mathematically represented by individual vector-components, denoted by index *k*, of the vector *S*). Ideally, we would get $S = \boldsymbol{W}X$, but in practice we might get $S' = \boldsymbol{W}X$ (e.g., comparing to *s*, vector-components of *S'* might be in different order and rescaled). Matrix ***W*** should separate the image in such a way that the independent codes *s* can encode as much information about image *X* as possible, but with as little redundancy as possible, so that we could reconstruct the initial image *X* as much as possible with available limited neural structure. Ideally, ***W*** is *inverse matrix* of ***A***, so that ***WA=I***, where ***I*** is the identical matrix which preserves or recovers the initial state (here *S*), although making a (composite) transformation. In such a case, it is thus *S'* = *S*, and the reconstruction of the image is perfect.

The *filters* ϕ, which extract the independent codes *S*, are hidden in *rows* of matrix ***W***. As long as the number of codes is equal to the number of distinct images and their vectors form an orthonormal basis[7], filters ϕ and independent components *Y* are equal. However, if the image-coding vectors are not orthogonal, their filters ϕ differ somewhat from corresponding *Y*, although filters recover the amplitudes of independent components *Y*. This is because such non-orthogonal independent components *Y* of the image *overlap* strongly, but filters ϕ concentrate their action to areas in spatial-(temporal)-frequency-space where the overlap of *Y* is minimal. With such filters, the amplitude of *Y* can be got without interference with other components *Y'*. Independent components *Y* can be considered as building blocks of the image *X*; their filters ϕ, on the other hand, are used for analyzing the image by determining how strongly each building block is present in it. Thus, it is more appropriate to say that the independent-component-*filters* (ϕ, not *Y*) correspond to V1-cells acting as "edge filters" (van Hateren & Ruderman, 1998).

In this infomax approximation, the role of primary visual system (from retina up to and including V1) is to find the proper *filter-matrix* ***W*** with the above-described properties. ***W*** is an extremely simplified and summarized mathematical description of results of the *cascade convolution processing weighted with receptive fields* (these are the filters!). They turn out to be Gabor-like in this infomax framework executed by the algorithms of Bell & Sejnowski or Olshausen & Field.

3.2 ICA BY BELL AND SEJNOWSKI

ICA-infomax. Bell & Sejnowski (1995, sec. 4) show precisely under what conditions an infomax computational algorithm minimizes the mutual information between resulting outputs called independent components. Namely, under specific conditions, mutual information $I(a,b)$ between output-components a and b is *minimized* while overall information is *maximized*[8] or equivalently joint-entropy $H(a,b)$ is maximized. This is evident from a well-known equation of information theory: $H(a,b) = H(a) + H(b) - I(a,b)$. Joint entropy $H(a,b)$ is maximal if individual entropies $H(a)$ and $H(b)$ are maximized and mutual information $I(a,b)$ is minimized. It was also proven in Bell & Sejnowski (1995, sec. 2, p. 1131) that[9] the mutual information $I(X,Y)$ between inputs $X = (a', b', ...)$ and outputs $Y = (a, b, ...)$ can be maximized by maximizing the entropy $H(Y)$ of the outputs $(a, b, ...)$ alone.

As long as a and b can interfere, the absolute minimum of $I(a,b)$ can not be achieved for sure, therefore the probability distribution *PD* cannot be decomposed into independent factors, i.e. can not be written as a product of individual probability distributions *PD(a)* and *PD(b)*. Unwanted interference is eliminated, thus giving absolute minimum of $I(a,b)$ and consequently realizing statistical independence where $PD = PD(a) \times PD(b)$, if the probability distributions of inputs are *super-gaussian*. Super-gaussian, as many natural signals are, means that the probability distribution has a sharper peak and longer tail than gaussian ("bell-shaped"). PCA gives optimal and complete results in the case of gaussian probability distributions (Intrator & Cooper in Arbib, 1995, p. 484), but super-gaussian probability distributions (where the bell-shaped function of statistics is thiner, except at the bottom where it is broader than the "normal bell") are optimal for ICA. In statistical language, the kurtosis (the fourth-order standardized cumulant measuring deviation from the normal gaussian distribution) must be positive (but see Baddeley, 1996). In the opposite case which is undesired: negative kurtosis, i.e. sub-gaussian distribution, means interference between image-components and so they will not get independent by applying an infomax-algorithm for image-processing. Such rare negative-kurtosis images would be problematic while attempting redundancy-reducing image-recognition using ICA.

Bell & Sejnowski (1995, 1996, 1997) present their version of an infomax-algorithm which attempts, usually successfully, to extract edges by evolving specific neuronal receptive fields selective to edge-like independent components. This means that the simulated receptive fields develop into Gabor-like wavelets which, seen from above, have stripes that are oriented like potential edges in the input-image are. So, Gabor stripes are parallel to edges. Not just orientation, but other subtle details of the edge-structure are also encoded in such a "Gabor filter" belonging to a simulated V1 simple cell and its tree-like connections. It will now be presented how the Bell & Sejnowski net incorporates one of many possible "hardwares and softwares" for computational modeling of evolution of edge-detecting V1 simple cells together with their receptive fields. Another option, the Olshausen & Field net, not based primarily on ICA, will be presented later.

ICA-algorithm by Bell & Sejnowski. The Bell & Sejnowski network maximizes information by maximizing the final joint entropy $H(Y) = H[g(\boldsymbol{W}X+B)]$; $Y = g(U)$; $U = \boldsymbol{W}X+B$. Input image is denoted by X, ICA-output is Y, $\boldsymbol{W}$ is the matrix of ICA-filters, B is the so-called bias. The activation function g is the sigmoid (tangens-hyperbolicus-like) function. It is very often used in ANN-models for non-linear summation of dendritic inputs in soma, because it enables optimal flow of information.[10] Joint entropy $H(Y)$ is maximized by performing stochastic gradient ascent, i.e. the elements of the filter-matrix $\boldsymbol{W}$ are changed in such a way that $H(Y)$ increases as much as possible. Bell & Sejnowski (1995, sec. 2) derive $\boldsymbol{W}$ for the gradient ascent with the sigmoid g. The so-called learning rule for the whole Bell & Sejnowski net is[11]

$$\Delta W \propto (W^T)^{-1} + (1 - 2\vec{Y})\vec{X}^T, \tag{3.3}$$

$$\Delta \vec{B} \propto 1 - 2\vec{Y}. \tag{3.4}$$

Here, output-vector Y is obtained from input-vector X, which is transformed by matrix $\boldsymbol{W}$, and then filtered by g after bias-vector B has been added[12]: $Y = g(\boldsymbol{W}X+B)$.

Such an input-output (X-to-Y) transformation is usual in ANN, taken from summation-process in soma. It has no direct relation to dendritic processing, although the results of dendritic processing shape the input X giving $\boldsymbol{W}X$ being transmitted to soma for g-weighted summation. *The Bell & Sejnowski algorithm parallels the holonomic theory with the infomax-result*, but *its implementation is*, it seems, *quite artificial*, or *it incorporates conventional neuronal computing rather than dendritic computing* advocated by the holonomic theory. Primarily, the learning rule (3.3) & (3.4) is *non-local*. This means that neurons must know information about synaptic weights (or, in other cases, activities) of many distant neurons without being directly connected to them.

Bell & Sejnowski (1997, p. 3330) improved the rule (3.3) & (3.4), making it less non-local, using an idea of Amari *et al.* (see *ibid.*) for a more local adaptation of filter-matrix $\boldsymbol{W}$:

$$\Delta W \propto \frac{\partial H(\vec{Y})}{\partial W} W^T W = (I + \hat{\vec{Y}} \vec{U}^T) W . \tag{3.5}$$

I is the identity-matrix (having 1s on diagonal and zeros elsewhere), and

$$\hat{y}_i = \frac{\partial}{\partial y_i} \frac{\partial y_i}{\partial u_i} = \frac{\partial}{\partial u_i} \ln \frac{\partial y_i}{\partial u_i} . \tag{3.6}$$

For this new Bell & Sejnowski algoritm, the above comment on implementation is still valid. After the entropy gradient-ascent[13] factor, the "auxiliary" $\boldsymbol{W}^T\boldsymbol{W}$-factor of (3.5) gives a positive-definite matrix which speeds up the gradient ascent and enables avoiding the problematic matrix-inversion (denoted by "exponent" –1 on ***W***) in rule (3.3).

Biological implementation of the ICA-net? Bell & Sejnowski, of course, recognize that the visual processing from analog retinal photoreceptors to spike-coding pyramidal cells in V1 is approximated very roughly by the filter-matrix ***W*** and its generalized or pre-processed ($\boldsymbol{W}_Z$-"whitened") versions $\boldsymbol{W}_I = \boldsymbol{W}\boldsymbol{W}_Z$ (for mathematical details see: Bell & Sejnowski, 1997, pp. 3330-3332). Bell & Sejnowski claim that their pre-processing of data has counterparts in retinal ganglion cells and LGN's relay-cells. The authors' simulated filters fit V1-filters in orientation and locality, but have somewhat too high spatial frequency.[14]

If the pre-processed filter-matrix ***W*** gives *relatively biologically-plausible outputs* (or *filters*, or *receptive fields*, respectively), this does *not* necessarily mean that a ***W***-like retino-geniculate-cortical transformation is *implemented in the living tissue*. The learning rules of Bell & Sejnowski, the new one in Equations (3.5) & (3.6) almost as much as the previous one in Equations (3.3) & (3.4), are both *non-local*. (Such) non-local processing is not necessarily biologically implausible, but has more danger of being so than local learning rules which get physiological support easier. Bell & Sejnowski (1997, p. 3336) write about their algorithm:

"…In both its feedforward and feedback versions, it involves a feedback of information from, or within, the output layer. […] Since it is difficult to identify the parameters of our static matrix, $\boldsymbol{W}_I$, with "true" biophysical parameters, we prefer to imagine that potentially real self-organizing processes […] occur in local spatial media where the feedforward and the feedback of information are tightly coupled, and where some microscopic and dynamic analogue of equation (13) may operate."[15]

Indeed, with present very limited and non-detailed knowledge on connectivity and physiology of the visual retino-cortical transformation, the ICA-algorithm of Bell & Sejnowski seems to have *no biologically-plausible implementation*. Future research might change this view or not. Nevertheless, their net, mathematically original and computationally very efficient, is a very important step forward and a basis for further computational and physiological research. There may be numerous different implementations of ICA-like processes[16] which might be found more biologically plausible for early vision, because of having more support from the infomax-view (of Pribram, among others) than Hebbian processes. Van Hateren (1992), Wainwright (1999) and Porrill *et al.* (1999), among others, studied optimal coding of natural images, comparing model with physiological (e.g., in flies) or psychophysical measurements, and found "infomax" ideas the most suitable for image processing.

3.3 SPARSENESS MAXIMIZATION BY OLSHAUSEN AND FIELD

Sparse coding is related to statistical independence. Bell & Sejnowski (1995, 1997) emphasize that *maximal preservation of information while reducing it to statistically-independent components* (their primary focus) *is very similar to maximizing sparseness of information-encoding* (primary focus of Olshausen & Field). Sparse coding means that a small percentage of neurons in the network is firing (is having non-zero activities), while the great majority of neurons is quiescent (has zero activity). The bulk of coding is thus realized by specialized minority of neurons, although the whole network participates in sorting the activity into such a restricted result or output (Olshausen & Field, 1996a,b). Sparse code is thus based on more distributed processing than localized codes like the "old" feature-detectors, but less than the so-called coarse (collective) codes. Olshausen & Field and Bell & Sejnowski achieve similar infomax-results (receptive fields like in V1) using different methods and algorithms. Sparseness is much related to the super-gaussian probability distribution of input-data with positive kurtosis. Very sparse probability distributions are more peaky and have longer tails than the gaussian ones. The difference between Olshausen & Field and Bell & Sejnowski in trying to *maximize information* is that Bell & Sejnowski force their net to *minimize mutual information of outputs*, but Olshausen & Field force their net to *produce sparse output-codes*.

Sparse-coding infomax by Olshausen & Field. Like those of Bell & Sejnowski, the Olshausen & Field method also decomposes the image X into a linear superposition of basis functions Y weighted by amplitude-coefficients s, just like in Equation (3.2). Image-codes s, which in the Olshausen & Field case do not necessarily represent *statistically*-independent image-components, but some basis-images of *very similar* type, are dynamic variables which change from one image to another (Olshausen & Field, 1996a,b).

Olashausen & Field's search for maximal sparseness of the codes *s* is done "globally and by force". They use minimization of the "cost function" *E* for regulation of "learning" as usual for ANN:

E = – *(preserve information)* – *λ (code-sparseness)*,

where *λ* is a positive constant determining the importance-rate of the sparseness-term relative to the first term. The first term *PI*, measuring preservation of information by optimal coding, incorporates the error or difference between the actual image *X* and the reconstructed image (sum of *s*-weighted components *Y*):

$$PI = -\sum_{x,y} [X(x,y) - \sum_{k=1}^{P} s_k Y_k(x,y)]^2 . \tag{3.7}$$

The difference should be as small as possible. If so, *E* is kept small, as desired.

The second term *CS*, sparseness of coding, "determines the cost" depending on how information is distributed among the codes *s* which are amplitude-coefficients of corresponding image-components *Y*. If distribution is limited to few coefficients *s*, then CS will be big, and its contribution to total "cost" *E* will be small, as desired.

$$CS = -\sum_{k} S\left(\frac{s_k}{\sigma}\right)^2 \tag{3.8}$$

This sparseness-term to be maximized was constructed by Olshausen & Field to concentrate encoding to as few components as possible, "globally and by force", to increase sparseness. *S* is a chosen nonlinear function, $log(1+x^2)$ or something similar, and *σ* is a scaling constant (details in: Olshausen & Field, 1997). Optimal *S* should promote such a set of components *Y* with equal variance, which would have maximal number of zero coefficients *s* (i.e., codes which could be eliminated to economize processing).

Olshausen & Field's algorithm. The learning procedure for the Olshausen & Field ANN goes as follows. After each image presentation, code-coefficients *s* are changed in such a way that *E* is minimized. After many image presentations, components *Y* evolve towards a convenient final form which ensures maximally-sparse coding directly and statistically-independent coding indirectly in trade-off with the requirement of minimal reconstruction-error.

Let us see details of iterative dynamics[17] of codes *s*, which are similar to Gabor coefficients, and of image-components *Y*, which are similar to Gabor wavelets. Olshausen & Field (1996a,b; 1997) derived, from the requirement of gradient-descent of E, the equation of iterative adaptation of amplitude-coefficients s_k:

$$\frac{\partial s_k(t)}{\partial t} = \eta \left[c_k - \sum_j C_{kj} s_j(t) - \frac{\lambda}{\sigma} S' \left(\frac{s_k(t)}{\sigma} \right) \right] \quad (3.9)$$

where η is the learning-constant (bigger η makes learning faster); the auxiliary scalar-product coefficients c_k and C_{kj} are:[18]

$$c_k = \sum_x \sum_y Y_k(x,y) X(x,y) \quad (3.10)$$

$$C_{kj} = \sum_x \sum_y Y_k(x,y) Y_j(x,y) \quad (3.11)$$

After several input-images X have been computed using Equation (3.9), Y_k are updated by making a stepwise decrease of E (gradient descent of E):

$$\Delta Y_k(x_m, y_n) = \eta_w \left\langle \left[X(x_m, y_n) - X'(x_m, y_n) \right] s_k \right\rangle \quad (3.12)$$

where η_w is the learning-rate, and X' is the reconstructed image: $X'(x_m, y_n) = \Sigma_k s_k Y_k(x_m, y_n)$.

Comparison Olshausen & Field — MacLennan. There are similarities of the Olshausen & Field algorithm and the *dendritic field computation model* by MacLennan (in Pribram, 1993, ch. 6, p. 179). MacLennan started from the model of Daugman (1988) and considered a possibility to implement it in a dendritic net. Daugman at an early stage used just the first left-side term in Equation (3.9), i.e. $\partial s_k(t)/\partial t = \eta c_k$. Following Daugman, MacLennan then made improvement by using the first and the second (with C_{kj}) left-side term of Equation (3.11). MacLennan uses, like Daugman (1988), the same gradient-descent process as in Equation (3.7).

This linear model, possibly implementable in dendritic nets (as shown in: MacLennan in Pribram, 1993, ch. 6), is *equivalent to the Olshausen & Field model with exception of the sparseness-term*. Olshausen & Field invented *CS*-term (3.8) which makes their algorithm "more infomax" than the Daugman and MacLennan algorithms, focused on *PI*-term (3.7), are. We should pay attention to a possibility that *the linear model with Olshausen & Field's extension, which uses the sparseness-maximization (CS) term, could be dendritically implemented as MacLennan has shown for the information-preservation (PI) term.*

Biological implementation of Olshausen & Field? Olshausen & Field claim that their learning algorithm, based on evolution of optimal parameters s and Y, can be implemented in a network with local connections. This is true, but the adaptation dynamics are regulated by a global requirement to minimize an abstract, artificial "cost"-parameter *E*. In the case of Hopfield ANN, *E* is similar to physical energy. In physical systems, energy is minimized naturally. In complex and biophysical systems, free energy is minimized naturally. Brain, however, seems such a special biosystem, that tries to maximize efficiency by maximization of information (infomax), or equivalently, by reducing redundancy and information-loss during net-processing which transforms inputs to outputs. As Olshausen & Field show, this may be done by sparseness maximization. This seems plausible and natural in principle, but how this is concretely implemented in the brain tissue is not clear. Minimization of Olshausen & Field's *E* is not so natural as in Hopfield's case. It will remain an open question, whether brain uses such a global regulation, until anatomical and physiological research reveals more details on visual-brain's structures and cooperation between them.

The fact that the needed connectivity and processing is not non-local raise the possibility of being biologically implementable. A special sort of lateral inhibition (maybe like in Kohonen's Self-Organizing Maps – see Appendix B) to enforce sparsification of encoding could be even sufficient, but this is not necessarily the case at all. Other options include top-down constraining of rather specific type, since the form of Olshausen & Field's *E* is quite specific, and this raises the question which higher visual centers could regulate it and how. Nevertheless, the Olshausen & Field algorithm seems to have more chance to be some day found to be biologically plausible than those by Bell & Sejnowski. Namely, the Bell & Sejnowski algorithm is nonlocal, more complicated and more artificial than Olshausen & Field's. Bell & Sejnowski derived their algorithm by purely mathematical means as abstract as factorization of probability distribution. Olshausen & Field, on the other hand, derived theirs by more natural and simple, it seems, means like sparse representation.

Overcomplete coding. The Olshausen & Field algorithm allows overcompleteness, i.e. output dimension is bigger than input dimension. Thus, the number of basis-vectors (memory-patterns or codes[19]) is greater than the effective dimensionality, or the so-called number of degrees of freedom, of input-images[20] (Olshausen & Millman, 2000). This is achievable by Olshausen & Field even with high degree of sparseness. Olshausen & Field (1997) note that sparse coding is not a general principle for finding statistically independent components in images; it functions only if the images really have sparse structure (as the natural scenes indeed usually do). Olshausen & Field's basis-vectors need not be linearly independent and are not orthogonal. Bell & Sejnowski (1997) say, in contrast, that they do not realize overcomplete representation.

Pribram (in Wang *et al*., 1998) and Lee (1996) emphasize physiological evidence on a redundant, tight-frame representation through *over*sampling, i.e. cortical sampling density is far greater than it would be minimally required for complete representation. This ensures flexibility and robustness of perception. Like Bell & Sejnowski, Olshausen & Field reproduced localized, oriented and band-pass receptive fields of V1 simple cells, but they did it by posing two global constraints on linear coding – information preservation and sparse representation. Overcompleteness, used by Olshausen & Field, serves better than completeness (equal input and output dimensions), used by Bell & Sejnowski, while constituting a sparse representation or a tight frame with non-orthogonal basis functions (Lewicki & Olshausen, 1999). Olshausen & Field's approach seems more biological in this respect also.

3.4 SPATIO-TEMPORAL AND EXTENDED INFOMAX

Natural image sequences make moving Gabors. Van Hateren & Ruderman (1998) studied ICA in spatio-*temporal* domain. They have successfully tried to computationally reproduce experimental evidence[21] that V1 simple cells have receptive fields localized in space *and time* (Artun, Shouval & Cooper, 1998), band-pass in spatial *and temporal* frequencies, tuned in orientation, and selective for *direction of movement*.

Simulated as well as natural receptive fields were found to span a spectrum of different values of these parameters. To achieve this, they have performed ICA on temporal sequences of natural images taken from television programs. A Cray-supercomputer has been used for running ICA-algorithm based on Bell & Sejnowski (1995, 1997), presented here in sec. 3.2, developed in the final form by Hyvärinen & Oja (1997).[22] Instead of detection of static edges, the simulation of van Hateren & Ruderman (1998) with input data-sets having spatial plus temporal dimensions reproduced spatio-temporal Gabor-like receptive fields responsive to *moving* edges. Corresponding edge-like, edge-detecting wavelets moved with fixed velocity perpendicularly to their own (or to edge's) orientation-axis. Bogacz *et al.* (2001) established a temporal convolution, i.e. $X^t = \Sigma_{f=0}^{T-1} A^f S^{f+t}$ (t is the time-sequence index), instead of Equation (3.1), and so achieved some biologically more plausible improvements.[23]

Comparisons and critiques. Beside adding temporal dimension to data-vectors and successfully reproducing V1-like "*moving* Gabor-wavelets" (i.e., moving sinusoids in a steady gaussian envelope), these simulation did not go essentially beyond algorithms by Bell & Sejnowski and by Hyvärinen & Oja (1997) — in the sense of satisfying the need for more biologically-plausible implementation, not merely output.[24] Filters which are focused on low spatial frequences tend to be tuned to faster movements in natural-scene movies, in contrast to filters centered at high spatial frequencies. Van Hateren & Ruderman (1998) somewhat reinterpreted Bell

& Sejnowski by emphasizing that the independent-component filters ϕ (in rows of *W*), and not the independent components *Y* (in columns of *A*) which differ somewhat (as in Fig. 2 in *ibid.*) from filters, are comparable to V1 simple cells with their receptive fields. Additionally to making this more precise specification, van Hateren & Ruderman (1998) write:

"This comparison should not be taken too far, however. First, the [Bell & Sejnowski] linear independent-component model for natural scenes can only be approximate [...], because objects in scenes do not superimpose linearly, but by occlusion [...]. Thus the ICA technique merely serves as a heuristic for maximizing independence. Nevertheless, it appears that even linear ICA is dominated by occurrence of occlusion: in preliminary ICA experiments on constructed static images we found that both occlusion and independent statistics (variances) of the occluding surfaces are necessary to get edge detectors. A second reason to be careful with comparing independent-component-filters with simple cells is that responses of simple cells are not expected to be completely independent of each other, because they appear to achieve a strongly overcomplete representation. This may be partly to the fact that single neurons are noisy, and have a limited information capacity. Some form of (local) ensemble coding may therefore be necessary. Thus the independent-component-filters should be viewed only as generic receptive fields, realizing a code that presumably underlies the actual neural implementation."

Beside successful agreement of computational results with experimental findings in the mentioned aspects, there were some disagreements in details. Let us mention two of them. First, ICA-filters are more symmetrical in time and more narrowly tuned to temporal frequency than V1 simple cells. Second, the finite spectrum of spatial coordinates is covered by a finite number of filter-cells which respond maximally to specific positions, but this is not the case for time dimension which is not finite. Stimuli can occur in different times which may span an unlimited range, so that the number of required neurons responsive to specific time-intervals could in principle become infinite. (Cf., Bogacz *et al.*, 2001.) A solution was proposed by combining ICA-filters which are identical in space, but centered at different times, into a single continuously-functioning filter which has to incorporate specific delays to obey the causality order. Finally, some technical difficulties also remained.

Extended infomax. T.-W. Lee made an extension of the Bell & Sejnowski algorithm and roughly reproduced the results of Bell & Sejnowski (1995, 1997). In the case of super-gaussian data, his[25] ICA learning rule is: $\Delta \boldsymbol{W} \propto (\boldsymbol{I} - th\ u\ u^T - u\ u^T)\boldsymbol{W}$. In the case of sub-gaussian data, the learning rule is the same, but the minus before *th* is replaced with a plus. Recall that sub-gaussian data were problematic for the Bell & Sejnowski net, but for Lee's extended ICA this is not so.

Lee[26] discusses biological plausibility of the preceeding and his ICA-algorithms as follows (literature mentioned in this quotation is in Lee's and not in my references):

"Linker's infomax principle and the sensory coding strategy proposed by Atick (1992) were biologically motivated. However, the learning rule in equation (2.56) [i.e., Lee's extended infomax, note by M.P.] in a single-layer feedforward neural network is non-local and would be more difficult to implement. An example of local learning rule using equation (2.56) is the extended exploratory projection pursuit network[27] with inhibitory lateral connections (Girolami & Fyfe, 1997b). A feedback architecture with local learning rules is presented by Cichocki et al. (1995). Recently Nadal & Parga (1997)[28] have suggested that the infomax learning rule can be related to the BCM theory of synaptic plasticity (Bienenstock et al., 1982)."

Lee (*ibid.*, ch. 8) reviews some other simulations. In simulations by Bartlett & Sejnowski (1997), ICA performs face-recognition tasks (first, in different poses; second, in different lighting conditions) better than PCA. In simulations following Gray, Movellan & Sejnowski (1997), where a word should be guessed by "reading it" from a temporal sequence of images of moving word-articulating lips, local algorithms were better than global ones, and specifically, local PCA slightly out-performed local ICA. So, local (i.e., individual inputs are image-patches, not whole images) PCA turned out to be the best in this case, since local features seem to be more suitable for lip-reading task, and since this is an associative and higher-order problem which transcends the edge-tracing by phase-alignments.

Other related models. It was also found that if data-bases are overcomplete to a higher degree, then input-data are processed more efficiently. Olshausen & Millman (2000) presented a general framework, based on *overcompleteness*-statistics, where the simulated model searches for sparse representations of data *without sparseness being imposed.* The sparse data-distribution is modeled by a mixture of gaussians. One gaussian captures non-active coefficients; the other one, two or more gaussians capture the active coefficients.

Blais *et al.* (1997) simulated formation of receptive fields by *single*-cell dynamics[29] after exposure to natural-scene input. They used *BCM learning rule* (in quadratic form), *maximization of kurtosis* (i.e., deviation from gaussian distribution) which turned out to operate similarly to BCM-net, *maximization of skewness* (i.e., deviation from symmetry of a distribution), and *single-cell ICA*. There are thus many infomax-ways, multicellular and "intracellular", to the same goal – reproduction of receptive fields.

Recent simulations by T.-W. Lee *et al.*[30] realized a sparse representation for the *color* natural scenes based on color-opponent processes (cf., Auxil. Sec. 6.11). They are reflected in red–green or yellow–blue pairings in resulting Gabor-like wavelet structures. ICA thus realizes efficient codes for color scenes like for black-white scenes, but redundancy is reduced even more in the color case.

3.5 SEARCH FOR BIOLOGICAL SUPPORT OF INFOMAX MODELS

Outputs plausible. Many different models with *different algorithms*, using different mathematical or statistical means and different computational "implementations"[31], have succeeded to reproduce satisfactorily (for present criteria) the receptive fields of V1 cells in their static, spatio-temporal and complex variants.[32] Information-theoretic meta-analyses of individual algorithms, attempting to provide a common framework for them, begin to emerge (Lewicki & Olshausen, 1999)[33], but still do not sufficiently indicate possible biological implementation. Successful models were usually variants and extensions of ICA which turned to be *clearly superior over PCA* in this task.[34] In some easier tasks, like modeling orientation-preference and ocular-dominance stripes (see Auxil. Sec. 6.10), models based on Self-Organizing Maps (Appendix B) and PCA (Appendix A) had some success also, so ICA is not overall winner in such tasks.

In a recent study by van Hateren & van der Schaaf (1998), receptive fields of simple cells in macaque's V1 were again compared with properties of ICA-generated (by the fixed-point algorithm by: Hyvärinen & Oja, 1997) independent-component filters. In the following quantities of the *amplitude spectrum*, agreement was good after processing of natural images has been executed:

1. *Spatial-frequency bandwidth* (defined as the full width at half maximum (FWHM) of each filter along the orientation of the peak)[35],
2. *orientation-tuning bandwidth* (defined as FWHM along a circle through the peak, with centre at zero spatial frequency),
3. *spatial frequency* and *orientation of the peak*,
4. *length* and *aspect-ratio* (defined as the quotient of length and width of the filter-envelope) *of the receptive fields / filters*.[36]

There is, however, no match for *peaks of spatial-frequency sensitivity*: ICA-filters are focused on the finest spatial scale, but simple cells process over many scales. Since natural images contain roughly the same amount of information on many scales (are multi-scale or even "fractal-like": compare with Olshausen & Field, 1996, fig. 2c), ICA performs poorly comparing to V1-neurons in this respect.

Implementation presently implausible. None of the models, in my present opinion based on present limited physiological evidence, can be considered as satisfactorily biological in the sense of implementation (except a part of the Olshausen & Field model). This is recognized by the authors of models themselves as well as their reviewers like Lee who calls for "reconsideration of algorithms for image processing issues in the visual cortex".

Van Hateren & van der Schaaf (1998) note (in "Introduction"): "The independent component model of the primary visual cortex should not be regarded as a full model of simple cells in primary cortex. As it is linear, non-adaptive model, many aspects of simple cells are ignored, such as contrast adaptation (e.g., Sclar et al., 1989), contrast normalisation (Heeger, 1992), nonlinearities involved in orientation tuning (Volgushev et al., 1996), adaptation to various stimulus statistics (Zipser et al., 1996), and so on. Nevertheless, the model has clear information theoretic interpretation, and is strictly based on stimulus statistics." [all references as in original, not cited in our bibliography]

ICA models are currently less supported by physiological evidence than Hebbian models. This might change if more experimental attention will be paid to infomax models in the future, since they have more support of psychophysical experiments and information-theoretic studies (e.g., Wainwright, 1999; van Hateren, 1992; Porrill *et al.*, 1999) than Hebbian models. Maybe infomax-processing is realized on a "bio-software" level more than on a "bio-hardware" level as Hebbian models seem to be (at least more than infomax models). This could be the reason why there is the unresolved problem of biological implementation of infomax-models like ICA.

Chances for biological support? Phase-Hebbian models, as I call them, can be a way for finding neural correlates of infomax-modulated image-processing. Phase-processing and reduction of redundancy seem enough biologically-plausible[37] and even unavoidable for perception modeling, therefore going beyond Hebbian-based models like PCA is necessary.

Phase-Hebbian learning rule, i.e. the Hebb correlation-rule with phase-differences (because complex-valued activities are correlated or convoluted), that's like in equation (1.4), seems to be a good starting candidate. Namely, it does not contradict current physiological and psychophysical experiments and can act as an executive element of infomax by processing higher-order correlations and dynamics. For early-vision problems like edge-detection, a "phase-Hebbian infomax" compromise-model is expected to have more chance in getting physiological support than for higher-order problems like recognizing occlusions of objects.

Edges are the first level of invariance or perceptual constancy and can be detected by linear transformations. But finding transformations that are invariant to shifting, scaling and rotation of object-patterns, are mainly an open problem for ICA (Lee after Bell & Sejnowski, 1997). These transformations were with certain success tackled with generalized Hebbian models using Fourier-preprocessing (e.g., Haken, 1991) and by other non-infomax specialized models. ICA seems to be a good model for image (pre)processing, but not necessarily for object perception (which is well distinguished from image processing by the holonomic theory and other models of vision, and this has psycho-physiological reasons). Object recognition, based on search for perceptual invariances, might need a combination of ICA and associative processing, probably in attractor-networks which manifest gestalt-like structures (Amit, 1989; Peruš, 1995a; Peruš & Ečimovič, 1998). Phase-Hebbian processing might be the searched-for biologically-implementable associative infomax-algorithm for higher image-processing *after natural images have been successfully preprocessed and abstracted (including extraction of edges) by an ICA-like infomax-algorithm*. So, the first stage would be ICA-like preprocessing of images, and then another (phase-Hebbian based?) associative / synergetic process would take over to recognize objects from images.[38] These intuitive speculations got some support by simulations of Bartlett & Sejnowski (1997) and seem to have chance for support of future experimental and computational research. In section 4.2, these indications will be tried to be integrated.

Weak points of ICA-models. Olshausen (in his website) recognizes that ICA-based image-processing models are *linear* and so do "not take into account the fact that the real causes of images (objects) occlude one another and also appear at varying positions, sizes, orientations, etc., in the retinal image". Because transformations like translation, rotation and scaling are linear, I do not see the reason why linearity of the model should be the main problem, at least theoretically. For instance, Haken (1991) simulated recognition of variously deformed, not merely translated, rotated or/and scaled, images in his Hebbian-based synergetic model without excessive non-linearity. I have simulated[39] response of a simple quasi-Hebbian linear associative ANN to input-cases, which were translated, rotated and scaled relative to the learned cases, with sufficient success — which was even perfect in the case of convenient data-structure (i.e., "cross-talk" eliminated)...

In practice, however, several other features (like edges and other invariances, for example) could well make problems which would *in*directly be related to limits of linearity. Namely, Olshausen (*ibid.*) continues: "In order to take into account such complexities in image structure [i.e., occlusions, and varying positions, sizes, orientations (note by M.P.)], it will be necessary to devise a significantly different and highly non-linear imaging model."

Indeed, individual objects, which are parts of an image, do not mix linearly, but they occlude each other. This needs nonlinear processing. Because of sparse coding, input-to-output transformations by Olshausen & Field actually *are* to some extent nonlinear (they suppress non-dominant responses and concentrate coding to dominant ones).

A real weakness of ICA-algorithms might be *non-local processing*, i.e. a neuron uses information which is coded in other neurons, but there is no direct connection to them. Or, in another view, (several) neurons need to process as if they would have a lot of information, which is encoded in many other sites, available at once. Quantum non-locality and coherence might help here, but not necessarily at all, because very specific information-access is needed in algorithms.

3.6 DENDRITIC FIELD COMPUTING

MacLennan's linear model. Bruce MacLennan (1994; 1999; in Pribram, 1993) has systematically demonstrated how could PDP be realized in more-or-less continuous physical media so that information would be processed in analog rather than digital manner. Such a *field computation* (mathematical foundations in MacLennan, 1990) could be implemented also in subcellular systems. Possible implementation of field computation in dendritic nets is systematically studied by MacLennan (in Pribram, 1993). The holonomic theory argues that function of axons is communication, but function of dendrites is computation. In spite of increasing awareness of importance of nonlinear processing, MacLennan (in Pribram, 1993, espec. pp. 170-172) shows precisely where linear approximation is not to be abandoned, especially not in dendritic nets. Namely, dendritic wave interactions (here is computing) are mainly linear, but nonlinear processing starts at axon-hillock where action potentials (here is communication) are generated. At low conductances, the postsynaptic potential depends linearly on the synaptic conductance induced by presynaptic potential. A synapse, especially on a dendritic spine, behaves approximately linearly as long as it does not get saturated at high rates.

MacLennan then demonstrates that for *global* subcellular field-computation, where numerous parameters (like concentrations of neurotransmitters and other charges, and resulting ionic currents and (de)polarizations in membranes) are involved, linear modeling is not to be underestimated in spite of many *local* nonlinear processing ingredients (like multiplication of signals, which may take place in addition to their summation or integration and differentiation).[40] Wolpert & MacLennan (1993) show that a linear field computer can even simulate a Turing machine which is the basis of all contemporary (von Neumann) computers.

In general, any deterministic state-transition of a system with orthogonal eigenstates v is given by a linear operator (matrix) $\boldsymbol{M}$ which is equivalent to a Hebbian sum of tensor-products (elements of the tensor product of eigenvectors are: $M_{ij} = v_i v_j$). Approximate global linearity of dendritic processing indicates that PDP models, like the models by Hopfield (1982), advanced for instance by Amit (1989), and by Haken (1991), are better implementable in dendritic nets than on neuronal level (meaning emphasis on processes in soma and axon). Furthermore, Bray (1995) argued that interaction-webs and chemical cooperation of proteins are a relatively good implementation of connectionist PDP models, and there are several such proposals for various subcellular levels (e.g., Dayhoff *et al.* in Pribram, 1993, ch. 12); and finally, I will show that quantum implementation of the linear Hopfield model with phase-Hebbian learning rule is the most direct and clear one (Auxil. Chapters 7 and 8).

Dendritic Olshausen & Field net? After presenting a general set of dynamic equations for dendritic-field processing, MacLennan (in Pribram, 1993, sec. 6.1) presents a concrete linear algorithm based on Gabor wavelets as eigenvectors (for our purposes: eigenimages). They are not orthogonal in general, but MacLennan (in Pribram, 1993, sec. 6.2) analyses an often-occurring case of optimal processing of "*nearly* orthogonal" data. This is discussed in detail in Auxil. Chapter 10, because "almost orthogonal" input-data, which are usually optimal for Hebbian-based associative processing and memory, are used, as proposed in Peruš & Dey (2000), also in the quantum associative net. Such a "fuzzification" of quantum-mechanics- or Hebb-based information processing on a so-called complete orthonormal set of eigenvectors is shown to be possible and useful in the case of an open system (i.e., having significant input from environment) — see Auxil. Chapter 10.

As shown in sec. 3.3, MacLennan's dendritic field model is equivalent to the Olshausen & Field algorithm without sparsification. This indicated that the Olshausen & Field algorithm can be implemented in the dendritic field. A "sparsification pressure" is then imposed on dendritic (and maybe also on neuronal) processing in order to get maximally-sparse codes. Biological origin of such a "pressure" is, for now, only to be guessed: It might arise from lateral inhibition, or from top-down influences, or from special feedbacks.[41] Such top-down influences might originate in virtual higher-level attractor structures (the "software" level), maybe in a similar way as in Haken (1991) (summarized in Peruš, 1995b). This is only a speculation as a result of unknown biological realization of sparsification. The inhibition hypothesis has support in Pribram's (1998a) words that "as the dendritic field can be described in terms of a spacetime constraint on a sinusoid – such as the Gabor elementary function, the constraint is embodied in the inhibitory surround of the field." See Auxil. Chapter 11 for some possibly-relevant models.

3.7 FROM INFOMAX-BASED IMAGE PROCESSING TO SECONDARY ASSOCIATIVE PROCESSING WITH ATTRACTOR NETS

Image processing plus object perception. Bartlett & Sejnowski (1997, 1998) have quite successfully (and better than with PCA) combined *ICA-based image processing* (V1-like) *and object-perception with attractor-networks* (ITC-like?) into an unified computational model. Specifically, results of ICA-preprocessing of input-images (in this case, faces from various views when a subject moves his head or we move around him) were put into attractor-ANN (e.g., Amit, 1989), after passing a low-pass temporal filter. As a result of inter-spatial and inter-temporal Hebbian associations, faces neighboring in temporal sequence landed into the same attractor. Thus, view-independent categories of individual objects (faces) were formed as corresponding attractors — an attractor for each (head-moving) person.

In the attractor-net, Hebbian associations between Gabor-filtered proximal time-sequences made the neurons' tuning to specific facial poses more broad. Addition of lateral connections, however, achieved invariance to pose of a face. This pose-invariance was manifested by all single-person's poses "falling into" the same attractor. One could speculate that *phase*-Hebbian memory-storage or "learning" would improve the results (cf., Spencer, 2001a,b) of such an encouraging integrative model which attempts to bridge the gap between image-perception, which is view-dependent, and object recognition which is view-invariant.

Gabor wavelet equivalent to quantum wave-packet. The Gabor wavelet was originally proposed by the "discoverer" of holography D. Gabor in 1946. The Gabor wavelet is a sequence of waves under a fixed gaussian envelope while the frequency of the wave inside the envelope varies for different cases (Fig. 2.2-right). Such a wavelet is *equivalent to a family of Weyl-Heisenberg coherent wave-packets used in quantum physics* (Lee, 1996). This observation allows me to relate infomax image-processing with quantum-implemented associative processing with attractors – in the quantum associative net.

ENDNOTES

1. PCA is Principal Component Analysis. It is presented in detail in Appendix A. ICA is Independent Component Analysis – to be presented soon.
2. *Second* order here means that it is based on a "learning rule" which is mathematically described by a multiplication of *two* variables describing state of the net, i.e. it is a *bi*linear rule.
3. (On phase, see footnote in sec. 1.3.) Exponential function is equal to a series (i.e., a long sum) of functions x, xx, xxx, $xxxx$, etc. *ad infinitum*, corresponding to "the first-, second-, third-, fourth-order statistics of multilateral auto-correlations" in an image. This indicates the relation between phase-information, higher-order statistics and multi-neuronal connections.
4. Complex synaptic processing is schematically illustrated in Figs. 1 and 2 of ch. 7 by MacLennan in Pribram (1993). It could, for example, be realized by a connection of a second synapse from neuron 3 onto (or onto axon 1 near to) the first synapse from neuron 1 to neuron 2. Another, much more "fine", possibility for multilateral coupling is the *dendritic web* where synaptic connections between many neurons can evolve in way which can be characterized as hardly-to-be-looked-over, immense, complex bundle.

5 Other such methods, although mainly based on second-order statistics, were: Harpur & Prager (1996) (Hebbian); P. Földiak (1990): *Biol. Cybern.* **64**, 165-170 (anti-Hebbian); N. Intrator (1992): *Neural Computat.* **4**, 98-197 (unsupervised ANN); G. Hinton *et al.* (1995): *Science* **268**, 1158-1161; and few others (Zemel; Saund; Lu *et al.*).

6 By comparing Fig. 1 with Fig. 4 in Olshausen & Field (1996a), and/or Fig. 1 with Fig. 3 in Olshausen & Field (1996b), and Fig. 3a with Fig. 3d (and Fig. 4) in Bell & Sejnowski (1997), it is clear that ICA is superior over PCA. As shown in Fig. 4 of Olshausen & Field (1996b), ICA develops independent components of natural scenes that are spatially localized, oriented and also band-pass (i.e., selective to structure at different spatial scales) like V1 simple cells are, therefore such independent components correspond to edge filters. PCA does not develop characteristics found in V1 simple cells, so principal components of natural scenes are not edge filters.

7 Orthonormal basis is a set of vectors which are orthogonal (have scalar-product zero, i.e. are perpendicular to each-other) and normalized (their length is 1). Orthonormal basis allows a complete topographic description of a (mathematical) space. E.g., "vectors of length, width/breadth and height" constitute such a basis for the 3-dimensional space.

8 The "infomax" details are very subtle. When speaking about maximal overall information, we mean maximal information-*preservation* while transforming input-data into memory-codes or outputs, but at the same time being as "economical" as possible. So, **input-output information should be preserved as much as possible ("infomax")**, but at the same time **all the redundant (multiple, unnecessary, common) data should be eliminated (i.e., mutual information in outputs should be minimized)**. Thus, when we try to remember many types of cars, we store in memory only the most important characteristics of all cars, then specific characteristics of particular types, and finally special characteristics of individual cars. *All other repeating features are not (separately) stored, but are represented in the condensed, abstracted code of essential characteristics* of the category "car", or sub-category "mercedes". *What remains in memory, are maximally* ***different*** *features only, therefore "infomax" and maximization of joint output-entropy are equivalent*, in this case.

9 Valid for invertible continuous deterministic mappings. (Note difference between mutual informations $I(X,Y)$ and $I(a,b)$, since they are of different sorts.)

10 Proof for maximum information transmission with sigmoid g (in gaussian case) see in Bell & Sejnowski (1995, p. 1132-3, espec. Fig. 1). Other "infomax"-rules with various activation-functions g are displayed on p. 1155 (Table 1). Their choice depends on data-structure. In general, the "infomax"-rules consist of an

anti-redundancy term and an anti-Hebbian term (i.e., Hebbian with negative sign) which prevents saturation of the neuron's activity.

11 ΔW means change of W; W^{-1} denotes inverse of matrix W; W^T denotes transposition, i.e. exchange of matrix-elements W_{ij} with W_{ji}.

12 In all vector-variables and equations in the text, vector-arrows are omitted for simplicity of typography.

13 Gradient-descent/ascent is a procedure of gradual decrease/increase of a quantity (here entropy H) because of changing (partial derivatives ∂/∂!) another quantity (here the synaptic "weights" W).

14 Details in Bell & Sejnowski (1997, p. 3336, upper right column).

15 Note by M.P.: Eq. (13) of Bell & Sejnowski is reproduced here in Equation (3.5); the mentioned "feedforward and feedback versions" are not reproduced here to avoid technical details which *presently* have no biological support.

16 The artificial neural net (ANN) model of Olshausen & Field, presented in the next section, is one of them!

17 Iterative processing means that the output of the operation described by the equation is fed back as the new input to the same equation. Such feedback dynamics usually turn out to be self-correcting and often lead to an optimal final rest-state. Adjustment of network's parameters by such dynamics is learning in the present case.

18 Coefficients c_k correspond to synergetic order-parameters by Haken (1991) and have quantum analogues. See Auxil. Sec. 7.1, analogy no. 2 and 4).

19 The activities of so-called hidden units are sometimes considered as the codes. In such cases (usually in multi-layer "perceptron" ANN), their number is greater than the number of input units if we have oversampling.

20 Here, input dimensionality means the number of discrete independent sources or causes. (Technically, it corresponds to the number of non-zero eigenvalues in the covariance-matrix of the inputs.) For example, if we watch a scene (e.g., a rider on a horse) for a while, many images of the couple enter the brain. They differ somewhat from each other, because the horse and the rider have both moved a little bit. However, the number of independent causes (objects) remained two (times the number of their subcomponent-units or pixels).

21 E.g.: G.C. DeAngelis, I. Ohzawa & R.D. Freeman (1993): Spatiotemporal organization of simple-cell receptive fields in the cat's striate cortex, I.; *J. Neurophysiol.* **69**, 1091-1117. R.L. DeValois, D.G. Albrecht & L.G. Thorell (1982): Spatial frequency selectivity of cells in macaque visual cortex; *Vision Res.* **22**, 545-559 (simple cells of V1-map of foveal area were measured in these monkeys). C.L. Baker Jr. (1990): Spatial- and temporal-frequency selectivity as a basis for velocity preference in cat striate cortex neurons; *Visual Neurosci.* **4**, 101-113. R.L. DeValois & K.K. DeValois (1991): Vernier acuity

with stationary moving Gabors; *Vision Res.* **31**, 1619-1626 (moving Gabor patterns presented to humans – a psychophysical study).

22 The fast algorithm of Hyvärinen & Oja (1997), which has been used in spatio-temporal image simulations, realizes the infomax ideas by special statistical procedures which can be considered as "fixed-point algorithm for maximum likelihood estimation of the ICA data model". It is different from the Bell & Sejnowski algorithm. The new versions of this extended ICA algorithm are in: Hyvärinen (1999) and, applied to modeling *complex* cells of V1, in Hyvärinen & Hoyer (2000).

23 Bogacz *et al.* (2001, sec. 5) compare their own ("our") model with the van Hateren & Ruderman (1998) ("their") model: "In their approach receptive fields are developed which encode different phases of the same movement. In contrast, in our approach the timing of activity encodes the phase of movement. [...] In their approach, where neurons encode particular phase of a movement, two sequences shifted in time, and therefore shifted in phase, will usually cause activity of completely different neurons (a situation unlikely to be used by the brain), while in our approach they result in the same pattern of neuronal activity shifted in time."

24 New versions, like Hyvärinen (1999) and Hyvärinen & Hoyer (2000), bring improvement of computational efficiency and speed only. In latter paper, however, interesting new relations to V1-modeling were proposed.

25 T.-W. Lee (1998): *Independent Component Analysis*; Boston: Kluwer. In this review book, equation (2.56) is the extended infomax learning rule of Lee, a version of ICA based on Bell & Sejnowski (1997).

26 From *ibidem* (Lee, 1998), sec. 2.9.3., p. 63-. Lee's simulated V1-filters are in Fig. 8.4 in *ibid.*, p. 171.

27 Note by M.P.: Projection pursuit learning is a method, related to BCM-theory, which attempts to extract as much information as possible from non-gaussian data and their higher-order statistics, especially third-order correlations of input-activities. See details in: Intrator & Cooper in Arbib (1995, p. 484-).

28 Note by M.P.: Also referenced in our bibliography. See their sec. 4.5 on links with BCM-theory. BCM-theory of synaptic plasticity (original in: E.L. **B**ienenstock, L.N. **C**ooper & P.W. **M**unro (1982): *J. Neurosci.* **2**, 32-48) bases on a nonlinear generalization of the Hebbian (covariance) rule with significant modifications. While the Hebbian rule prescribes strengthening of synaptic connection if presynaptic and postsynaptic activity are coactivated, BCM-model only does so if the coactivity *exceeds a threshold*, else the synapse is weakened. The threshold is variable and depends on the mean-activity level of the neuron. For details and physiological support of BCM-theory see: Intrator

& Cooper in Arbib (1995, p. 153-). Intrator & Gould (1992) used a BCM-network for 3-dimensional object recognition.

29 Interactions between neurons were eliminated in order to study statistics-based learning-rules on a neuron which receives external input only.

30 NIPS 2000 in Denver, and in T. Wachtler, T.-W. Lee & T. Sejnowski (2001): *J. Opt. Soc. Am. A*18(1), 1-13.

31 A comprehensive collection of web-links concerning ICA-research with various applications and codes is available at: http://www.media.mit.edu/~paris/ica.html

32 See T.S. Lee (1996) for maximally precise mathematical description of the receptive field of a V1 simple cell which gives a *response that is a projection of an image onto* the physiologically-supported *Gabor wavelets*. This projection is given by Lee's equation (4), and fully-detailed Gabor wavelets are given by Lee's equation (3).

33 Beside the "maximum likelihood" model, another meta-model is Independent Factor Analysis, which is a generalized combination of ICA, PCA and factor analysis, by: H. Attias (1999): *Neural Comput.* **11**, (4)803-851.

34 In addition to receptive-field modeling, ICA was superior over PCA in search for independent sources in a mixed signal, e.g. in identification of independent neuronal groups of correlated firing, each with common neuronal input. So, correlated firing of neurons of an independent group was reflected in higher-order correlations between neurons, not only in their (second-order) covariance. Many other "demixing" studies gave similar results like these reported in: M. Laubach (1999): *J. Neurosci. Methods***94**, 141-154.

35 Spatial-frequency bandwidth is expressed in octaves, i.e. factors of 2 in frequency.

36 The length is defined as FWHM of the envelope along the orientation where the filter is low-pass. The width is defined as FWHM along the orientation where the filter is band-pass. The low-pass and band-pass orientation-directions are usually orthogonal.

37 Infomax and reduction of redundancy are very similar approaches (comparison in: D. Obradović & G. Deco (1998): *Neural Comput.* **10**, (8)2085-2101), both related to exploitation of higher-order correlations which have to be handled to achieve statistical decomposition (i.e., factorization of the probability distribution). Higher-order spatial correlations in visual textures were reflected in measured single V1-neurons, is reported in: K. Purpura, J. Victor & E. Katz (1994): *Proc. Natl. Acad. Sci. USA***91**, 8482-8486. They also write that second-order correlations capture spatial frequencies of an image, and higher-order correlations handle object contours.

38 *Spectral* associative memory, based on the Hopfield model, has been proposed by Spencer (2001a,b). It could be added among phase-Hebbian models. It is not (yet) quantum-implemented, but in special chip-electronics.

39 M. Peruš: M.Sci. work (physics), and: M. Peruš & B. Borštnik (1996): "The study of hard-wired neural networks by solving traveling salesman problem" (poster); and "Computer simulations of biological evolution of hard-wired neural nets"; in *Consciousness Research Abstracts*, Tucson II conference-abstract no. 236.

40 J. Shawe-Taylor argued in handbook *Phenomenology and Cognitive Science* (see AuxLit 29) and lectures based on MacLennan's works, that sensory cortices first make "hard-work" nonlinear preprocessing, so that higher association-cortices can, after that, do their integrative work more elegantly – in a linear manner (and so "conscious process could also be roughly linear"). This agrees with our hypothesis of essential involvement of quantum processes in consciousness, since they are linear (at least until not perturbed).

41 Pribram in Dubois (2000b, p. 55) writes about "top-down" modulations of inhibitory influences in receptive fields: "At both the LGN and the cortical level, the inhibitory surrounds and flanks of receptive fields could be augmented or diminished: ITC and putamen stimulation increased the inhibitory surround; anterior frontal lobe and caudate nucleus stimulation practically wiped out the surrounds [after experiments of: D. Spinelli & K. Pribram (1967): *EEG Clin. Neurophysiol.* **22**, 143-149. M. Lassonde, M. Ptito & K. Pribram (1981): *Exp. Brain Res.* **43**, 131-144]. [...]

[...] When the surround was maximized, processing was multiplexed and proceeded in the spacetime domain; when the surround was minimized, processing emphasized the spectral domain and was more holographic-like."

Chapter 4
Images, Associations and Conscious Experience

This chapter describes our model of "emergence" of conscious experience, as correlated with neural and quantum-(like) processes, by taking neuroscience and the holonomic theory optimally into account. We present arguments why neural nets might not be sufficient to "explain" consciousenss, and that sub-cellular and quantum processes within dendritic webs might be necessary. Presentation of characteristics of (visual) consciousness should make evident that contemporary models can hardly be satisfactory, since consciousness is an ego-centered qualitative experience.

4.1 CONVOLUTION, CORRELATION, AND MATRIX-PROCESSING

Pribram (1991, Lecture 10) presents indications that the far-frontal cortex prefers continuous processing and thus convolution / correlation processing. On the other hand, the posterior cerebral convexity prefers discrete processing and thus matrix (interference) processing. The model of section 1.2 is convolutional, as is evident from equation (1.1). My quantum associative net of section 1.4 and all the related models presented in detail in Auxil. Chapters (HNeT, the Hopfield and Haken ANN

DOI: 10.4018/978-1-61520-785-5.ch004

models) have a matrix-core, as in equation (1.4) where the elements of matrix ***G*** (similar to ***J***) are given, but they take part in a sort of convolutional process – as in equation (1.5). Convolutional and matrix operations are thus often combined, and both can involve correlation-processing.

Convolution / correlation processing is more appropriate for configural perception, while matrix (interference) processing is more convenient for modeling associative cognitive processes like formation of prototypes.

Inner versus outer product. The key-difference is that convolutional computing incorporates the *scalar or inner products* of pattern-vectors, and matrix-computing uses a Hebb-based *tensor or outer product* of pattern-vectors. Computing scalar (inner) products gives an "enfolded" code — the result is a single number, i.e. an activity-parameter or probability-coefficient. Computing tensor (outer) products gives a distributed, "unfolded" encoding — the result is a whole matrix, i.e. an array of numbers–weights (which is like description of a hologram).[1]

The convolutional processing, as executed along the retino-geniculo-striate visual pathway[2], is more economic and up-to-date than the matrix-processing is. Namely, a single coefficient, i.e. the scalar product which describes *global* correlation of images, is a much more economic code than the matrix of many elements where each of them describes a *local* correlation of image-patches (and the matrix as a whole represent the combined, overall correlation). Economic refers to a lower amount of used processing elements, and this brings abstraction and condensation of information. Up-to-date refers to *serial* convolving of the current input with the memory (encoded in weights which are often matrix-elements).

In a brief summary: convolutions are used in early abstraction and perception (which is configural and serial); tensor-product matrix-computations are used for description of associative memory and top-down- or memory-based operations including cognitive associations and categorization.[3]

The holonomic theory views imaging as based on convolving the various stages of primary visual processing. In contrast, "[…] Object perception […] depends on correlations among patterns in which centers of symmetry are determined by operations performed in the superior colliculus and the visuomotor system." (Pribram, 1991, p. 261)

4.2 ATTEMPT OF AN INTEGRATED MODEL OF IMAGE PROCESSING IN V1 AND BEYOND

Synopsis. The holonomic theory of the retino-geniculo-striate image processing was presented in section 1.2. It uses Gabor wavelets, eq. (1.2.), as "weighting"- or "filtering"-functions while performing convolution, eq. (1.1), with the retinal image. The result of this Gabor transform is a spectral image-representation in V1. This, roughly hologram-like, representation in V1 is then reconstructed by an inverse Gabor-transform into the spatial representation in V2, probably. Thus, the topologically-correct "image" is recovered in inverted form in V2.[4]

The overlapped Gabor wavelets, which are used in image processing in V1, describe the receptive-field profiles of V1 neurons which realize the Gabor transform using their dendritic trees. Gabor wavelets were shaped by an ICA-like process. I prefer the Olshausen & Field sparseness-maximization process over the ICA-variant of Bell & Sejnowski for the implementation-model of shaping the Gabor wavelets g (i.e., the independent components Y[5]) and especially determining their coefficients s (i.e., the amplitudes or sparse codes of the independent components of input-images). The Gabor coefficients are updated, approximately according to equation (3.9), much more rapidly than the Gabor wavelets. Coefficients change with each new input image, but the Gabor receptive fields adapt in a longer term – after a lot of different images have been presented. In fact, they adapt slowly all the time, approximately according to equation (3.12), but substantial change is seen after a while.

From Gabor receptive fields to wavelets. A Gabor elementary function g has two roles in modeling vision which seem to be somewhat different. First, it is used as a description of the receptive-field profile, i.e. of the "weighting" which the synapto-dendritic net imposes on all the inputs to a neuron. Second, it is used as a Gabor wavelet which represents the encoding of an independent component of input-images. Of course, both roles have the same origin (a sort of ICA) and "are two sides of the same coin". However, the consequence of infomax-processing *manifests in different places*: in the receptive-field profile which "lies hidden" in the synapto-dendritic net, *and* in the Gabor wavelet which propagates to other brain areas and gets involved in further holography-like processing there. Namely, if the receptive field is Gabor-shaped, then it gives Gabor-shaped outputs, or at least something similar or generalized, on a later stage. These Gabor wavelets might be of another sort – e.g., time-dependent or spectral.[6]

This effect (i.e., Gabor wavelets produced because/out of Gabor receptive fields) is more clearly evident if the retinal input is made uniform (i.e., "white noise" or Ganzfeld). This is related to well-known observations that an uniform stimulation triggers a system's response which is a sort of internal "expectation" (or a "hallucination", respectively) of the filtering system (e.g., MacLennan in Pribram, 1993, p. 189).

Sorts of representations. According to our model, there are two sorts of representations in V1 available for further brain processing: the spectral "compound of images", or equivalently, the sparse assembly of codes (the so-called sparse codes) representing or "weighting" the independent components extracted from a collection of images. The first representation (independent component) is hidden in V1 dendritic fields; the second representation (sparse coding) is encoded in activities of V1-neurons. After the spectral representation of V1 has been inverse-Gabor-transformed in the connections between V1 and V2, the retinal image re-emerges in V2. Thus, the usual image, as once originally fallen on the retina, should be reconstructed (turned upside-down, maybe also somewhat "deformed") in V2 or nearby. This "image" is then the third available representation.

Why three representations? (Cf., Kirvelis, 2000.) We could suppose that the spectral (Gabor wavelet) representation is for perceptual image processing in V1. The sparse-code (Gabor coefficient) representation is for robust, rough encoding needed for automatic, immediate, reflex actions — they are unconscious and probably realized by neural circuits alone (dendrites just transmit the signals, do not process them). The "image" representation in V2 is used for the usual phenomenal conscious experience.

Let me explain. When a person is, say, involved in conversation, (s)he sees another person and at the same time processes a lot of information — e.g., "decodes" the other-person's body-language, not to mention more multi-modal and symbolic cognitive processes like understanding of words and thinking about the topic. The person sees the other one in full phenomenal integrity and quality all the time, without interruptions of the information processing (understanding body-language and spoken language, thoughts, etc.) going on in the background. For the seemingly-direct "realistic" experience of the environmental image, as conscious process offers it, the V2 space-time "image" is needed. For the abundant accompanying apperceptual processing, which is unconscious and abstract, the spectral representation is needed. I will suppose that the "image" is also needed for additional processing, mainly limited to visual cognition, which uses associative processing that is more similar to holography than the perceptual spectral processing is.

Phase-Hebbian associations take over. I think that visual associative processes of V1, after perceptual preprocessing has been finished (ICA has generated Gabor wavelets and Gabor coefficients, and the Gabor transform using convolution has produced spectral responses of V1 simple cells), could be well realized by a Hebbian or phase-Hebbian mechanism. The second one has much more chances for good performance. In the second part of this book, examples of Hebbian second-order statistics models are presented: PCA in Appendix A, Hopfield-like associative neural nets in Auxil. Section 8.2. Examples of phase-Hebbian models, which process more or less higher-order statistics, are also presented in detail in Auxiliary Chapters: HNeT in 8.3, the quantum associative net in 8.4 (and 10). These phase-processing models are those which are the most similar to holography!

These models may in some respect be less efficient than other phase-processing models, which are not phase-Hebbian, like ICA and MacLennan's dendritic model as far as it has similarities with the Olshausen & Field model. But the phase-Hebbian models have a peculiar symmetry which makes them fundamental and close to physics. So, I believe, there is a division of labor. Other models, like ICA, convolution-models, Kohonen's Self-Organizing Maps (Appendix B), "do the hard job" first. After their processing with nonlinear moments (like sparsification) is finished, the phase-Hebbian associative dynamics start the "fine job". Phase-Hebbian models have an ability to construct rich multi-level attractor-structures. In this, they can go beyond Hebbian models which are already successful in flexible attractor-dynamics (my example in Auxil. Sec. 8.5; full details in: Peruš, 1995a, 2000a; Peruš & Ečimovič, 1998).

Attractor processing. For secondary visual processing[7], processing with attractors is unavoidable (as indicated in sec. 3.7). Pribram also says that cortical processing uses largely parallel-distributed and redundant representations. The model, which realizes this most directly and is also the best ANN embodiment of holography, is sketched as follows. In simple words, the whole network of units with their connections encodes numerous "images" simultaneously: In the weight-matrix (encoded in the array of connections / junctions), there is the content-addressable associative memory. In the configuration-vector (encoded in the set of units), there is the "image" which is currently processed (which "we are *conscious* of"). Each vector of activity-configurations, which represents an "image", acts as an attractor of network's dynamics, because it is at the minimum of its potential well — as in the model of Hopfield (1982).

In the matrix of connections (or "hologram"), not the whole patterns or images are stored. Merely pair-wise ***(auto)correlations*** *of all previously-input images are stored.* With other words, *condensed information about (dis)agreements among all image-parts of all the watched images is encoded so that it is parallelly-distributed across the whole matrix.* This is sufficient for reconstruction of an image from the memorized image-traces *if* a recall-key (i.e., a new, similar input pattern) is presented to the array of connectionist units (formal neurons) of the network. The interaction of these units across the connections is modeled by multiplication of the vector of units' states with the matrix of inter-unit connections.

The processing of this model has been extensively presented for experts and non-experts in Peruš (1995a, 2000a) and Peruš & Ečimovič (1998), and, of course, also in Amit (1989) and numerous physical literature. Let this rough sketch therefore be sufficient, although the subtle flexibility of attractor dynamics, so much necessary for pictorial cognition, has been neglected fully here.

Quantum net as the inner processor. The essential point is that processing of the *Quantum Associative Network*, as sketched in section 1.4, as derived from ANN in Auxil. Ch. 8 based on analogies of Auxil. Ch. 7, and as presented as a sort of mathematical / physical "recipe" in Auxil. Ch. 10, realizes the attractor-dynamics, associative processing and image recognition in a "compact" and effective way. It is progressive that Hebbian processing is enriched with phase-processing. *Because this model can be quantum-implemented in a natural way, it is, for now, hard to imagine anything more fundamental, more holographic, more effective, and hypothetically more directly linked to conscious experience, in the sense of associative processing.* Processing similar to that of quantum associative net might take place in V1 and partially also maybe in V2.

(Dis)advantages of the quantum net. Certainly, much more sophisticated quantum information processing is, in principle, possible in deeper quantum levels than those of *Quantum Associative Network*. The quantum associative net is based on the usual quantum-mechanics formalism in the Feynman path-integral interpretation. The deeper levels are, for example, those of quantum field theory, including multiparticle systems and collective modes (Vitiello, 2001), and quantum inseparability called entanglement.[8] However, the quantum associative net is the *basic* quantum model which efficiently enough realizes image processing and all the mentioned features of associative information processing. It is, so far, the only natural and purely quantum model with a direct and complete information-processing "mechanism" or "algorithm" specified in sufficient, but not superfluous, mathematical detail. Its simplicity is advantage rather than disadvantage in spite of not being an universal model (as no model is). And, the quantum associative net is more akin to holography than any of the other presented quantum models.

The quantum associative net can process, by interference, various kinds of eigen-wave-functions (eigen-images). They can be Gabor wavelets. Gabor wavelets are very similar to the natural quantum wave-packets (Lee, 1996; MacLennan in Pribram, 1993). In sec. 1.4 and in Auxil. Chapters, I present interference-processing in the quantum associative net using the plane-waves for image-bearing eigen-wave-functions — for simplicity, although efficient. This is the most usual / basic quantum-type holography. *Interfering Gabor wavelets* could enable more sophisticated, maximally information-preserving processing, in accord with the holonomic theory. This has not yet been realized theoretically (analytically) or computationally in such a way[9], as far as I know. Namely, one would need, beside a plentitude of original natural images used for ICA-learning, the carefully-produced wavelets from the preceding compatible ICA – to use them as inputs for processing and memory-storage in the quantum associative net.[10] (I am starting to work on it.)

Associative basis for visual cognition. In Auxil. Section 6.5, the uniformity of (neo)cortical structure (Ebdon, 1993; Oakley, 1985, after Mountcastle) was described. This allows the use of phase-Hebbian associative models for a rough (but maybe the best available) approximation of global (neo)cortical processing which is at roots of visual cognition. The proper modeling combination, I suppose, would be: *ICA-constrained convolutional preprocessing up to V1, followed by fractal-based* *associative processing in (neo)cortical neural, dendritic and quantum attractor networks – one within another inside V1.*

Of course, I am probably entering pure speculation if I try to be more specific now. The idea is that the multi-level phase-Hebbian associative processing, having the quantum associative net as the most deep/inner level (for now), is currently the most convenient one for *cognitive manipulations of images or*, rather, *object-forms*, as performed probably in the inferior temporal cortex (ITC). ITC is specialized for prototype-converging[11] recognition or *comprehension* of objects, including discriminations and choices resulting from it (Pribram, 1971). In accord with Rainer & Miller (2000), Riesenhuber & Poggio (2000) argue that the prefrontal cortex finishes the object-recognition started by ITC. They write on p. 1202: "In anterior ITC, invariances to object-based transformations, such as rotation in depth, illumination and so forth, is achieved by pooling together the appropriate view-tuned cells for each object." Then Riesenhuber & Poggio (2000) continue in caption to Figure 3: "The stages up to the object-centered units probably encompass V1 to anterior ITC. The last stage of task-dependent modules[12] may be located in the prefrontal cortex." (I would agree, but would avoid the use of mechanistic terms like "units" and "cells" for carriers of coding.)

Experiments of Rainer & Miller (2000) on object recognition in the prefrontal cortex showed that familiar objects activate fewer neurons than novel objects do, but these neurons are more narrowly tuned. Such a sparse representation of a familiar object is also more robust to degradation (made after the learning period) of a newly-posed stimulus-object. Based on ITC inputs (in vision), the prefrontal cortex is the region most important for the so-called working memory used in cognition. Present models[13], based on works like Amit (1989), use Hopfield-net-produced Hebbian attractors for working-memory representations and attractor dynamics for (visual) cognition. I believe, generalization of these models (which were used in: Peruš, 1995a; Peruš & Ečimovič, 1998) by incorporating phase-processing (i.e., using the phase-Hebb rule) and implementing it in dendritic or/and quantum networks would be more appropriate.

It should be emphasized once again that image processing and subsequent object recognition could be possible only because of "the hard job done" by ICA and the perceptual convolutional cascade. They provided Gabor wavelets and spectral representations of images which are still used in many higher cortical areas for more abstract processing (i.e., processing without the topologically-correct pictorial representation – the usual image). Volition-, "I"-based control- and symbolic processing are examples of abstract processing. On the other hand, operations of visual cognition — like imagery, mental manipulations of objects (end of Auxil. Sec. 6.7), visual modeling or planning (e.g., vividly imagining how to drive from A to B) — could have cortical implementation based on the quantum associative net, especially if these processes are performed consciously. This fits the Crick and Koch (in Hameroff *et al.*, 1996) hypothesis that we begin to be conscious of visual processing in V2 and beyond (encompassing visual cognition based on cooperation of ITC and the prefrontal cortex).

4.3 ADDITION OF CONSCIOUS EXPERIENCE AND QUANTUM PROCESSES INTO CONSIDERATION

From dendrites to conscious experience. The following citate from Pribram (1971, p. 105) summarizes the view which has been later elaborated by the holonomic theory and will be discussed in my following sections:

"Neural impulses and slow potentials are two kinds of processes that could function reciprocally. A simple hypothesis would state that the more efficient the processing of arrival patterns into departure patterns, the shorter the duration of the design formed by the slow potential junctional microstructure. Once habit and habituation have occurred, behavior becomes "reflex" — meanwhile the more or less persistent designs of slow potential patterns are coordinate with awareness. This view carries a corollary, *viz.* that nerve impulse patterns per se and the behavior they generate are unavailable to immediate awareness. [...]

In short, nerve impulses arriving at junctions generate a slow potential microstructure. The design of this microstructure interacts with that already present by virtue of the spontaneous activity of the nervous system and its previous "experience". The interaction is enhanced by inhibitory processes and the whole procedure produces effects akin to the interference patterns resulting from the interaction of simultaneously occurring wave fronts. The slow potential microstructures act thus as analogue cross-correlation devices to produce new figures from which the patterns of departure of nerve impulses are iniciated. The rapidly paced changes in awareness could well reflect the duration of the correlation process."

Discussion on sorts of representation in section 4.2 seems to fit this citate. Sparse-coding assemblies of neurons (i.e., just few neurons fire, and this is enough for encoding entire images) serve in reflexes without awareness. The second representations of images, the Gabor wavelets, interfere in the (sub)dendritic microstructure. The correlation process, hidden in subcellular or quantum (as I prefer) interference, could also be accompanied by awareness. The final result of interference processing, the conscious image, would be reconstructed after the collapse of quantum (or at least quantum-like) wave-function.

In support for his hypothesis on junctional electric activity[14] as the substrate for awareness, Pribram (1971, 1995) mentions that using biofeedback subjects can discriminate α-EEG waves in their brain by "feeling them as pleasantly relaxed awareness". He also cites Libet's findings that stimulation-produced awareness occurs in patients 0.5–5 seconds after the relevant brain-area has been stimulated — as if some electrical state would have to be built before the patients can experience anything.[15]

Neural and quantum "sides" of dendrites. Infomax-processing is probably based in dendritic-*fiber* networks or/and neural circuits – on the (sub)cellular level, not quantum. V1 image processing and subsequent visual associations are probably realized by quantum-based dendritic *micro*processes. Dendritic processing thus combines two levels. Its *macroscopic fiber-part* is involved, under some top-down influences probably, in shaping the Gabor receptive-field profiles by specific collective dynamics of dendritic trees of many neurons that criss-cross. Its *microscopic membrane–"bioplasma" part* (in the patches or "holes" in-between the criss-crossed dendritic fibers beneath their membranes) implements the holography-like image-processing, as will be described in sec. 4.6, but probably by interfering a sort of Gabor wavelets instead of plane-waves. Wavelets could be naturally rooted in quantum background.

Why fractal-like multi-level processing. Systematic observations show that brain structures repeat roughly on many levels or scales like in a fractal. Why such a (seeming?) redundancy? The answer is probably: flexibility, adaptability and universality.

Pribram (1971, 1991) observes that patterns shaped or learned in one part (area, level) can be transferred to another part (area, level) of the brain. One perceives an image, can recall it, manipulate it in imagination, one can use it to guide and control motor action directed toward its object (the image of achievement). In Pribram's example, one can draw a circle with a pencil on a paper or wall using fingers of a hand or even of a foot, or one can put the pencil in the mouth. *The same pattern* (circle) *can be produced* (drawn) *in different circumstances using different body-actions and different brain areas.* Even different levels of the tissue are needed: microscopic for processing, macroscopic for execution of action. To mobilize a muscle, amplification of (sub)neural signals is necessary – that's why neurons are needed also, not just dendrites and quantum systems. Neurons are cells – like the muscle cells are also. Since Nature is multi-scale, body and brain also have to be multi-scale to handle it. The many levels therefore have to cooperate fluently, so they must be compatible in information exchange. Patterns as global information therefore have to be able to travel from one level to another. This is realizable by *fractal-like* dynamics that is intrinsic to complex (bio)systems anyway. How the *inter-level or inter-scale transfer of patterns* or images is realized is much harder to find out than to realize that this is indeed happening (J. Anderson).

Attractors are very probably those emergent virtual structures which can "travel across the brain". They are the bioinformational or PDP correlate of gestalts – each represents an invariant information-unit (percept). An attractor, a primitive "ghost in the machine", is rooted in a network-state, but *changes its substrate-elements* (Peruš, 1995a, 2000a) like the electric current changes its underlying crystal structure or like (water) waves change (water) molecules while propagating.

4.4 NON-MATHEMATICAL DESCRIPTION OF Q.A.N. MODEL

Essentials. A verbal (partially metaphoric) description of processing in the quantum associative net, in comparison to holography, will now be given following the same list of steps (1-4) as in section 1.4. The processing of quantum associative net *is* a sort of holography, if one is allowed to use the term outside classical optics, since the net interferes quantum waves. In fact, the quantum associative net is a quantum-mechanics-based mathematical model which can be computationally simulated. No reasons have up to now been found why the model could not be implemented in a real quantum-physical system. The model also needs *specific input–output transformations*, therefore it is an informational model as much as it is a physical one.

Interpretation of states. The quantum associative net is the core of basic quantum mechanics (in Feynman's interpretation) *which is put into an intelligible interaction* with the environment (visual field).[16] This is new: the input–output dynamics. Another essential new thing is that eigen-wave-functions (i.e., the basic, natural quantum states[17] – they are often particles–waves, but not necessary) are harnessed to encode some information like an image. An intelligent being must be there which interacts with the system in such a way that the input-, output- and internal (memory) states represent some *meaningful information for the being*. His *interpretation "transforms" an ordinary quantum system into an information-processing system* as soon as he is satisfied with the input-to-output transformations. Let us assume so. (It is like in the case of a round piece of wood "becoming" a wheel if put in a proper context – the axis, other wheels, upper plate, etc.)

Inputs. Image processing can be done during the holographic process (presented in sec. 2.3). It works perfectly and simply, as all physicists and opticians know (Hariharan, 1996). It is natural in holography (as well as photography and any other optics) that the light encodes the 3-dimensional form of an object by specific modulation (i.e., shaping) of amplitudes, frequencies and phases of its waves (rays). So, it is possible to encode complicated object-forms into usual electro-magnetic waves — even with perfect resolution when the code is being reconstructed. We thus have: objects, their codes or representations (in a medium), and we need object-to-code transformations (encoding) and, finally, code-to-object transformations (decoding or reconstruction).

Because holography works with all sorts of waves, the information-carrying waves can be quantum waves. This might bring new capabilities, but not eliminate the classical ones. Hence, the input-waves can be plane-waves (like the sea-waves toward the beach)[18], mathematically described by equation (1.3), or Gabor wavelets (made of "increasing and decreasing waves under a gaussian, i.e. bell-shaped, roof or envelope"). (But mathematics shall not be important in this section.)

We merely need means for proper manipulations of waves. Even for quantum waves, technology is so far today. Brain might also be able to do it. Auxil. Chapters 8, 9 and 10, along with some sections of the main text, provide numerous details and evidence.

We thus perform STEP 1 by illuminating an object so that light-rays (or, in brain's complicated version, ICA-produced wavelets) "fall into" the quantum associative net.

Interference constitutes memory. We do that with different objects and let the waves, each belonging to an object, "mix together" while falling on a medium (the hologram-plate). This is interference of waves (like when two water-waves criss-cross) produced so that it leaves a trace (the hologram) on the medium. The wave-parts add or suppress each other (the constructive or destructive interference), and a criss-cross matrix of their relationships is recorded on the hologram. This hologram is a "frozen" *content-addressable associative memory* which becomes active when light is sent through it![19]

Holography is a very straight-forward process, but it is in detail very subtle and complicated (and fundamental!), therefore it is nearly impossible to describe its "almost-omnipotence" if applied in various circumstances.

The quantum interference and quantum holography are when the waves and the hologram (but not necessarily the object) are quantum. The quantum hologram is the interference-pattern of quantum waves which leaves traces *in the quantum medium itself*. In quantum world, "parts are virtually a whole", so the waves and the hologram "inter-penetrate". Matrix $\boldsymbol{G}$ of equation (1.4) describes the quantum hologram and it is at the same time the carrier and transformer of waves. $\boldsymbol{G}$ is the quantum-holographic memory (STEP 2) which is active – performs associations "through itself".

In quantum mechanics, we sometimes say (if we don't just write equations which avoids the traps of language) that $\boldsymbol{G}$ describes the "self-organizing internal restructuring" of the quantum system by "internal interactions between its (seeming) parts", i.e. by *self*-interference. It should be emphasized here that this is not an interaction in the sense of chemical or quantum-particle (nuclear, subnuclear) reaction, but in the sense of mutual mechanical (or electrodynamic / optical) influence or *re-arrangement* ("mixing"[20]) on a quantum level — that's why quantum mechanics. (In quantum fields, the "mechanism" becomes much more "holistic".) In the language of quantum informatics, $\boldsymbol{G}$ describes a *compound*. (The "deeper, holistic" quantum fields incorporate *entanglements*, where parts which have once interacted cannot be really separated any more, but just seemingly.) Compounds can be "un-mixed" like the images can be reconstructed from the hologram (memory).[21]

Associative processing. The matrix / hologram / propagator ***G*** describes phase-relationships between "infinitely"-small parts of the waves which were "mixed". This is associative memory, which also acts like a "turbine" for associative "computing" (STEP 3). Each quantum wave Ψ "flows through the ***G***-turbine", and this changes both the ***G*** and the wave. This is mathematically described in eq. (1.5). It shows, because eq. (1.4) should be inserted into eq. (1.5) to replace ***G***, that (and how, why) waves Ψ change ***G*** and ***G*** changes waves Ψ. This is called the coupled dynamics of the quantum system — it is a "self-holography" triggered by our inputs. We call it associative processing, because it is realized by "projecting" the quantum eigen-state or -wave "through the associative memory or hologram" ***G***. Initial quantum-encoded informational state (Ψ_{in}) is thus transformed into an *associated* quantum-encoded informational state (Ψ_{out}).

Image recognition by wave-function collapse. If we want to recall a memory, or to reconstruct a stored image out of ***G***, respectively, we have to present a part of the image or a similar image (the memory-"key") to the system (Ψ_{in}). The similarity activates matching of relations, encoded in phases, and thus *selectively associates* the "key" with the most similar stored image which then "comes out of the mixture" (i.e., ***G***) in a clearly-reconstructed form.

I can claim that this quantum process, called *"wave-function collapse"*, is *typically holographic* in the framework of quantum associative nets (details in Auxil. Ch. 8). It is also essential for all quantum measurements, where one "chosen" eigen-state Ψ_k is realized in the quantum state Ψ, all the other eigen-states "retreat" (into the implicate order). Thus, the input-triggered wave-function "collapse" is the *memory-to-consciousness transition.* An image is reconstructed from memory and simultaneously "appears in consciousness", because it has been associated with all the relevant contexts during this very process! Therefore, the image is also (consciously) *recognized* at the same time! That's STEP 4.

For such an associative memory-based and input-triggered image recognition, careful combination of informational relations (i.e., associations) are needed. The "key"-image must be proper in the sense that it is more similar to the desired stored image than to other stored images. In contrary, a "mixture" comes out and "we do not understand". Therefore, the user of the system has to know approximately what is stored in the system, if he wants to get eigen-states (images) out. (A conscious being has much more chance here than another physical system, therefore many say that consciousness is necessary to "collapse" the wave-function.) These conditions for image recognition are mathematically precisely described in / after eq. (1.6), and in detail in Auxil. Chapter 10, where also the complete "recipe" for processing of the quantum associative net is given.

Remarks. Memory associations are encoded in correlations of wave-amplitudes A and additionally in differences of wave-phases φ. The latter encoding turns out to be more important and more fundamental, although both encodings are complementary.

In sum, the significance of the quantum associative net is in the fact that all the elements or aspects of an input-image are compared with all the elements or aspects of all the images, condensely stored in the system (as described by $\boldsymbol{G}$ or, alternatively, by the so-called *probability-density matrix* $\boldsymbol{\rho}$). An optimal, "compromise" output-image is then given as the result.

In the following sections, some related quantum or similar models will be presented, and they will be, together with the quantum associative net, discussed in the context of consciousness.

4.5 HOLOGRAPHIC PERCEPTUAL OUT-TO-SPACE BACK-PROJECTION AND OBJECT-IMAGE MATCH

Mathematical-physical description of holography. A hologram is a complex linear superposition of collective stationary interference fringe patterns. Storage of information (i.e., object-image) is usually made by global interaction (mixing) of a coherent information-bearing object-wave (reflected from the object) and a no-information-bearing coherent reference-wave under a particular angle. Information can be later retrieved by illuminating the hologram (no object needed any more) with the anti-wave of the original reference-wave used at storage. The anti-wave is an original-like information-bearing phase-conjugate reference-wave in the opposite direction of the original wave.

The phase-conjugate wave ($-k$) has, in the case of all local fields having the same frequency, an unique property to propagate back, in real or virtual form, along the path of the original wave (k).[22] The advanced wave k and the retarded wave $-k$, which is *as-if* time-reversed, get superposed (giving $e^{2\pi i v t} - e^{-2\pi i v t} = 2\ cos\ 2\pi v t$), due to precise timing.

Hologram's parallel-distributed organization is globally regulated by the local relative-phase variable implemented by the infinite-dimensional irreducible unitary linear representation of the Heisenberg nilpotent Lie group. The virtual "slices" (pages) of the hologram are frequency-organized, selective by incident angle of the page-oriented retrieval scanning reference-wave. Pattern/page-selection is executed by phase-conjugate adaptive resonance. The fractal self-identity is encoded in a hologram enabling reconstructional resolution proportional to any hologram's fraction where the total information is enfolded from. (Schempp, 1993)

Implementation of (bio)holograms. Hologram is realizable in brain tissue[23] as well as in fundamental (quantum)-physical media. Neural holography could be realized by dendritic transmembrane-potential oscillations characterized by microwave-frequency coherence of the non-thermal excitation-states of biomolecules with high permanent dipole moments. The needed coherent "wave" could be *internally* incident. Holograms can be made also with *partly coherent* waves (Marcer & Schempp in Fedorec & Marcer, 1996; Hariharan, 1996), although usually coherent waves are used. Information between neurons is exchanged also independently of synaptic connections via glia-cells or non-anatomic coherent resonance coupling.

Phase-Hebbian holographic associative memory has many concrete implementations, e.g. in photorefractive media — see Anderson (in Zornetzer, Davis & Lau, 1990, ch. 18). A variety of holographic processes including single-state (e.g., single-photon) holograms are possible (why not, at least in a generalized sense, also in the brain?).

Fundamental (sub)quantum holography could be realized with coherent overall wave-functions of dynamic quantum-vacuum. The fundamental Berry phase or geometric phase (Berry, 1988; Anandan, 1992) of the quantum system is promising for quantum memory and holographic (bio)information processing. [24]

Holographic perceptual projections. We have an impression that an object we see is located "outside". The naive view is as follows: There is an object in the environment. Perception of it is produced in our brains in such a way that we see the object as it "really is – in external space". The perception appears to be somehow projected from the brain out to the original location. Pribram (1998b) mentions several cases of such perceptual or even cognitive projections: For example, one feels the paper, on which one is writing, at the tip of his/her pen, not at the tip of fingers holding the pen. A well-known case are also the so-called phantom-limbs — a patient feels the amputated limb which he has just lost. The pain is felt outside the remaining part of the limb — at the location where the former complete limb was or should be. In experiments, cited by Trstenjak (1996), subjects spontaneously write on their own foreheads a letter (e.g., F) oriented as if they would read it from inside out (with their "mind's eye"), or as if they would write it on the internal side of their own foreheads. Projective nature and use of percepts are thus a part of human performance.

Let us illustrate how we could start to model holography-like perception and memory-recall. First, in stage 1, a subject faces an external, illuminated object. Light-waves reflect from object's surface toward subject's eyes. Image of the object is processed in his visual cortex, and gets memorized in a holography-like manner.

Later, in stage 2, the subject faces a similar object or the same one. Its light-wave indirectly interacts with holographic memory of the (original) object. A memory-image is remembered when the corresponding hologram-page is selectively reconstructed (as detailed in Auxil. Ch. 8 and in Peruš, 2000c). By a *phase-conjugate wave*[25], the perception (a compromise of the stimulus and memory) is experientially projected back (from brain) into the surrounding space to the location of the original object. *The virtual (holographically back-projected) image of the remembered or perceived object coincidences in space with the original object.* This important and plausible hypothesis originates from: Marcer & Schempp (1998, 2000), Marcer in AuxLit 27.[26] The idea of holographic back-projection by the (quantum) phase-conjugate wave is not yet finally proven, but it fits our *feeling* that the object and its brain-made and out-projected virtual image coincide out-there as if they are one!

The *perceptual* projection has been proven, for example, by the G. von Bekesy experiments (Pribram, 1971, pp. 168-171; 1991, pp. 90-91), but the *quantum-optical* phase-conjugate back-wave remains a question of quantum "reality" (i.e., theoretically it exists and is useful, experimentally there are indications, but there are different interpretations about their "reality"). The hypothetical back-projection waves would be *quantum* – not classical electro-magnetic waves like the input-waves to the retina. They would *not* exist *really* in the ordinary, i.e. classical-physical, sense.[27]

To repeat: In stage 1, an original object is seen — a representation (a virtual "image") of it is produced in the brain. In stage 2a, a similar or the same object triggers remembering the original object of stage 1. In stage 2b, which follows stage 2a immediately, a joint perceptual image (usually a "compromise" of images of stages 1 and 2a) is projected back to the location of 1 or 2a, respectively. These stages of the dynamic holography-like process usually *iterate* and possibly converge to a maximal agreement of perception with phenomenal reality. Beside pragmatic advantages to the organism, this is also necessary for unambiguous, consistent communication between image-making subjects, since they share the perceived objects in positions relative to one another in the 3-dimensional "Cartesian theatre" they co-create (Marcer & Schempp, 2000).

Iterative matching loop. Thus, the image, which brain/mind creates, is perceptually coincident, maybe also quantumly coincident, i.e. phase-conjugate[28], in external space with the object imaged. One can ask: what is (or even: is there any) difference between the perceived object and the back-projected image of it (i.e., the image in the original location in the environment, not the image/code in the brain, say in V1). Disagreements lead to errors or misperceptions. The iterative matching process can also be led to creative generalizations and associations.

In *imagery*, or more plastically in hallucinations, the (possibly modified) back-projected image of the object replaces the non-existent object (which was present in stage 1, but not in stage 2 when the reconstructed reference-wave has some internal sources). In *perception*, the back-projected image phenomenally fits the real object.

Why (quantum) holography is necessary for spatial perception. Phase-conjugate projection of the image back into the space-location of the original object is an *exclusive characteristic of holography* (or at least of optics, if the image would perhaps be processed in another way, not holographically). Namely, *neural networks* *or other hard-wired subcellular networks, without having electro-magnetic or quantum embedding,* ***can definitely not back-project*** *their images into the external space on their own.* But we experience that perceptual projection is happening. Because holography (not photography) is involved, it is not directly the object-image that is back-projected (as in photography), but it is the *wave (-k)*, which carries the encoded object-image, that is back-projected by phase-conjugation during/after holographic retrieval (Marcer & Schempp, 1998, 2000; Marcer in AuxLit 27). So, *the external medium must be of the same or at least similar nature than the medium of the brain-hologram.* Hence, the *common medium can only be* ***quantum field****!*

Philosophical questions. Actually, there can be no plastic, geometrically-/topologically-correct 3-dimensional perception of the object, which we experience and call "the real/true perception", without that fitting of the object with the back-projected image of it (which has been a moment earlier reconstructed from the "brain hologram"). This iterative fitting seems to need time[29], but in fact space-time is co-created by ("in") our conscious experience "during" this visual processing. Objects and brain-states seem to be located in space and time, but conscious experience "has been / is / will be there all the time – as long as the Cartesian theatre is in the play".[30]

Nothing is perceived outside mind: There are no perceptions, which we are aware of, without consciousness (i.e., conscious experience), and there are no phenomenal unconscious or subconscious sensations or "detections" (i.e., perceptions we are not aware of) without mind. (Detailed studies, including special cases, see in Gennaro, 1995, and in Peruš, 1998c, 2000a.) A Kant's *Ding-an-Sich* is not perceived; just the co-created thing is, i.e. a "deformed shadow" of the hypothetical *Ding-an-Sich*. Thus, a perception is a Plato's "shadow" created by consciousness in cooperation with Nature, or "deformed" by brain-processing.

Various back-projections, like visual, tactile, auditory ones, match with the object and with each-other! Our vision "quantum-touches" the object successfully as well as our hands mechanically touch it simultaneously. Indeed a peculiar space-time fit. (But maybe this "resonance", and space-time incorporating objects, and conscious experience including objects, is the *same* thing/process… However, this symmetry or harmony (fit) may be broken, e.g. in hallucinations… So, what is what?…)

4.6 DENDRITIC HOLOGRAPHY-LIKE IMAGE PROCESSING

Bioplasma dynamics. How can a dendritic net implement holography-like image processing of V1? Stimulus-specific waves of the dendritic field are produced, and these information-encoding waves interfere. A phase-Hebbian PDP is realized.

There are (at least, roughly) two related kinds of collective oscillations accompanying the dendrites cris-crossed in networks: first, the oscillations of dendritic membrane potential, and second, the oscillations of dendritic ionic "bioplasma", or of the electric polarizations within, respectively. The "bioplasma"[31] "flows" and "waves" near the membrane-surface of the dendrite. It depends on the biomolecules (proteins, lipids, etc.) of high dipole moments located on (beneath, in, near, along, respectively) the dendritic membrane, and the ionic charge travels through it.[32]

This membrane–"bioplasma" system of numerous coupled electric dipoles exhibits dynamic ordering which is determined by the distribution of phase-differences in oscillations of the corresponding dendritic potentials. The isophase-contours of oscillations in the polarization-field, extended over dendrites (especially their membranes) and the accompanying "bioplasma", determine the collective wave- and fluid-phenomena that are the *correlates of image processing at the subcellular level of V1* (after Appendix A of Pribram, 1991).

The Jibu, Yasue and Pribram model. The density (or concentration) of the ionic "bioplasma" $\rho(x,t)$ changes as a result of the dynamic pattern of hyperpolarizations and depolarizations across the space inside and outside dendrites. Yasue, Jibu & Pribram (in Appendix A of Pribram, 1991) defined a wave-function $\Psi = \sqrt{(\rho)}\, e^{i\varphi}$ (φ is the phase). They derived (*ibid.*, pp. 282-286)[33] a wave-equation for the membrane–"bioplasma" system:

$$i\nu \frac{\partial \psi}{\partial t} = \left(-\frac{\nu^2}{2} \nabla^2 + U_{ex} \right) \psi \tag{4.1}$$

which has the same mathematical form as the Schrödinger quantum wave-equation, but with different interpretations of variables.[34] U_{ex} is the external static energy / potential, i.e. the external electric influence (stimulus). ν is a constant (a "relative" of the quantum Planck constant) equal to a quotient of "flow"-velocity v and the length of the so-called wave-vector k which is equivalent to spatial frequency, related to the change of phase: $k = \nabla\varphi$. $\nabla = \partial/\partial x$ is the spatial derivative, i.e. the rate of change across the space (x); $\nabla^2 = \nabla\nabla = \partial(\partial/\partial x)/\partial x = \partial^2/\partial x^2$ is the second / double derivative, i.e. the change of the rate of change.

Successful derivation of such an equation, having a characteristic form for wave phenomena, for an idealized dendritic network / field shows that global polarization-waves, described here by Ψ, emerge in the subcellular medium. As a consequence of these waves, "flows" and interference are also produced in the dendritic net. There are, of course, many variations of oscillations / waves / interferences in that complex medium. Ψ (with different interpretations) could be chosen to approximate (m)any of them.

Phase-Hebb-like memory-storage in bioplasma. Wave-equation (4.1) is alone not enough for image processing. From wavelets

$$\psi(x,t) = \sum_n c_n \left(\frac{1}{L}\right) \exp\left[\frac{i}{\nu}\left(\frac{2\pi}{L}\nu x - \lambda_n t\right)\right] \tag{4.2}$$

we obtain (details in Pribram, 1991, A, pp. 288-291) the density $\rho(x,t)$ of charge-distribution in "bioplasma":

$$\rho(x,t) = \left|\psi(x,t)\right|^2 = \psi(x,t)^* \psi(x,t) =$$
$$= \sum_n \left|c_n(t)\right|^2 \left(\frac{1}{L}\right)^2 + \sum_{n \neq n'} c^*_{n'}(t) c_n(t) \left(\frac{1}{L}\right)^2 \exp\left[i\frac{2\pi}{L}\left(n - n'\right)x\right] \cdot \tag{4.3}$$

c_n are the Fourier coefficients; L is a characteristic spatial extent of the dendritic system along the spatial axis x; λ_n is a constant; n, n' are integer numbers. The last term manifests interference, which is essential for holographic memory, since it has a phase-Hebb-like structure: $c_{n'}{}^*(t)c_n(t)$ represents "interference of amplitudes" and $exp[(2\pi i/L)(n-n')x]$ "interference of oscillations" – in the exponent one finds the phase-differences.

The flow of dendritic perimembranous "bioplasma" is driven by the phase-differences among isophase-contours (i.e., curves connecting all synchronized oscillations). The "density-based flow" toward an attractor at the centre of the concentric contour-family is regulated by the gradient (i.e., maximal change-rate) of phase ($\nabla\varphi$). (Recall secs. 2.1, 2.3.) Exactly:

$$\frac{\partial \rho}{\partial t} = -\nu \nabla \cdot (\rho \nabla \phi) \,. \tag{4.4}$$

The wave-equation (4.1) and Equation (4.4) describe the subcellular "fluid-dynamical" correlate of associative image processing in the V1 dendritic net.

Papers like Berg *et al.* (1996) and those in Fedorec & Marcer (1996) support the possibility of biomolecular realization of holographic processing at the subcellular level (Nobili, 1985). These dendritic dynamics (Yasue, Jibu & Pribram inYasue, Jibu & Pribram inYasue, Jibu & Pribram inYasue, Jibu & Pribram in: Pribram, 1991, Appendix A) are in general principles equivalent to processing of the quantum associative net, but incorporate some sophisticated constraints imposed by characteristics of the subcellular tissue. The difference also is that the dendritic wave dynamics are quantum-like, but quantum associative nets are purely-quantum.[35]

4.7 MICROTUBULES, COHERENT SUBCELLULAR AND QUANTUM PROCESSES, AND CONSCIOUSNESS

Microtubules. *Microtubules*, cilindric / filamentary tubes, are the most important ingredient of the *cytoskeleton* which is a protein-made intracellular network.[36] Microtubules extend along dendrites, come together at soma, and extend further along axons. They consist of oriented assemblies of electrical dipoles, or permanent electric polarization systems (electrets), respectively, which globally act as mega-dipoles.

A hypothesis (by Hameroff, 1994) with increasing support (AuxLit 20) has been presented that cytoskeletal microtubules, constituting a network in cooperation with MAP (microtubule associated proteins), realize sub-cellular information processing based on coupled oscillatory collective dynamics (roughly similar to ANN).[37] Since Hameroff & Penrose (1995) emphasize that mainly *dendritic* microtubules act such a role, this hypothesis might not be entirely incompatible with the holonomic theory, but complementary. Such dynamics emerges from conformational transitions[38] of tubulin electric-dipolar molecules, which act as "bit flips", and from soliton-based[39], almost loss-less transfer of energy and information (phase!) along the paracrystalline microtubules. Tubulin states might encode the pixels of patterns which are processed (Hameroff, 1994; Nanopoulos, 1995). Globally, the processing is manifested in changes of concentration of electric polarization (polarization density), and moving of the concentration peaks from one side of the tubulin-web to another (AuxLit 20).

Long-range quantum coherence and related laser-like, thermal-noise-free (and information-loss-free) ordering phenomena, like super-radiance and self-induced transparency may take place in microtubules.[40] The hypothesis that quantum coherence subserves binding of conscious perceptions is supported by an increasing number of authors.[41] Microtubules are viewed by Jibu & Yasue (1997) as information-encoders into a coherent subcellular optical PDP network. Fröhlich (1968) has shown the first signatures of interdependence of biological and quantum oscillatory dynamics.

Fröhlich coherence. As proposed by Fröhlich (1968) and succesors (AuxLit 19), the so-called Fröhlich (long-range, microwave) coherence emerges from interacting oscillating (10^{11}-10^{12} Hz) dipoles of biomolecules. Electric polarization density serves as the biological order-parameter. Fröhlich coherent oscillations may lead to two sorts of extreme collective states: to the Bose-Einstein condensate, where all dipole-elements act as if they are one, or to loss-less solitonic polarization-waves (proposed by the Davydov model), where the dipole order propagates as one traveling condensate (Denschlag *et al.*, 2000). This is analogous to superconductivity. Indeed, it was proposed by Jibu & Yasue (1995, 1996, 1997) that the experimentally-supported Fröhlich waves along the protein filaments can propagate without resistance, thermal loss and damping. Such superconductivity hypothetically occurs even at body temperature.

Subcellular automata. Many nanobiological systems could be represented as assemblies of dipoles: 1. cell membrane as a double sheet full of dipoles; 2. cytoplasmatic and extracellular water; 3. microtubules as systems of tubulins; etc. Systems of dipoles or spins can be arranged: 1. randomly; 2. ferroelectrically (i.e., aligned in parallel); 3. as spin-glass (i.e., in domains of frozen (dis)order, each with its own parallel alignment) (Mezard *et al.*, 1987; AuxLit 14). The membrane bi-layer of dipoles might incorporate sandwich-like *Josephson junctions*, over which superconducting electrical currents with special effects would flow. They might be connected into a peculiar PDP "Josephson net-computer" (Rein in Pribram, 1993; Jibu & Yasue, 1995).

Quantum effects in synapse. Eccles (in Pribram, 1993)[42] pioneered the idea that conscious mind, using attention, could influence the probability of discrete (quantal) release (*exocytosis*) of vesicles full with neurotransmitter-molecules at the hexagonal-paracrystalline presynaptic vesicular grid. Conscious mind would impose effect on probabilistic quantum processes (e.g., the wave-function collapse) underlying the probabilistic exocytosis in synapses (AuxLit 21). So, conscious process would selectively modulate, through quantum fields, the essential ingredients of memory-storage and associative processes – synaptic efficiacies.[43] To be precise, quantum influences should trigger electronic rearrangements resulting in movement of hydrogen-bridges which would affect vesicle-release from the presynaptic hexagonal grid (Hameroff, 1994).

Collapse and consciousess. The hypothesis of Hameroff & Penrose (1995, 1996, 1997) advocates microtubules and their nets as the main subneuronal substrate of consciousness. They are flexible, fast-changing and might allow retrograde signaling, thus mediating bi-directional subneuronal links between synapses. Hameroff (1994) argues, based on observations, that general anaesthetics cancel conscious experience[44] by operating mainly on specific microtubular ingredients. He writes that an anaesthetized brain usually remains quite active (as persistent EEG, evoked potentials and autonomic functions show), but this activity is neural, not microscopic quantum. So, quantum coherence, which gets disrupted by anaesthetics, should be essential for conscious experience.

In contrast to the environment-induced wave-function collapse of quantum theory (and of the quantum associative net), the wave-function collapse in microtubules is supposed by Hameroff & Penrose (1995, 1996) to depend on quantum gravity: Condensates which become larger than a threshold-size should cause their common wave-function to collapse "under their own mass". This would be thus a self-collapse called *orchestrated objective* collapse. Each such collapse is considered by Penrose and Hameroff to be a single conscious event. A temporal sequence of such conscious "nows" would constitute the "flow" of conscious experience, by this hypothesis.

According to the Hameroff & Penrose (1995, 1996) sketch, preconscious net-computing, when the classical (sub)neuronal PDP is gradually replaced / complemented with quantum PDP, leads to emergence of a quantum coherent superposition. Each of its superposed alternative states has its own "competing" space-time geometry. When an instability-threshold is exceeded, the time-irreversible orchestrated objective self-collapse occurs, and this *is* the conscious experience (the "now"). This moment of maximal coherence "illuminates" the (results of) preconscious network-processing of images etc. (executed till the collapse) "by making it conscious" at the very moment of collapse which "chooses" one from many alternatives. The selected information-state (e.g., a recognized image) is further-on processed unconsciously in a classical way — until a new quantum coherence "consciously illuminates" the new mental state to make a qualitative experience of it, says the hypothesis. [45]

Subcellular "holography". A number of oscillatory network-structures were mentioned: electric-dipole systems, microtubular nets, "bioplasma", extracellular matrix, protein nets, Josephson-junction nets, etc. They individually or in cooperation (as is usual) are able to exhibit holography-like interference processing.[46] However, molecular vibrational fields in these nets are just a sort of *interface between quantum networks and neural networks*. They all are very probably influencing, directly or indirectly, the synaptic efficiacies (e.g., whether they are inhibitory or excitatory, and how much) (e.g., Nanopoulos, 1995).

Subcellular coherence. What follows, is based on speculations by Jibu & Yasue (1995, 1996, 1997), derived from quantum field theory of Umezawa.

Beneath many levels of cell's biomolecular structure, many levels of sub-atomic or inter-atomic quantum particles, and their ensembles or condensates, can manifest collective dynamics capable of coherence and interference processing of holography-like sorts. They should mainly "live" in the intra- and inter-cellular *water* which composes more than 70% of the material composition of brain-cells like neurons and glia. These particles should collectively take part in water's rotational fields (or spinor-fields, emerging from spins of particles or from molecular spinning dipoles) and their interactions with the electro-magnetic field (i.e., with its quanta – photons). (Cf., Dimitrovski *et al.*, 1998.)

Of special importance is supposed to be the Nambu-Goldstone boson (a sort of dipole phonon) which is a mass-less quantum of a long-range correlation-wave of the water rotational field created in an ordered vacuum-state.

Quantum binding. Macroscopic condensates of Nambu-Goldstone bosons are, after the hypothesis of Jibu & Yasue (1995, 1996, 1997), the fundamental carriers of perceptual memory and cues for reconstruction of the original stimulus-perception. Since they depend on the interaction of radiation (photons) and dipoles, they lead to evanescent (i.e., virtual, tunneling) photons[47] which may collectively produce *Bose-Einstein condensates*. In such a condensate, many particles merge into a collective, unified, macrosopic quantum state. Particles (e.g., photons), which are able to unite into such a coherent condensate, are called bosons; particles (e.g., electrons), which never occupy the same quantum state, are called fermions. A Bose-Einstein condensate of evanescent photons is proposed to be the ultimate neurophysical correlate of an *unified conscious experience*. In Hameroff (1998, Box 1), Jibu & Yasue (1995, 1996, 1997) and AuxLit 20, suggestions are given how the Bose-Einstein condensates could be shielded enough by special biomolecular structures against the destroying thermal fluctuations.[48]

The coherent dipole-field (i.e., having dipoles oriented in the same direction) of water might extend over the whole brain tissue or even whole body, not just over several cells.[49] Such ordered water with presumably laser-type processes is assumed to enable photonic holography in and around microtubules and in extracellular matrices (Jibu & Yasue, 1995; Hameroff, 1994).

Qualia unexplained. I can agree (Peruš, 1995-2000) with Hameroff and Penrose that the wave-function collapse seems essential for transitions from subconsciousness (or preconsciousness, or unconscious memory) to conscious experience. It also illustrates the classical–quantum (neuro–quantum, macro–micro) transitions. But saying (Hameroff & Penrose, 1996) that only and merely the "orchestrated collapse" (not any usual stimulus-induced collapse) provides the non-computable element necessary for consciousness, does not give any explanation of the qualitative experience (Peruš, 1998b,c; AuxLit 32).

Anyway, one might speculate that the usual stimulus-induced collapse is related to conscious perception of the stimulus, and that the orchestrated objective (if indeed induced by quantum gravity) collapse is rather related to (introspective) awareness. Although this hypothesis provides relations of conscious process to the origin of space-time[50], the problem of qualia[51] remains (Flanagan, 1992; Davies & Humphreys, 1993; AuxLit 33). Qualia are only (with justice, I think) transferred to the most fundamental level (also). This could be concluded also for the suggestions of Jibu and Yasue: They might "explain" the origin of the unity of conscious experience, but not its qualitative phenomenal character.

4.8 CONSCIOUS EXPERIENCE

"Definitions." Although all such attempts are disputable, let us try to "define" conscious experience[52] and related notions. Two levels of brain-process "incorporating" conscious experience can be distinguished:

1. *Information-processing background* emerging from the so-called neural / physical correlates of conscious process (the third person aspect);
2. Subjective, qualitative, *phenomenal experience* of the irreducible *I* (the first person perspective).[53]

The first aspect incorporates roots of conscious experience in a "sea" of unconscious and subconscious background processes. Together with conscious experience they constitute the mind, i.e. mental processes.[54]

A special ("second-order cybernetics") case of the first-person perspective is *awareness*. It is self-reflective (recursive) and, since it is Self-based and entails self-consciousness, it is also Self-reflective, i.e. self-aware. *Introspective* awareness is a deeper or active (self-monitoring) case of self-awareness ("third-order cybernetics").

Neural correlates of consciousness. One should distinguish the following aspects of conscious experience: *conscious process itself* (e.g., conscious thought, attention, intention), *contents of consciousness* (percepts, or mental representations of, usually, external objects), and *neural correlates of consciousness* (neuro-chemically determined states or patterns of neural activity that, presumably necessarily, accompany conscious processes) (e.g., Pribram, 1998a; cf., Davies & Humphreys, 1993; ASSC, 1998; AuxLit 8). They are based on physiological *arousal* which is, at least partially, controlled by the *reticular activating system* of the middle part of the brainstem. In this context, locus coeruleus is important for vigilance, and damage to intralaminal nuclei of the thalamus can cause irreversible coma and vegetative states (Frith *et al.*, 1999).

Both wakefulness and dreaming are manifested by similar activity-patterns in the thalamo-cortical circuit (Jasper *et al.*, 1998), and both include subjective experience, but in REM-periods of dreaming we are usually not aware of it. Dreams are characterized by vivid, but unreflected, imagery, similar to a sort of delirium (rather than psychosis), in spite of lower arousal. (Experiments show that higher animals also have such dreams.) (Frith *et al.*, 1999)

States of conscious experience are in general accompanied by increased neural and metabolic activity comparing to unconscious states. It is the same with superliminal versus subliminal stimulation and with paying attention (to an object encoded in a neural state) versus being non-attentive. Attention regulates access to conscious experience, but is, in turn, also often controlled by it. (Baars, 1997)

Theories. There are several rival groups of hypothetical theories of conscious experience (reviews in: Marcel & Bisiach, 1988; Flanagan, 1992; Hameroff *et al.*, 1996; Newmann, 1997; Rakić *et al.*, 1997; Železnikar & Peruš, 1998; AuxLit 28-31):

- Conscious process *emerges* from collective dynamics in specific networks (e.g., neural nets);

- Conscious process is a result of *attentional scanning circuits* (e.g., thalamo-cortical, reticular);
- Conscious process arises from *coherent 40(-80) Hz firing* of far-apart neurons belonging to the same brain area (e.g., V1) or to different areas;
- Conscious process is essentially a *quantum* phenomenon modulated by (sub) neuronal PDP (e.g., in dendritic, or/and microtubular, or/and biomolecular networks);
- Conscious process is entirely non-material or even non-natural, or it is entirely mysterious.

Ad 1. Emergence theories are usually bottom-up physicalist theories emphasizing hierarchical structures of assemblies of assemblies (attractors of attractors) of neurons or other units. I agree that such processes take place, but they operate top-down as well as bottom-up, and are not exclusively responsible for consciousness.

Ad 2. Conscious process is identified, usually by proponents of the main-stream neuroscience, with (an emergent process arising from):

2.1. ERTAS[55]: *thalamo-cortical-loop*-driven (Jasper *et al.*, 1998) voting-like competition, modulated by *nucleus reticularis thalami*, of modules for attention (Baars, 1997; Newmann, 1997);
2.2. Processes of the *intralaminar complex* of the thalamus in cooperation with the cortical areas interconnected with it (but also the reticular formation of the brain-stem);
2.3. Activation of the *theta-system* in the hippocampus (O'Keefe in Oakley, 1985)[56].

These views base on findings that, in a rough sketch, reticular activating system regulates wakefulness, the nucleus reticularis of the thalamus mediates processes of directing attention to various brain-areas (Bickle *et al.*, 1999), and the superior colliculus of the tectum[57] mediates integration of sensory information.

A critique (probably in accord with Pribram) of these hypotheses can be given: They say more about the physiological origins of arousal, alertness and attention than about conscious experience. These arousal centers are, of course, necessary for conscious process, but are not sufficient to describe the conscious processing of mental *contents*. For the latter, neocortex is needed — although coma happens with damage to the intralaminar complex and (usually) not with damage to neocortex. The intralaminar nuclei trigger the cortical EEG characteristics. (Baars, 1997)

Ad 3. Global processing and unification of information-"parts" *that belong together* is realized by coherent oscillations in neural units that encode the "parts". Cooperative phase-synchronization[58] was found, by many independent multi-electrode recordings[59], 1. among assemblies of neurons in the same cortical column where similar features are processed, 2. among distant assemblies belonging to different columns of the same cortical area, and 3. also among assemblies in different areas of the visual cortex when they simultaneously process some *common feature* of the stimulus (Gray *et al.*, 1989, 1990).

However, Gray *et al.* (1990, p. 94) acknowledge that there is a question how can distant cortical columns become coherent often even with *zero phase-delay* in spite of considerable transmission-delays in the synchronizing connections. (Sub)quantum non-local processing with *instantaneous*, noise-less intercommunication[60] within a coherent (bio)substrate might well provide solution to this problem (Bohm, 1980; Peruš, 1997a,b,c; AuxLit 26). Thus, coherent neuronal oscillations realize merely a subset of perceptual binding phenomena, but relation with conscious experience remains unclear.

Ad 4. There are two branches of the quantum idea:

4.1. Conscious experience, or (more precisely) at least its physical correlates, are ontologically of quantum nature, albeit not exclusively. This means that conscious experience is essentially related to quantum phenomena, but might not be reduced to them. It might have additional, unknown features. (Lockwood, 1989; AuxLit 27, 26, 21; Peruš, 1997a,b,c, 1998b, 2000a). [61]
4.2. Quantum theory can help in modeling (sub)neuronal complex-system PDP by analogical reasoning. So, it can help us epistemologically, indirectly in understanding (sub)neuronal correlates of consciousness (e.g., Barahona da Fonseca *et al.*, 1999; AuxLit 22, 23, 28, 30).

The holonomic theory advocates both possibilities as either–or: Conscious experience is either related to quantum (4.1) or to quantum-like (4.2) PDP, or both.

For me, 4.2 is true for sure[62], and 4.1 is very likely, but not definitely proved, except by taking parapsychology seriously. Here I mean especially the non-local (transpersonal) and transcendental experiences. Although many individual reports are not (entirely) reliable, the enormous amount of parapsychological reports is not to be rejected *a priori*. On the other hand, for instance, a single clear case of telepathy (cf.: Wackermann *et al.*, 2003) (not to mention more exotic phenomena) is enough to reject the theory of exclusively classical-physical (e.g., exclusively skull-limited neural-net) nature of consciousness.

It has been definitely theoretically (AuxLit 18, 26; Bohm & Hiley, 1993; Gould, 1989) and experimentally (Aspect *et al.*, 1982) demonstrated that the (sub)quantum world is essentially *non-locally inter-connected* (AuxLit 26) and *entangled* (AuxLit 18). These results suggest parallels with parapsychology, i.e. with altered states of consciousness with possibly non-local manifestations. However, if it will turn out definitely that conscious experience has quantum roots (Stapp, 1993; AuxLit 27) and that it might have in exceptional cases also transpersonal manifestations, this does not yet mean that the problem of consciousness is solved. (Sub)quantum unity (coherence) or re-unification can solve the binding problem at the ultimate level, but cannot explain qualia.

This is my opinion in spite of the fact that Penrose and Hameroff (1998, p. 125) considered qualia as fundamental as the most microscopic[63] *networks of quantum spins*: "Qualia might be particular patterns in fundamental space-time geometry [...] encoded in Planck-scale spin networks." To say again, this might well turn out to be true, but it does not decipher the enigma of the essential nature of qualitative appearances within phenomenal consciousness.

Ad 5. I think, dualist or idealist hypotheses neglect too much the natural phenomena, especially the numerous physio-psychological signatures or brain–mind parallels, inter-dependence, interactions, complements.

None of the present theories explains qualia, although there is an increasing number of scientists who take them as essential and irreducible. The "quantitative skeleton of qualia" (AuxLit 8, 30, 31) is increasingly traced with PET-scans etc., but the qualitative features (AuxLit 32, 33) still remain encoded in subjects' adjectives used in verbal reports.

Pribram on conscious experience. The holonomic theory distinguishes the surface-structure of memory and conscious experience, which correlates with processing in neural systems, and the deep structure which correlates with synapto-dendritic field dynamics (Pribram, 1997b). Pribram (in Hameroff *et al.*, 1996, ch. 10) writes that the *electrochemical synapto-dendritic states are coordinate with the states of conscious experience* as well as with the results of (conscious) learning including memory-storage. Varieties of conscious states are produced by various organizations of the polarization field within the same synapto-dendritic substrate (Pribram, 1971).

Conscious experience is attenuated when actions cohere with their guides and so actions become automatic (Pribram in Hameroff *et al.*, 1996). Similar attenuation happens when perceptions match with expectations, or when no novelty is perceived, respectively. In such cases, neural circuits do the job. Unconscious processing in neural circuits is sufficient also for reflex-actions. For a reflex-move of a hand, neural circuits guide the muscles more directly and thus more quickly than dendritic nets — because neurons are closer in size and power to muscle-cells than dendrites are. So, the conscious processing in dendritic nets is excluded during reflex-actions — in order to cause no dendritic-processing delays, to be as fast as possible.

Pribram (1998a, Summary) writes: "Weiskrantz's, Stoerig's and Cowey's distinction between reflexive, phenomenal and conscious by accessible processing[64] becomes [...] a distinction between automatic, referential (semantic) and episodic executive processing. Automatic, reflexive, processing occurs whenever a neural circuit has become thoroughly established with a minimal synapto-dendritic delay. Referential, phenomenal, processing is semi-automatic but easily accessible to monitoring when shifts among reference frames[65] are initiated within circuits, either "spontaneously" or by some more organized sensory or central input.

Executive monitoring comes about when frontolimbic processes become involved in processing delays allowing shifts among reference frames in the systems of posterior convexity."

Pribram on delays and consciousness. Pribram (1998a) emphasizes that *conscious experience depends on delays because of processing in modifiable dendritic patches between synaptic arrival signals and axonal departure signals*. Shifts of reference-frames and monitoring of such shifts occur during this dendritic processing. Namely, dendritic processing introduces delay in overall neuronal input-to-output processing (resulting in the axonal output-signal) and this "gives opportunity" for the *consciously*-monitored and controlled shifts. Delays are unavoidable, if signals get distributed over the dendritic trees and their junctions and if they are extensively processed there. So, delays are a sign of extensive dendritic processing accompanied by structural subcellular changes (which are induced by transit of signals and thus increase delays). Reflex-actions, as it is necessary, avoid (dendritic) delays as long as they are executed by neuronal-circuit processing only. Dendrites probably just transmit signals in such cases.

Delays also "give time" for communication with other brain-areas and their cooperation in processing. Other systems, especially the frontal cortex, seem even to facilitate the delays (Pribram, 1998a, after Fuster) in order to intervene and consequently, to appropriately change the dendritic and global structure of the circuits. The rapidity of changes in awareness would so be proportional to the delay, i.e. to the duration of interference-processing in synapto-dendritic polarization- or quantum fields (Pribram, 1995). I interpret such interference-processing as the phase-Hebbian memory-storage, but strongly top-down modulated, or shaped by other processing layers, respectively.

To make it clear: (Conscious) monitoring processes are: attention which monitors sensory input, intention which monitors action, and thought which monitors memory (recall) (Pribram, 1998a).

To summarize Pribram's view, *processing delays, dendritic processing and consciousness are coincident*, and seem to be thus *actively* correlated. Processing in circuits of neurons which by-passes consciousness, is possible in the cases of automatic behavior, subliminal perception, priming and unconscious learning. On the other hand, reference-frames, which configure sub-neuronal processes, become accessible to consciousness (a precondition for introspection) during dendritic-processing delay — according to Pribram.

Awareness. Consciousness, or better to say conscious experience, is representational; awareness is *re*-representational: Conscious experience is about a content (it is, by a philosophical word, intentional); awareness is (more or less) about consciousness (of that content). The division is not strict, of course, since representational levels may often "inter-penetrate" or even "merge" (Gennaro, 1995).

Awareness is needed for many specific changes in learning and behavior. Namely, in experiments by McIntosh *et al.* (1999), such changes[66] occurred only in aware subjects. Awareness was correlated mainly with the (left) prefrontal cortex, but only if its activity was integrated with many other (higher) distributed, large-scale networks.

The PDP background or correlates of awareness are very probably *self-interaction or self-interference processes* of collective states (configurations or assemblies that constitute patterns–attractors) in numerous bionetworks we discussed (details in: Peruš, 1995a, 2000a; Peruš & Ečimovič, 1998). In short, a content of our awareness bases in the "virtual action" of a specific attractor upon or within itself, or within its own basin of attraction, respectively.

Quantum basis? As long as mental processes are not conscious, they might be (to a large extent) realizable by classical physics and neural networks. When they become conscious, deeper processes may start to operate significantly. The holonomic theory supposes that dendritic field dynamics, significantly quantum-influenced, are needed for conscious experience.

I have a hypothesis, which expands Pribram's view to the quantum domain, that purely-quantum processes also are *necessarily* involved in conscious processes. It is based, additionally to the numerous unrepeatable phenomenological indications of parapsychology, on mathematical-physical observations and considerations, described in Auxil. Chapters 7—10. My hypothesis is not yet experimentally provable or testable *more than* parapsychology can provide, but Auxil. Chapters 7—10 allow establishing such a hypothesis which has many relatives by other authors.

Multi-level binding. The mainstream- versus the holonomic view differ in neuronal versus sub-neuronal binding suggestions. Main-stream cognitive neuroscience has problems with binding all the specialized, partial perceptions of "feature detectors" and similar sorts of cardinal neurons into an unified perception.

Roelfsema (1998) proposed two possibilities for assembly-labeling and binding: first, the *synchrony of neuronal discharges*, and second, the *enhanced firing-rates* (evoked by *attention*) of the neurons encoding features which belong to a common object and should be bound together. Precise timing of action potentials could also be needed as a label of such features. Responses of lower visual centers, which are triggered through an eye that participates in conscious visual perception, are different from responses which are evoked through an eye with suppressed vision, only in the higher rate of synchrony (Roelfsema, 1998). This stronger synchrony might cause higher activities in related subsequent processing areas.

According to Pribram, the binding problem, and the problem of needing extremely many cardinal neurons for all the special features, are artificial problems which emerge from a wrong theory.

The neuronal coherence is probably, at least indirectly, bi-directionally related to coherence-phenomena at subcellular (e.g., dendritic) and quantum levels. The ultimate origin of inter-connectedness and coherence of (neuro)physical states is undoubtly at the (sub)quantum level.[67] The neuronal and dendritic coherence is a higher-order stimuli-induced *re*-establishment of the fundamental (sub)quantum coherence or its fractal-like replica.

So, the "binding solutions" of Roelfsema (1998), Baars (1997), and of other scientists who do not want to involve quantum physics, are just auxiliary *re*-binding suggestions, since conscious states have always been bound quantumly and can re-bind quantumly if re-arrangements of conscious states are needed, I believe. Quantum binding (at least sometimes manifested as the so-called quantum entanglement, i.e. inseparability) is always more perfect than (sub)neuronal binding. [68]

4.9 VISUAL CONSCIOUS EXPERIENCE

Neural correlates of conscious vision. Experiments by Logothetis (1999) and many others show that we are (almost) always unaware of (the great majority of) neural processes subserving visual conscious experience. It was evident from fMRI-images that monkeys had active visual areas although they were anaesthetized. Logothetis (1999) concludes that only a small percentage of the whole community of neurons seems to represent the direct "neural correlates" of visual consciousness, and such neurons are probably distributed all over the visual pathway, with some concentration in higher areas like ITC.

Clinical data on selective loss. Conscious experience appears to normal subjects to be integrated, unified, holistic and stable, but this is so only "in the final product" or "on the level of envelope" of a heterogeneous multi-level process. Its internal structure becomes evident after selective loss of performance in lesions:

In visual agnosia, conscious object-perception functions well, but patients find no meaning (or purpose) of the object with recognized form and color (Baars, 1997). Prosopagnosia, where a patient cannot recognize faces of known-to-him persons, for instance, is shown (Young & de Haan in Davies & Humphreys, 1993, p. 64) to be a failure of being conscious about the recognized rather than a loss of the recognition mechanism.

In color-agnosia, patients describe their world as "black-white, gray, dirty, erased".

Reticular formation is necessary for wakefulness; intralaminar nuclei are necessary for waking consciousness, but cortex is not. Cortex shapes the contents of conscious experience (Baars, 1997). Damage to visual association-cortex does not lead to any overall loss of consciousness, but usually to a specific, partial loss (e.g., failure to consciously recognize known faces). Some patients can be blind to specific parts of the external visual field only (e.g., to the left half-space); the other part they can see.

Blindsight. So-called blind-sight is such a case, presumably a result of lesion in V1 and almost always limited to one hemisphere. Such a damaged V1 gives no (conscious) visual responses, and although the extrastriate cortex is active, the patient is cortically blind in the contralateral visual half-field.[69] Blindsight patients can visually discriminate objects in the blind half-field, although without phenomenal visual experience. That is, they are not conscious of seeing anything as if they would not see at all, but if they are "forced" to guess, they give correct location and description of presented objects or patterns. Some of blindsight patients "have conscious residual vision in response to fast moving stimuli or sudden changes in light flux, characterized by a contentless kind of awareness, a feeling of something happening, albeit not normal seeing".[70]

Blindsight may reveal something about the role of V1 in *conscious* image processing. Koch (in Hameroff *et al.*, 1996) argues that we are *not vividly aware of* the majority of processing in V1 (e.g., not of the eye-dominance in imaging, and not of the high spatial frequencies which V1 responds to). Koch and Crick (in *ibid.*) reasoned that "we are visually aware of certain [results of [71]] neural activity in V4 that could be triggered by activity in V1." It appears that V1 is crucial for vision including conscious vision[72], but our visual awareness resulting from neural correlates of visual processes starts with higher extrastriate areas (V4, ITC?) etc.

Visual consciousness. Mainly all the characteristics of consciousness and awareness are valid specifically for their visual branches if the content is a pictorial representation (e.g., a "real", externally-induced image in imaging, a "virtual", internally-induced image in imagery and in so-called lucid dreaming). Conscious image processing is the processing based on some visual scene where we are conscious of its, usually colorful, pictorial content – image. The processing *preceding that* is mainly not conscious. Conscious object perception and manipulation is actually more clear and frequent than conscious image processing and its result – conscious image recognition. So, with "conscious image processing" I mean being conscious (that's a process) of an image — e.g., just stare, gaze, not yet categorizing. Conscious image recognition is the final and clearest fenomenal level of context-dependent image processing, correlated to so-called collapse of the image-modulated state-vector, probably quantum.[73] The flow of visual conscious experience seems to be a sequence of such events. Conscious object perception and categorization (by implicitly comparing with stored objects) are advanced topics for other books (for start: Pribram, 1991, chs. 5, 7).

Visual awareness is awareness using visual resources. It is strengthened, often, by seeing and recognizing one's own image in the mirror. Since visual attention is being attentive to visual stimuli, all characteristics of attention are valid for its visual branch also, except the physiological pathways of attentional influence onto vision which are much specific / limited to visual cortex (in contrast to, say, auditory attention).

Pictorial experience (my model). Visual patterns–attractors, i.e. carriers of mental images[74], move from one physiological and virtual level to another, because of fractal-like self-similarity and "mirror-like reflectivity" of networks and their holography-like processing at various levels, and because of universality of attractors irrespectively of the substrate. When an image-attractor is projected to the quantum level, the Bose-Einstein condensation or/and quantum coherence provide *the unified conscious experience of the image*, I propose (after Marshall and others). The quantum associative net may realize *associative processing of images, associative storage and conscious image recognition simultaneously*. Binding is provided by special coherence phenomena within the quantum associative net and its deeper levels.

Wavelength- and color-responsive neurons. Zeki (after Lockwood, 1989) has found three types of neurons in V4 and V1:[75]

1. *Wavelength-selective neurons* respond to any color-area which reflects enough light of that wavelength;
2. *Color-coded neurons* respond preferentially when a colored patch would *look to us* to be, say, red (remarkable!);
3. *Wavelength-selective opponent neurons* have a base-level of firing which increases in response to light of one wavelength and decreases in response to light of a complementary wavelength. Their hypothetical relatives, "color-coded opponent neurons", could take part in producing colored after-images — if a patch *looks* red, then it produces a green after-image.

Color as secondary quality. Extension or form, for example, are in philosophy called primary qualities of an object, because they are "more objective" (i.e., depend on object's own characteristics) than secondary qualities. Color, for example, is called secondary quality, since it is relatively "more subjective" (i.e., depends on mentality). Here we shall not enter the philosophical debate how much is color a (co)product of characteristics of the object or illumination and how much a (co) product of conscious experience (Peruš, 2000a; AuxLit 3, 32, 33). By the way, dogs and cats are color-blind (Trstenjak, 1996; Baars, 1997). In other words, colors do not exist for them apart of different rates of grayness. Indeed, colors are subjective (i.e., species-specific or even subject-specific). Secondary qualities like color are not so much something in the objects themselves, but rather "potential powers" to trigger various sensations and perceptions in our brain and conscious experience — based on the objects' primary qualities like extension, form, or wavelength of the object-reflected light.

Phenomenology of color qualia. Color is phenomenally equal, if the responses of cones are equal, to a corresponding superposition (i.e., weighted sum) of the principal colors. This is a consequence of sensational equivalence like, for example, in the case of monochromatic yellow with wavelength 580 nm versus a mixture of yellows with 540 nm and 620 nm. The binding of color is so perfect that the original colors "poured together" cannot be phenomenally traced any more (in contrast to auditory tone-mixing). Perception of the optically-mixed color is thus completely unified or unitary (Trstenjak, 1996).

Basic "dimensions" of color are brightness (Helligkeit), hue (Farbton) and saturation (Sättigung). Trstenjak (1996) phenomenologically describes hue as the "classical color", brightness as the "similarity of color to whiteness", and saturation as the "rate of clearness, distinctness, (unique richness) of a non-gray color" or as the "rate of difference from the grayness". About brightness Trstenjak (1996) says that it is not simply a result of the intensity of light-effect, but is also related to "intrusiveness" of color. The rate of intrusiveness is not an absolute function of illumination, but it is rather a relative function of illumination-difference between the observed color and its immediate external context or background. This direct dependence on background distinguishes intrusiveness from brightness which is conditioned by constancy laws.

Theory of color qualia? Color qualia incorporate aspects which remain, in their very phenomenological essence, non-quantifiable. However, colors also have quantifiable backgrounds in perceptual processing and its stimuli, i.e. in unique wavelengths. This allows a theory of relations (comparisons, contrasts) and combinations (constructive and destructive mixing) of protoqualia like "elementary" colors (MacLennan in Železnikar & Peruš, 1998). Visual conscious experience is a global superposition of more elementary experiences. Flashing of a red light for 20 ms, followed immediately by 20 ms of green light gives impression of a transient yellow light corresponding to the mixture of red and green (Koch in Hameroff *et al.*, 1996, after Efron).

The last but not least remark that I share with Pribram (1998a, Epilogue): "The so-called hard problem[76] is common to all epistemological endeavor. Scientific and other types of knowledge always *begin* with our conscious experience. We can no more ontologically "know" a quark or a table than we ontologically "know" a color or a pain. We relate our experience to one another and match these communications to our own experience. [...]"

ENDNOTES

1 The use of terms "enfolded" and "unfolded" is inspired here by their use by Bohm (1980).

2 The convolution of equation (1.1) is a scalar product in Hilbert space.

3 Pribram wrote (pers. e-mail): "The posterior extrasensory cortex begets objects and categories by enhancing the inhibitory surrounds and flanks of receptive fields in the primary sensory cortex, thus making the process use outer products (i.e. the processing patches become separated from each other [these patches then become symbolized by separate vector-components or 'units']). The anterior frontal cortex works in the opposite fashion. [I.e.,] by eliminating the inhibitory surrounds [in LGN], the processing patches act more in unison, thus the convolutional inner product type of processing prevails [in sensory processing]." Compare, including my note(s) in [...], with the last footnote of sec. 3.6 for justification.

4 Kirvelis (2000, p. 282) writes: "We suggest that the layer $\mathbf{IV}_C$ [of V1] is the 'sensory screen' which represents the retinal projection of the visual scene. Neurons of the layer $\mathbf{III}_{B\beta}$ (a.k.a. $\mathbf{III}_A$) function as the 'screen' for reproducing images stored in memory. The existence of such 'mental screen' explains mental visualization and memory. [... ...] We propose that the neuronal structure implementing the quasiholographic analysis-by-synthesis ought to possess at least three layers: (i) the sensory layer containing the projection of the retinal image – the visual scene; (ii) quasiholographic memory layer, which carries the transformed image $\mathbf{Q}^{-1}$ and stores this transformed memory trace; (iii) reconstruction layer, onto which the back-transformed image is mapped; this reconstructed image possesses the topology of the image in layer (i)."

5 To be precise, independent components are good approximations of Gabor wavelets (but are not the same). The Olshausen & Field model is similar enough, and gives equivalent results, as the ICA-net of Bell & Sejnowski, therefore I will talk about ICA as the process generating the Gabor striate receptive fields γ composed as a superposition of Gabor wavelets g weighted by their Gabor coefficients (sparse codes) s. A single Gabor wavelet g, after being convoluted with the retinal image f, determines the role of an individual V1 neuron in the collective encoding of the analyzed image in the striate cortex (i.e.,V1).

6 The time-dependent Gabor wavelet see in MacLennan (1991, fig. 10). The Gabor wavelet in spectral representation see in Daugman & Downing (in Arbib, 1995). This Fourier transform of the original Gabor profile is expressed in the same functional form as the original spatial Gabor "wavelet", but with the spectral (u) and spatial (x) variables interchanged.

7 I.e., e.g., object perception, from V1 to V2, and beyond it.

8 One could continue to list more "exotic" quantum and quantum-relativity / quantum-gravity models, quantum-spin/spinor models, superstrings, etc. Possibilities are, in principle, enormous… (Nanopoulos, 1995; Kaminski & Linkevich, 1997/8; Van der Wal *et al.*, 2000; AuxLit 26, 27; cf., Mahler, Pitkänen, Kherennikov, Amoroso, Rakovic, Dugic,…)

9 Pötzsch *et al.* (1996) made a non-quantum Gabor-based model – a potential base for associative processing… Jibu, Yasue and Pribram (in Pribram, 1991, Appendix A) also used plane-wave approximation – see sec. 4.6.

10 Modularity is still the main obstacle for integrative computational models.

11 I would add that global associations and context-searches are necessary during search of the right prototype.

12 These are needed for tasks like object-identification, -discrimination and -categorization, they say before.

13 See, for example: D. Durstewitz, J. Seamans & T. Sejnowski (2000): *Nature Neurosci. (Suppl.)***3**, 1184-1191.

14 In the central nervous system.

15 Details on experimental indications that a critical *duration* of cortical activation might separate conscious and unconscious operations see in: B. Libet *et al.* (1983): *Brain***106**, 623. (Also inAlso in: Rakić *et al.*, 1997/II, p. 149.)

16 Cf., Grössing (1988), but also Cahill (2001) and Aharonov & Vaidman (1994).

17 Eigen-state (from German "eigen" = own) is a state which the system stabilizes in.

18 Quantum (plane) waves importantly differ from classical ones not in form and dynamics, but in their essence: They are a result of neither the oscillations of matter-pieces nor oscillations of the magnitude of electric / magnetic field, but of oscillations of the probability for measuring a quantum state (particle) at a location.

19 Content-addressable memory means that partial information triggers reconstruction of the whole memory. This is done by finding relations (associations) between the part(s) and the whole – that's why associative CAM.

20 "Mixing" is used here as an approximate descriptive word. It must not be understood as the "mixture-state" (i.e., the "mixture" of "pure" quantum states) used in terminology of quantum statistical physics.

21 More on these issues in: Blanchard & Jadczyk (1999); Cerf & Adami (1997); Wheeler & Zurek (1983); AuxLit 16.

22 Wave-vectors with opposite signs ($k=2\pi/\lambda$ and $-k$) indicate wave-propagation in opposite directions, but with the same wavelength λ.

23 O'Keefe (in Oakley, 1985, p. 88-89) gives a concrete proposal of holographic processing in the hippocampus, and Nobili (1985) using damping-constant variations of local oscillators in the cortical glia-tissue.

24 A quantum system, which evolves so that it returns to its initial state, acquires a "memory" of this trajectory. This quantum "memory" is encoded in the Berry phase which is added to the phase of system's wave-function.

25 Phase-conjugation refers to the change of sign (direction) of the phase (in exponent) and thus of the wave-vector: k to $-k$. This means that the phase-conjugate wave ($-k$) propagates in the direction opposite to that of the original wave (k), similarly to propagation of the original wave *backwards in time* (as well as in space).

26 Cf., A. Freitas da Rocha *et al.* (2001) emphasizing importance of the coherence between incoming and reentrant signals in conscious perception.

27 These waves could "propagate backwards in time", "symmetrically" to the original input-wave (as in quantum field theory). See also further notes on the Cramer interpretation of quantum mechanics.

28 I.e., there is a coincidence and annihilation ($\Psi\Psi^*$) of positive phases of forward-propagating waves (Ψ, having wave-vector k) with negative phases of backward-propagating waves (phase-conjugate Ψ^* with $-k$). Forward waves encode the original object and backward waves its perceptual image. When they meet and match ($\Psi\Psi^*$) on the path they share, the perceptual transaction is completed, hence the wave-function collapse occurs (Cramer, pers. commun.) and so the image becomes conscious. This adaptive-resonance hypothesis is best understood with the *transactional interpretation of quantum mechanics* by Cramer (1986).

29 Unless the Cramer (1986) transactional interpretation with "quantum waves backward in time" is adopted.

30 E.g., Cramer (1986) even says that his interpretation, or a quantum transaction, respectively, is *a*temporal.

31 Plasma is an ionized gas – a collection of electric charges (+,–) moving around. It is a special state of matter.

32 Details on dendritic dynamics, extensively described in popular manner, see in Jibu & Yasue (1995).

33 Introductory explanatory description of these mathematics see in Jibu & Yasue (1995, pp. 124-129). Studies of *analogies of the dendritic wave dynamics and the Bohm interpretation of quantum theory* can be found in *ibid.*, pp. 127-129, in Gould (1995), and in Gould (in Pylkkänen & Pylkkö, 1995). Bohm's quantum formalism is briefly and popularly described in: L.I. Gould (1993): *Arkhimedes* **2**, 144-157; Jibu & Yasue (1995, pp. 32-35); M. Škarja in *Proc. IMC Info. Soc. '99*, Ljubljana.

34 See paragraph 18 in Auxil. Ch. 7 and compare equations (24) and (25).

35 There is a different interpretation of the wave-function Ψ which here corresponds to the "bioplasma-density" ρ – in contrast to Peruš, Bohm and the

Copenhagen quantum interpretations. Namely, the "bioplasma-density" ρ does not exactly correspond to the quantum density matrix ρ (as used in my Auxil. Chapters 7—10 and in AuxLit 16), because the amplitude of the "bioplasma"-Ψ is defined as $\sqrt{(\rho)}$ (not quantum!) to get the "quantum" $\Psi^*\Psi = \rho$.

36 Actin and myosin are examples of proteins which constitute filaments connected to a dense protein net (Bray, 1995). Membrane proteins link the cytoskeleton also with the so-called extracellular matrix.

37 Evidence on relations of cytoskeletal processes with learning and cognition is listed by Hameroff *et al.* in Pribram (1993, ch. 10, pp. 330-331).

38 I.e., a molecule (tubulin) changes its form or 3-dimensional structure.

39 Soliton is a special solitary conglomerate of waves added together into a big, strong wave.

40 Theoretical predictions by: M. Jibu, S. Hagan, S. Hameroff, K. Pribram & K. Yasue (1994): *BioSys.* 32, 3195-3209. Super-radiance is coherent, dissipation-free electromagnetic emission. Cf., Matsuno (2002).

41 E.g., Hameroff, Penrose, Jibu, Yasue, Pribram, Stapp, Walker, Marshall – see Hameroff *et al.* (1996), and find such ideas in: Nanopoulos (1995); M. Gemine (1991): *Med. Hypotheses* 36, 3277-3283.

42 Originally in: J.C. Eccles (1986): *Proc. Roy. Soc. London B* 227, 411-428. Later: J. Eccles & F. Beck.

43 Experimental modulation of neurotransmitter-function by quantum fields see in Rein (in Pribram, 1993).

44 They also inhibit 40-Hz neuronal coherence (Nanopoulos, 1995, after: Madler & Poppel, Plourde & Picton).

45 One of the sceptics is Tegmark (2000). His calculation result says that the decoherence time-scales ($\sim 10^{-13}$–10^{-20} s) are usually much shorter than the relevant dynamical time-scales ($\sim 10^{-3}$–10^{-1} s) for regular neuron firing as well as for kink-like polarization excitation in microtubules. This might have a negative effect on the microtubular hypothesis by Hameroff and Penrose (however, they rejected this possibility), and maybe also on *some* views of Jibu and Yasue, but *not* also on the *more fundamental* model of quantum associative net (sec. 4.4). Peruš's model (Auxil. Chapters 7—10) belongs to levels below the levels explored by Tegmark (2000).

46 E.g., cytoplasmatic calcium-ion waves, emitted from excited microtubules, might interfere to produce holography-like processing (Hameroff *et al.* in Pribram, 1993 — after Lechleiter *et al.* in *Science*).

47 By the so-called Higgs mechanism which creates mass of these damping-field quanta.

48 Some experimental realizations of noise-resistant coherent and macroscopic quantum states, Bose-Einstein condensates and entanglement reported in Kwiat *et al.* (2000), Denschlag *et al.* (2000), van der Wal *et al.* (2000).

49 The coherence-length, i.e. a "diameter" of the region of coherent oscillations of all net-elements like dipoles, is calculated to be in the case of outer perimembraneous water about 20 to 50 μm (Jibu & Yasue, 1997) (more than cell-dimensions).

50 After Einstein, gravity is equivalent to the (mathematical) relativistic curvature of space-time.

51 Qualia are subjective, qualitative, phenomenal experiences ("how things seem to be to us"). Examples are experiencing yellowness of a lemon, feeling pain in own elbow, and in general also *what it is like to be* a person, etc. Qualia are felt in first person only, not in third person. A blind person cannot imagine precisely how it is to see; person A does not know precisely how person B feels.

52 Conscious experience will here be used as equivalent to the more vague, reified expression – consciousness.

53 Pribram (1998a) considers the *Self* having two aspects: the *I* which is a first-person self-perspective (e.g., self-visualizational, -episodic, -narrative – as in telling about own actions: e.g., "*I* have taken…"), and the *Me* which is a third-person self-perspective (intentional – as in referring to oneself, i.e. to "*my*self": e.g., "This belongs to *me*." "This is *me*, photographed thirty years ago.").

54 Let us list, in a broad context, some mental functions that can be mediated by conscious experience: emotion, motivation, intentionality, planning, goal-achieving, I / self-schema, self-reflective thought, judgements, control in novel circumstances, volition, creativity, spiritual "dimensions", but also ordinary feelings and (felt) needs, and qualia (details in: AuxLit 3, 4, 6; ASSC, 1998; Marcel & Bisiach, 1988; Pribram, 1998a; Peruš, 1998c).

55 ERTAS = Extended Reticulo-Thalamic Activation System.

56 This might be related to Ad 3. In general, these suggestions (1-4) may partially be complementary.

57 Multiple inputs to the superior colliculus show that it should even be needed to generate a meaningful and coherent sense of selfness and its relations to experience – as argued by: B. Strehler (1991): *Synapse* **7**, 144-191.

58 I.e., oscillations "go up and down" *simultaneously* or always with the same delay.

59 Multiple micro-electrode recording is necessary for measurement of phase-dynamics.

60 Ho in Fedorec & Marcer (1996); cf., A. Freitas da Rocha *et al.* (2001)

[61] These indications are faced with skepticism (e.g., Tegmark, 2000), but experimental evidence is missing to decide between the pro- and contra-quantum camps. The latter abundantly rejected Tegmark's objections.

[62] Consider numerous analogies between quantum systems and other brain's complex systems on my list of Auxil. Ch. 7 (associative neural nets) and in sec. 4.6 (dendritic "bioplasma").

[63] Planck-scale dimension is 10^{-35} m. Spins are the most fundamental "binary" variables attributed to particles.

[64] Note by M.P.: This divison is presented in: P. Stoerig & A. Cowey (1995): *Behav. Brain Res.* 71, 147-156. They conclude that phenomenality is prerequisite for the so-called conscious accessibility (a term for conscious representations available for verbal reports, judgements, planning, guiding of action, and the like) to the present or previous visual processing (i.e., for "usable" visual introspection) which needs higher cortical functions.

[65] Note by M.P.: Pribram (1991) defines a frame as the immediate local context influencing an input-perception.

[66] Subjects became aware or not that a tone predicted a visual event and another tone did not.

[67] Sub-quantum dynamics (in "vacuum" or "holomovement") are defined in Bohm (1980). See also AuxLit 26.

[68] Cf., A. Freitas da Rocha *et al.* (2001), Fidelman (2002).

[69] From this, it was concluded that "blindsight does not depend on functional islands of tissue preserved within the deafferented striate cortex" – in the abstract of: P. Stoerig *et al.* (1998): *Neuroreport,* **9**, 121-125.

[70] From the abstract of: L. Weiskrantz, J. Barbur, A. Sahraie (1995): *Proc. Natl. Acad. Sci. USA* 92, 6122-6126.

[71] Note added by M.P., as also Pribram would say. Namely, it is not clear what do Koch and Crick mean by "awareness of neural activity". The note is *our* interpretation – after personal communication with Prof. Pribram.

[72] Cognitive psychologist Baars (1997, p. 67) writes: "V1 is the only region whose loss abolishes our ability to consciously see objects, events, people, dramatic gestures, delicate textures, the bold shapes of a Picasso painting. But the cells in V1 respond only to a sort of pointillist level of visual perception, like the dots of light on a television screen. Thus it seems that area V1 is needed for such higher-level experiences, even though it does not contain higher-level elements! [... ...] Damage to V1 'blocks out' *all* higher-level visual consciousness, while damage to object perception [areas] causes only a local loss." (Emphasis his.)

[73] To make it clear, we are interested here in how a human *becomes conscious of the recalled image* (which is a result of, mainly *non*-conscious, image process-

ing). For this, quantum implementation is, by my hypothesis, needed. Since image processing has come to the stage where intra-image and inter-image context information has already been processed, one can talk about image recognition, being aware that higher levels of recognition (of objects and their categories, etc.) are following (I leave them to other books).

74 A pattern of neural-units' activities which acts as an attractor represents a gestalt — it is the neural substrate of the mental image or gestalt. The "mentality" itself emerges from the "virtual influence" of the attractor's potential well and from consciousness (which one cannot explain, but just see its relation to the "virtual action").

75 Partially such or similar neurons, especially no. 3, could be found already in retina.

76 I.e., how to explain qualia, which is beyond capabilities of contemporary science — note by M.P.

Chapter 5
Computer Simulations and Applications of Quantum Associative Network

5.1 CENTRAL QUANTUM HOLOGRAPHIC MODEL

Quantum holography (Peruš M. et al,2004-2005), is a new paradigm of quantum information processing that constructs the Hopfield-like associative neural networks in an extremely natural way by exploiting quantum physical system. Instead of utilizing complicated and complex quantum-based devices abiding to quantum mechanics, Quantum Holography harnesses the fact that an open system can be manipulated in a specific measurement-like way by triggering non-linear, non-unitary and irreversible collapse like phenomena instead of the Schrodinger evolution of the quantum state. It is a parallel-distributed process at quantum level which operates similarly to (oscillatory) associative neural nets. Since the natural fundamental quantum-wave dynamics is harnessed, it allows much easier and cheaper physical realization with

DOI: 10.4018/978-1-61520-785-5.ch005

bigger sizes and resolutions of images than the mainstream quantum-computing approaches like Turing-machine based quantum computing using quantum logic gates. The birth of the approach is due to big experimental success on holography followed by fast development of quantum optics. With quantum realm resolution, it promises to implement the Hopfield model and its generalizations in an original way into a new framework where its former obstacles such as memory capacity limitations and crosstalk due to non-orthogonal patterns are substantially weakened.

The implementation is based on translation on encoding of intensity of pixels to their encoding by wave-phases, that is, a parallel-distributed encoding of a N-pixel image into a front of waves ($A_1e^{i\phi 1}, A_2e^{i\phi 2}, A3e^{i\phi 3}......$). The encoding has 2 special cases, which is, encoding in amplitudes A only, and encoding in oscillatory φ phases only. The two cases enable effectively the same image processing as far as the following variable exchange can be made in the mathematics of the model/ algorithm: $A \leftrightarrow e^{i\varphi}$. It will be shown that wave-based model is equivalent to intensity-based model.

The simplest Hopfield neural network incorporates Hebbian storage of patterns $\overset{\rightarrow k}{v}$ into correlation matrix J, that is, $J = \sum_{k=1}^{p} \overset{\rightarrow k}{v} \otimes \overset{\rightarrow k}{v}$, and memory-influenced transformation of patterns $\vec{v}$: $\overset{\rightarrow output}{v} = J \overset{\rightarrow input}{v}$, with each pattern stored denoted by superscript k. Preferably orthonormal patterns, which become Hopfield net Eigen states (attractors), can be complex-valued and can be quantum-encoded into the net-state $\vec{q}(= \sum_k c^k \vec{v}^k)$. For the quantum implementation, we will henceforth use the quantum notation, i.e. Ψ corresponds to $\vec{q}$, and Ψ^k corresponds to $\vec{v}^k$ (= A^k if non-oscillatory). Wave function Ψ describes the whole state of the quantum system/ net; k describes the kth of its Eigen states. Thus, images are assumed to be encoded into quantum Eigen-wave-functions k. It is: $\Psi = \sum_k c^k \psi^k$. The images (if huge and random) and quantum Eigen states share the orthonormality property.

For quantum-physical realization, like neural nets and holography, there is a storage and preparation phase governed by (5.1), and a "collapse"-like retrieval or measurement phase governed by (5.2). Associative memory matrix J consists of interaction-weights J_{hj} which are correlations (from quantum perspective, junctions in wave-interference array) of unit-states Ψ_j (j,h=1,...,N, for N units; N is huge):

$$J_{hj} = \sum_{k=1}^{p} \psi_h^k (\psi_j^k)^*, \tag{5.1}$$

Where the asterisk denotes complex conjugation (or, optically, phase conjugation). (5.1) describes a multiple-interference array, i.e. a "quantum hologram". The most natural physical realization is possible using single-photon self-interference. Each laser-illuminated object / image is encoded into a specific photonic mode (index k). Each such quantum light-mode Ψ^k is left to interfere with itself in the same quantum-optical medium simultaneously (or sequentially) to the other modes. We believe that this is a fundamental natural process, since J corresponds to the Green-function propagator G governing dynamics of any quantum system. However, it can be artificially realized by laser beams where quantum states of light are assumed, entering some type of beam-splitter/interferometer, which is quite similar to classical holography. However, if quantum waves are used, the process can be identified as the simplest case of quantum holography. A generalized Hebb learning rule can be found in chapter 7.7.

The net-state transforms, in dependence on the "hologram-points/weights" J_{hj} according to

$$\Psi_h^{output} = \sum_{j=1}^{N} J_{hj} \Psi_j^{input} \tag{5.2}$$

Combining (5.1) and (5.2), we obtain the Eigen state/image retrieval process using an equivalent of relation $(\vec{c} \otimes \vec{a})\vec{b} = \langle \vec{a}, \vec{b} \rangle \vec{c}$,

$$\begin{aligned} \Psi_h^{output} &= \sum_{j=1}^{N} \left(\sum_{K=1}^{P} \psi_h^k (\psi_j^k)^* \right) \Psi_j^{input} \\ &= \sum_{k=1}^{P} \left(\sum_{j=1}^{N} (\psi_j^k)^* \Psi_j^{input} \right) \psi_h^k = \sum_{k=1}^{P} c^k \psi_h^k = \psi_h^{k0} \end{aligned} \tag{5.3}$$

This corresponds to input-triggered "collapse of the wave-function" to a "selected" Eigen state Ψ^{k0}. Usually, exclusively this coefficient is close to 1 which belongs to the memorized image that is most similar to the input image. Consequently, all others are close to zero. In this case, the quantum associative net recognizes the input-state/image. Refer to auxiliary chapter 7.8 for more details on wave-collapse.

Let us now show mathematically how the quantum-holographic procedure realizes the Hopfield net dynamics. Quantum plane waves, i.e. photon modes $\psi^k(\vec{r}, t) = A^k(\vec{r}, t) e^{i\phi^k(\vec{r}, t)}$, are the most suitable. Here the phase is

$$\psi^k = \vec{k}^k \vec{r} - w^k t = \frac{\vec{p}^k}{\hbar} \vec{r} - \frac{E^k}{\hbar} t$$

For the kth mode of the photon having momentum $\vec{p}^{k}$ and energy E^k. Advanced alternatives are: (a) sum of plane-waves (modes), (b) quantum wave-packets nearly identical to Gabor wavelets.

Holography shows at least for non-quantum waves, how one can parallel-distributively encode images k into a web of waves ($A_1^k e^{i\phi_1^k}, A_2^k e^{i\phi_2^k}, A_3^k e^{i\phi_3^k}....$). The amplitude A_j^k and the oscillatory phase ϕ_j^k have the same lower index j, since they belong to the same "waving" point, which is our "unit" (encoding a point of the image). Plane waves can be set to constant amplitude at $A = 1/\sqrt{N}$ to ensure proper quantum normalization. For phase information holography, we replace all Ψ-variables of the above equations with $e^{i\varphi}$ instead of Hopfield's A. This ($A \leftrightarrow e^{i\varphi}$) is usual mathematical exchange of variables, a legal mathematical step, giving $\psi_j^k = e^{i\phi_j^k}$ instead of $\psi_j^k = A_j^k$. All the simulation tested mathematics remains valid for sinusoidal encoded images. So, Hopfield algorithm works with sinusoidal inputs equivalently as with real-valued inputs.

In order to prove quantum-wave image-recognition with the system of (5.1) and (5.2), it suffices to execute the exchange, $\psi_j^k \leftrightarrow e^{i\phi_j^k}$, first in the Hebb rule (5.1), using $(e^{i\varphi})^* = e^{-i\varphi}$:

$$J_{hj} = \sum_{k=1}^{p} e^{i\phi_h^k} e^{-i\phi_j^k} = \sum_{k=1}^{p} e^{i(\phi_h^k - \phi_j^k)} \tag{5.4}$$

And secondly in (5.2), which we obtain:

$$e^{i\phi_h^{output}} = \sum_{k=1}^{P} (\sum_{j=1}^{N} e^{i\phi_j^{input}} e^{-i\phi_j^k}) e^{i\phi_h^k} = e^{i\phi_h^{k0}} \tag{5.5}$$

The right most expression of (5.5) gives an image-like output, where if a wave has a similar phase to one of the memorized waves, then those waves will be reconstructed. If those conditions are not satisfied by the data correlation-structure, then destructive interferences (cross-talk) lead to a mixed or averaged output. For details on relevance of mathematical formalism of associative neural network and quantum theory, refer to auxiliary chapter 7.

Classical Hopfield network has two major limitations which are storage limitation and cross-talk. Storage capacity of the network is limited by P=0.14N where N is the number of processing nodes while P is the number of patterns stored. An example is that it can only store only 14 10X10 pattern images. Cross-talk is due to interferences from other stored patterns due to being non-orthogonal between train-

Figure 1. (a) Original image. (b) Original image (a) with 80% salt-and-pepper noise. (c)-(g) Image restored from memory of 10 different simultaneously-stored fingerprints after presentation of the query-image which is: (c) whole original image (a); (d) 25% (first quadrant) occluded image (a); (e) 50% (bottom half) occluded (a); (f) 75% (2nd - 4th quadrants) occluded (a); (g) noisy image (b). (from Loo Chu Kiong, Mitja Perus, Horst Bischof, 2004b, p.284)

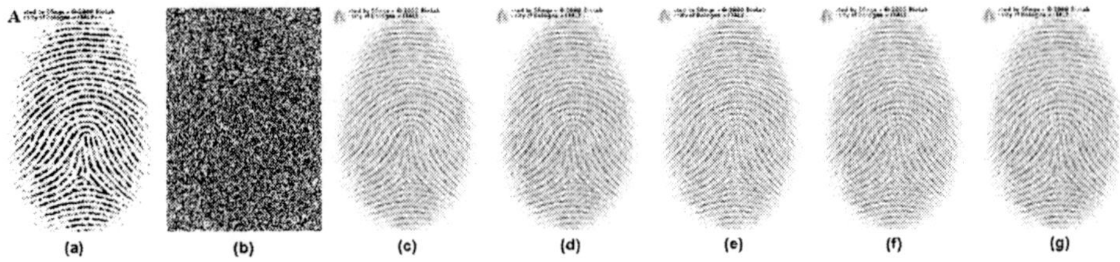

ing images, with Figure 8(a) as an example. Because reconstruction depends on the covariant component of vectors, which is a summation of stored patterns with their respective covariant coefficient, non-orthogonality will cause a projection to other vectors, thus, non-zero coefficient even with 100% perfect images (meaning, same as the stored images). These problems are solved by quantum holography due to its immense size. Working in quantum realm means astronomically high resolution or processing nodes or N. So the problem of the limitation P=0.14N is much reduced. Cross-talk is reduced also since images with huge number of pixels are with high probability almost orthogonal. A summary of the proof: The probability distribution of values of the scalar product converges to Gaussian distribution, owing to the Central Limit Theorem, which can be applied for scalar product $\langle\Psi^a,\Psi^b\rangle$ if the components of Ψ are random. The Gaussian distribution has zero mean, so $\langle\Psi^a,\Psi^b\rangle=0$ and $\sigma^2=R/N\rightarrow 0$ if N approaches infinity. Orthogonality is therefore most probable.

5.2 FROM RECOGNIZING IMAGES TO RECOGNIZING OBJECTS

From image recognition, Hopfield network can be extended to view-invariant object recognition. Same object but with different poses is to be recognized as one object. It is the attractor net which is responsible for mapping of many views into common attractor, so that any of the views triggers reconstruction of the same attractor

Figure 2. (a)-(e) some training images for mixed-set experiment. (f)& (g) Query images for mixed-set experiment. ((f) is occluded (a); (g) is occluded Fig. 1(a)). (from Loo Chu Kiong, Mitja Perus, Horst Bischof, 2004b, p.284)

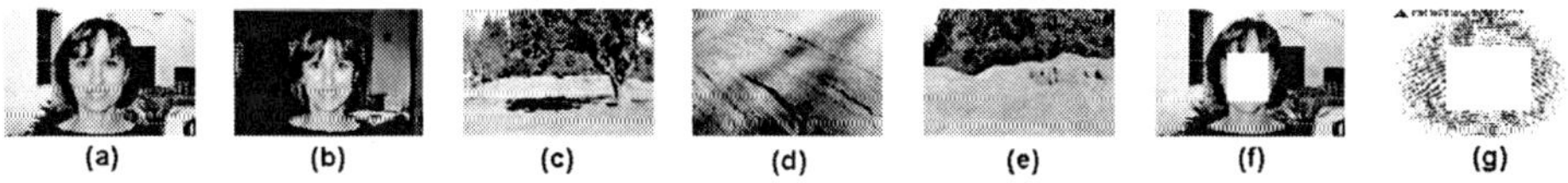

5.3 ORTHOGONALIZATION PREPROCESSING

For highly view-invariant object recognition task, images are not as orthogonal as the random-image case as discussed above. In this case, images are normalized and orthogonalized. This is achieved by deriving a vector ψ_k^+ where $\psi_k^+\psi_{k'} = \delta_{kk'}$ (Kronecker). So, ψ_k^+ replaces ψ_k^*. They were obtained from $\psi_k^+ = \sum_{k'} a_{kk'}\psi_{k'}^T$, where the matrix (optical mask/SLM) $[a_{kk}]$ had been calculated as the Moore-Penrose pseudo-inverse of $v_k v_{k'}^T$ (interference term) due to the fact that the matrix is singular caused by normalization, which is a preprocess to "standardize" the images. In geometrical point of view, the approach converts the covariant components to the contra-variant counterparts. Image reconstruction then follows the parallelogram summation rule of the contra-variant components. Reconstructed image can be seen

Figure 3. (a, b) Selective reconstruction from a memory of 30 simultaneously-stored images (10 different natural scenes like Figures 2(c)-(e) and 10 different face-poses like Figures 2(a)&(b) and 10 different fingerprints): (a) Reconstruction from query image Figure 2(f); (b) Reconstruction from Figure 2(g). (c) Reconstruction from another 30 simultaneously-stored images (10 different Chinese pictograms and 10 different face-poses as before and 10 different fingerprints as before) triggered by 50% (upper half) occluded "query-pictogram". (from Loo Chu Kiong, Mitja Perus, Horst Bischof, 2004b, p.285)

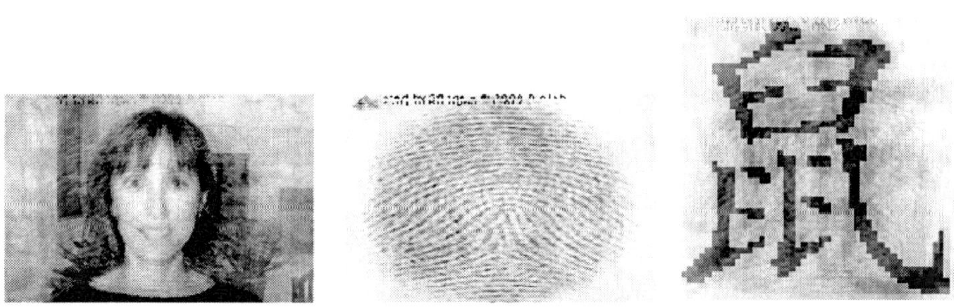

Figure 4. Plots of Peak Signal-to-Noise Ratio of reconstructed image from query image versus number of simultaneously-stored images of (a) fingerprints, (b) Chinese pictograms, where (a1) query is a fingerprint with salt-and-pepper noise, (b1) query is a Chinese pictogram with salt-and-pepper noise, (a2) query is an occluded fingerprint, (b2) query is an occluded Chinese pictogram.(from Loo Chu Kiong, Mitja Perus, Horst Bischof, 2004b, p.283)

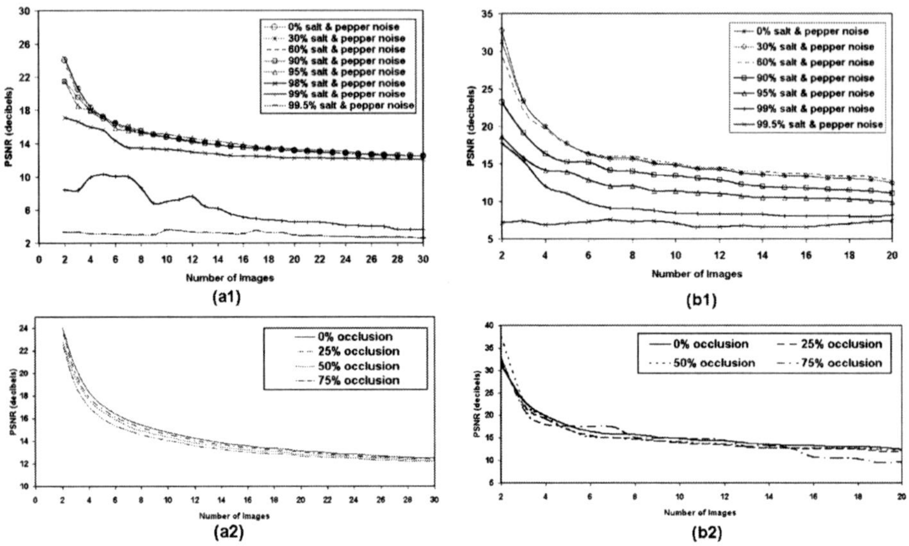

from Figure 8(b). Orthogonalization algorithm is not quantum-implementable, so computing with classical electronic or optical devices is assumed.

But instead of plane waves, images can be encoded in Gabor wavelets, which are similar to quantum wave packets. This can be a natural alternative to the present orthogonalization procedure. Namely, nearly orthogonal ("tight-framed") Gabor wavelets may be constructed.

5.4 BIO-COMPUTATIONAL MODEL OF OBJECT RECOGNITION

It is one of the rare detailed agreements of natural vision research (Palmer, 1999; Buser & Imbert, 1992) and computer-vision practice that Gabor wavelets (Navarro et al., 1996) are the best descriptors of V1 receptive fields and computationally efficient (cf., Wundrich et al., 2002), which are approximately invariant to affine transformation. This means Gabor-encoded object is recognized also after translation, rotation, dilation etc. Lee(1996) derives conditions under which Gabor wavelets which are in general non-orthogonal, behave as if they are (one says that they form

Figure 5. Retinal image (top) is convoluted with Gabor wavelets (middle row). The bottom image is the output of the whole simulated procedure.(from Mijta Perus, Horst Bischof, Chu Kiong Loo, 2005a, p.118)

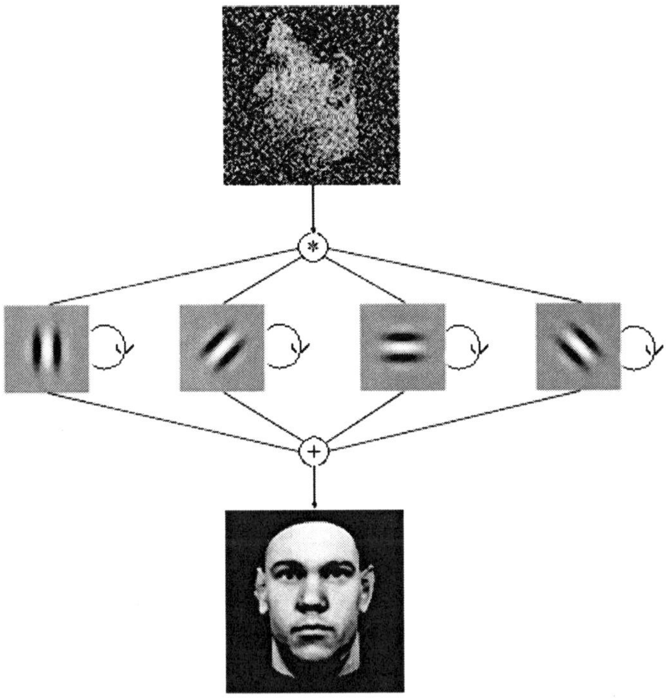

tight frame). This allows good reconstruction of high resolution images from low-precision neural codes. There is physiological evidence (listed in Lee, 1996, sec. 6 end; R. & K. De Valois, 1990; Pribram) on an *overcomplete* (much redundant) and even *almost tight-frame* Gabor image-representation obtained through significant *over*sampling by the primate visual system. Non-orthogonality and non-compact support of Gabor wavelets is thus no obstacle neither for natural nor artificial object recognition, since they can be made almost / effectively orthogonal and mainly compact (i.e., with negligible "tails") (e.g., Krüger & Sommer, 2002, sec. 2.2; Resconi & van der Wal, 2002, sec. 5.3).

A receptive field of a neuron is everything that arrives into its dendritic tree and affects its output. Retinal image *I* gets convoluted (*) with the Gabor receptive-field profile *G* of V1 simple cells: *I*G* (Figure 5). Convolution is, roughly, integration (multiplication) of all image points with all Gabor-wavelet patches. This is a very rough model of the primate image processing (Pribram, 1991) imitated successfully by computer vision (e.g., Pötzsch et al., 1996; Krüger & Sommer, 2002). According to Pribram (1991) the spectral (holographic-like) image representation of V1 is then reconstructed by an inverse Gabor transform into a spatial topologically-correct representation probably in V2. Wavelet representation as well as (phenomenal) pictorial representation could be processed in a Hebbian-like associative way – quantum-implemented or classically. Explanation is provided in accordance to contemporary main-stream neuroscience on visual processing described in auxiliary chapter 6.

5.5 QUANTUM HEBBIAN PROCESSING WITH NEURALLY SHAPED GABOR WAVELETS

For our simulation we have chosen the following family of elementary Gabor wavelets (with j=√-1) as described in Lades et al. (1993) (orig. Gabor, 1946).

$$g_i(\mathbf{x}) = \frac{\|\mathbf{K}_i\|}{\sigma^2} \exp\left(-\frac{\|\mathbf{K}_i\|^2 \|\mathbf{x}\|^2}{2\sigma^2}\right)\left[\exp\left(j\mathbf{K}_i\mathbf{x}\right) - \exp\left(-\frac{\sigma^2}{2}\right)\right] \tag{5.6}$$

Each g_i is an elementary Gabor wavelet, characterized by the vector **K**, enveloped by a Gaussian function. Its parameter σ=2π determines the ratio of window width to wavelength, i.e. the number of oscillations under the Gaussian envelope. The first term in the square brackets determines the essential oscillatory part of the wavelet; the second term makes the wavelet response independent of the absolute intensity of the image (Lades et al., 1993). The vector $\mathbf{K}_i$ is defined as:

$$\mathbf{K}_i = \begin{pmatrix} k_\nu \cos\varphi_\mu \\ k_\nu \sin\varphi_\mu \end{pmatrix} \tag{5.7}$$

where $k_\nu = \frac{k_{\max}}{f_\nu}$, $f_\nu = 2^{-\frac{1+\nu}{2}}$, $\varphi_\mu = \frac{\mu\pi}{8}$, $k_{\max} = \frac{\pi}{2}$, $\sigma = 2\pi$

The parameters ν and μ define the frequency and orientation of the wavelets. We used five frequencies (ν=0,1,2,3,4) and eight orientations (μ=0,1,2,3,...,7) in the final representation following the methods in Lades et al. (1993). The parameter f_ν is the spacing factor between wavelets in the frequency domain.

Let us define Gabor-wavelet field G_k as a linear combination of N elementary Gabor wavelets g_i:

$$G_k = \sum_{i=1}^{N} a_i g_i \tag{5.8}$$

For image reconstruction, it is sufficient to consider the real part of g_i:

$$G_k = \sum_{i=1}^{N} a_i \operatorname{Re}\left(g_i\right). \tag{5.9}$$

Since elementary Gabor wavelets g_i are not orthogonal, it is not possible to calculate the Gabor coefficients a_i by a simple projection of wavelet g_i onto the image Ψ. However, the dual wavelets can compensate for the non-orthogonality of the Gabor wavelets (Krüger & Sommer, 2002). The wavelet $\mathbf{g}^{+}$ is the dual wavelet to $\mathbf{g}$ iff $\mathbf{g}^{T}\mathbf{g}^{+}=\mathbf{I}$ where $\mathbf{g}=\operatorname{Re}(g_1,g_2,\ldots,g_N)$ and $\mathbf{I}$ is an identity matrix (Krüger & Sommer, 2002). Therefore, $\mathbf{g}^{+}$ is the pseudo-inverse of $\mathbf{g}$, i.e. $\mathbf{g}^{+}=\mathbf{g}^{-1}$. The dual wavelet allows an orthogonal projection of an image Ψ onto the closely linear span of g_i, i.e.

$$G_k=(\Psi\mathbf{g}^{+})=(\Psi\mathbf{g}^{-1})\mathbf{g} \tag{5.10}$$

Equating Equation (5.9) and Eq. (5.10), we obtain: $\mathbf{a}=\Psi\mathbf{g}^{+}=\Psi\mathbf{g}^{-1}$. Gabor coefficients a_i were thus calculated from Ψg^{-1}. Then, Gabor-wavelet fields G_k were obtained from Equation (5.9). After such classical pre-processing for input-dependent Gabor-field shaping, the quantum Hebb process $\Sigma_k|G_k\rangle\langle G_k|$, or the quantum-implementable Griniasty et al. (1993) process:

$$\Sigma_k|G_k\rangle\langle G_k| +\gamma(\Sigma_k|G_{k+1}\rangle\langle G_k| +\Sigma_k|G_k\rangle\langle G_{k+1}|) \tag{5.11}$$

was executed. The shaping of Gabor-wavelet fields by input-dependent setting of Gabor coefficients, each corresponding to an elementary Gabor wavelet, (Equation (5.9) with $a=\Psi g^{-1}$) is remarkably similar to models of processing in the visual cortex, e.g. by K.H. Pribram (1991).

In the main text, Gabor-wavelet fields G_k will be called just Gabor wavelets, to be short. Namely, a superposition of wave(let)s is also a wave(let).

5.6 SIMULATED BIO-HOLOGRAPHY WITH GABOR WAVELETS ENCODING

Basic Quantum implementable model- We can abstract findings from previous sections into a schematic computational model (core) which is a gross simplification but functions itself sufficiently well. It seems that images and representations of segregated objects are processed in many different ways, brain regions and levels, architectures, "algorithms" – depending on the task and on circumstances or contexts. However, the basic (core) image- and object-recognition is probably realized in V1 (wavelet interference) and in its connections to V2 (inverse Gabor transform) (Pribram, 1991). A V1 net of neurons can, in principle, be effective for this task after it has received inputs which had been suitably pre-processed (including possibly ortho-normalized) on the way from retinas through LGN (lateral geniculate nuclei) to V1. However, direct, attentive, clear *phenomenal* perception probably requires that a quantum net (pervading neural nets with dendrites and accompanying electro-magnetic fields) is also activated to increase resolution and effectiveness, and hypothetically to support conscious experience (Peruš, 2001b, 2004; Woolf & Hameroff, 2001). Perception of images is phenomenal (incorporating qualia); recognition of images requires also support of unconscious processing in a quantum and/or neural attractor net (the "core"); recognition of objects, on the other hand, requires other non-phenomenal pre- and post-processing (relative to the "core") up to cognitive levels based in ITC and PFC.

Let us discuss the quantum core of the model. Previously, we emphasized that quantum wave-packets can be shaped like Gabor wavelets. Let us use the Dirac quantum notation: $|.\rangle$ is quantum eigenvector; $\langle.|$ is its transpose and complex-conjugate or phase-conjugate. The first vector describes the state of a quantum system (incorporating pixel-values, in our case); the second vector corresponds to the same state but with the opposite direction of wave-propagation. Recognition of an object, *re*presented in V2, is realized by a Gabor-filtered quantum-encoded retinal input ($I=|\Psi\rangle$):

$$|\Psi_{v2}\rangle \cong \sum_k |G_k^{v1}\rangle\langle G_k^{v1}|\Psi_{retina}\rangle \tag{5.12}$$

where the scalar product (Gabor coefficient) $\langle G_k^{V1}|\Psi_{retina}\rangle \equiv G_k^{V1} * \Psi_{retina}$ (convolution); Ψ is a (quantum) plane-wave, i.e. sinusoid, encoding a retinotopic image; G_k is the kth Gabor wavelet, i.e. a Gaussian-enveloped sinusoid. Various such wavelets are either rotated (for angle φ), or translated (for shift x), or dilated, or scaled (λ), or combined, versions of the mother wavelet g, e.g. $G_k = S_\lambda R_\varphi T_x g$. They contribute to view-invariant object-recognition. Figure 5 illustrates the process of Equation (5.12).

Peruš, Bischof, Caulfield & Loo (2004) proved that *quantum neural net* by Peruš (2000a,b), executing $|\Psi_{out}\rangle \cong \Sigma_k|\Psi_k\rangle\langle\Psi_k|\Psi_{in}\rangle$, can *recognize images,* encoded into quantum eigen-states $|\Psi_k\rangle$, *like an attractor neural net does.* (Such a net takes part in Bartlett & Sejnowski, 1997, 1998.) One can experimentally (quantum-optically) produce $|\Psi_k\rangle=|G_k\rangle$, thus realizing Equation (5.12).

Since Peruš et al. (2004) have shown with computer simulations that their quantum-wave net using a quantum-holographic procedure can emulate the Hopfield attractor neural net, it is important to note the proof by Maass & Natschläger (1997) that a network of spiking neurons can emulate arbitrary Hopfield nets in temporal coding. There, time-delay between two spikes has a memorizing role analogous to the phase-difference between two waves used by Peruš et al. (2004). See Figure 6. There are certain analogies or "isomorphisms" between associative nets of quantum, oscillating, spiking, "real-valued", "binary" and spin/dipole (cf., Shuai et al., 1995) neuron-like units. Such nets can recognize images by their attractor dynamics.

Let us describe, turning to Figure 6, such "coding isomorphisms" specifically. We transform the usual associative memory, which uses the correlation of pixel intensities $v_1 v_2$, into a hologram made of wave-interference $v_1 v_2 = exp[i\ (\varphi_1 - \varphi_2)] = exp[i\omega(t_1 - t_2)]$. (In this paragraph only: i = √(-1).) The wave-phase φ is proportional to time t: $\varphi = \omega t$, where ω is the angular frequency. Note that $v_1 = cos\ \varphi_1 =$

Figure 6. Our memory is encoded in phase-differences ($\varphi_1 - \varphi_2$) of waves. This is analogous to an encoding in time-differences or delays ($t_1 - t_2$) of spikes. To demonstrate the analogy, a spike is illustrated by a horizontal line reaching the maximum of a wave. Instead of the signal intensity or wave amplitude (classical code), the wave's phase-angle is the relevant code in our model (or time of firing in the "isomorphous" spike model). (from Mijta Perus, Horst Bischof, Chu Kiong Loo, 2005, p.121)

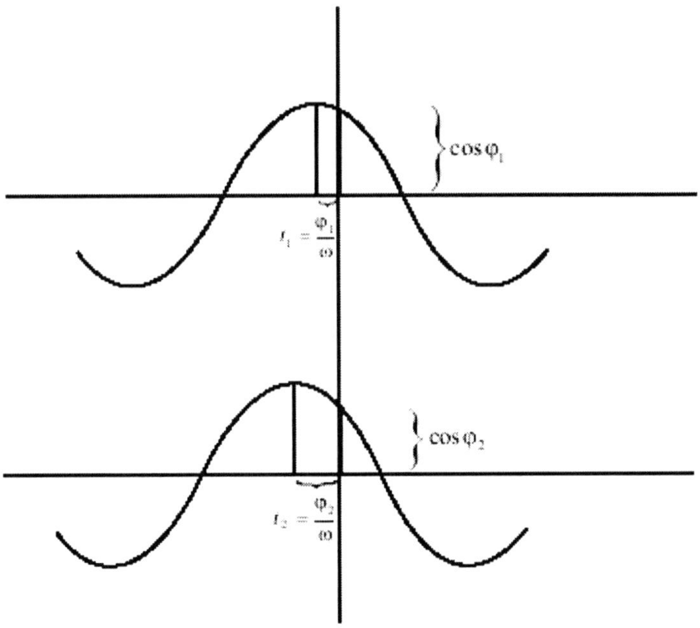

$Re[exp(i\varphi_1)]$ and $v_2 = cos\ \varphi_2 = Re[exp(i\varphi_2)]$, where *Re* means the real value. Pixel-intensities ("line heights" in Figure 6) are encoded by wave-heights which are actually determined by the current position of the wave-peaks, or by the wave's time evolution, respectively. Consequently, our memory is realized by phase-differences $\varphi_1 - \varphi_2$ (hologram). A mathematically analogous possibility is an encoding by spike-firing delays $t_1 - t_2$ ("coincidence memory"). Just imagine in Figure 6 that the vertical lines denoting wave-peaks represent spikes.

It is possible that there is a subcellular mechanism in brain which shapes the underlying quantum state into a Gabor-wavepacket form according to a neural Gabor-wavelet "template." Such a "template" is neurally stored as the receptive-field profile which has been produced along the retino-geniculo-striate pathway (details in: Peruš, 2001a,b). In the next section, we summarize our computer simulations containing essential parts of this process.

Bio-Holography with Gabor wavelets encoding - Gabor wavelets of different orientations and frequencies were implemented as defined previously. Images of objects, photographed from different viewpoints, as in Figure 7, have been successfully encoded into them. These Gabor wavelets were then interfered in a Hebbian process for associative storage, i.e. $\Sigma_k |G_k\rangle\langle G_k|$ was formed. Interference corresponds to calculation of correlation. Retrieval procedure, resembling that of Hopfield neural nets or quantum holography, was finally executed to recognize a selected object. Results (Loo, Peruš & Bischof, 2005b) were always clearly better than those with the conventional quantum Hebb-like storage, i.e. with self-interference of quantum plane-waves (instead of Gabor wavelets), respectively (Figure 8a). The Hebb storage with Gabor wavelets is a significant advance regarding biological plausibility while retaining quantum implementability. However, our object-recognition results (Loo et al., 2005b) were good merely for a small number of simultaneously stored poses.

This is the cost of our demand for biological plausibility and quantum implementability. Namely, Loo, Peruš & Bischof (2005a) have already associatively stored many neurally ortho-normalized object-images and recognized versions of the memorized objects by the quantum neural net of Peruš (2000). Results were almost perfect (e.g., Figure 8b), but the used orthogonalization procedure, i.e. the Moore-Penrose pseudo-inverse matrix method, is hardly realizable in biological

Figure 7. Example of training images of various poses of objects

Figure 8. Object-recognition results: (a) using Hebbian storage of image-carrying plane waves (5 simultaneously stored images of Figure 6; recall-query was that image occluded as in (d)). (b) using Hebbian storage of plane waves after orthogonalization pre-processing (35 simultaneously stored images of 7 different poses of 5 persons; recall-query was that image with 30% salt-and-pepper noise). (c) using the Griniasty et al. storage of image-encoding Gabor wavelets (5 stored images of Figure 6, $\gamma = 0.1$, recall-query was (d)). (d) Occluded image as the recall-query input which triggered outputs (a) and (c)

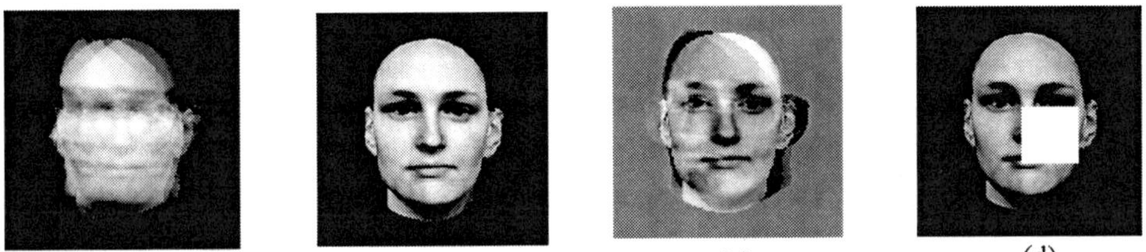

(a) (b) (c) (d)

and/or quantum systems. Therefore, we later incorporated the learning rule from Griniasty et al. (1993), Equation (5.11), instead of the Hebb learning rule $\Sigma_k|G_k\rangle\langle G_k|$ (γ is a coefficient between 0 and 1).

It is possible to implement the Griniasty et al. storage with quantum interference optics — easier than so to implement the Bartlett & Sejnowski (1997, 1998) processing. However, it is *harder* to implement the Griniasty et al. storage than to implement the Hebb storage by quantum optics (done by Peruš et al., 2004). On the other hand, the Griniasty et al. net is found to be more efficient for object recognition than the Hebb net (Figure 8). Both learning rules are relatively biologically plausible (sec. 5; Griniasty et al., 1993; Bartlett & Sejnowski, 1998, 1997; Miyashita, 1988).

For quantum implementation of the Griniasty et al. storage, three superposed multiple quantum-wave interference processes would have to be done – one for each term of the above expression – i.e., a self-interference and two cross-interferences of Gabor wavelets encoding neighbouring poses. Neural or (bio)electromagnetic implementation, too, are realizable (easily, in principle).

The new Storkey & Valabregue (1999) learning rule is also to some extent relevant for bio-like and neuro-quantum processing models.

Although the Griniasty et al. net reconstructs the correct object merely in the case of a small number of stored images (Figure 8c), the result is nevertheless much better than the non-selective output of the Hebb net (Figure 8a).

After training, i.e. inputting images as in Figure 7 and storing them using Equation (5.13) (Griniasty et al.), a testing recall-image is presented (Figure 8d) which triggers associative reconstruction of a selected (most similar to query) object-image (Figure 8c): $|\Psi_{output}\rangle = J\,|\Psi_{input}\rangle$. In the Hebb case we obtained Figure 8a.

Why is the Griniasty et al. rule more successful for object recognition than the Hebb rule? The Griniasty et al. (1993) rule for the memory-matrix (i.e., the hologram, quantum-implemented as the Green-function propagator), i.e. $J = \Sigma_k |G_k\rangle\langle G_k|$, is:

$$J_{ij} = \Sigma_k\, G_i^k\, G_j^k + \gamma\, (\Sigma_k\, G_i^{k+1}\, G_j^k + \Sigma_k\, G_i^k\, G_j^{k+1}) \qquad (5.13)$$

where *i* and *j* are pixel indices, *k* is the index of an image, G_k is the Gabor wavelet encoding it.

For object-recognition, it is essential that indices *k* correspond to time-sequence of moments capturing poses of a rotated object. Thus, the net is trained by a sequence of view-points which has to be temporally ordered. The second and the third term of the Griniasty et al. rule then store temporal correlations between object-views, additionally to Hebbian self-correlations of each view (the first term). This produces correlated attractors in the sense that poses, neighbouring in the sequence, fall into the same attractor. This leads to appearance-based viewpoint-invariant object-recognition if the attractors are flexible enough to allow some pose-interpolation.

Conclusion:
Review of Book's Specific Contributions

To provide an integrative model of conscious image processing, a large body of neuropsychological, neurophysiological, psychophysical, biocybernetical and other theoretical, experimental and simulation-based literature on vision has been systematically reviewed, compared, and an synthesis of it has been proposed. These data as well as several relevant models have been considered in the context of the Holonomic Brain Theory. Many informative complementarities were found. Additionally to a general model of image processing in brain, the following specific results and new contributions were presented.

It was realized[1] that the infomax models, like ICA by Bell & Sejnowski and sparseness-maximization net by Olshausen & Field, outperform the classical Hebbian or PCA models, because they incorporate phase information, or higher-order statistics, respectively. Infomax models were shown to give much more biologically-plausible outputs (receptive-field profiles), but a biologically-plausible implementation on the "hardware"-level is possible (for now) only for the Olshausen & Field net, not for the Bell & Sejnowski net. Relations between the Olshausen & Field net and MacLennan's dendritic field computation model were found, which indicate a possibility of dendritic implementation of the Olshausen & Field net. However, dendritic processing "following the Olshausen & Field algorithm" would be strongly constrained by sparseness-maximization process which could originate from the lateral inhibition or from top-down influences.

Since the oscillatory-dynamic phase-processing is experimentally supported, a question arose whether ICA infomax-processing, or at least the sparsification process, might be realized virtually, i.e. on a "software"-level (higher-order attractor dynamics). ICA-like infomax processing shapes the receptive-field profiles into Gabor wavelets, and these are then convoluted with the sensory inputs. The infomax processing is thus viewed as an information-saving preprocessing procedure for optimal encoding into Gabor wavelets. Infomax-based (appropriately weighted) Gabor wavelets are spectral image-representations which are involved in convolution (during perceptual processing), or in interference, or in other phase-Hebbian processes (during pictorial cognitive processing and associations). The *Quantum Associative Network*, my original computational model, was presented as a possible core-model for holonomic associative image processing. This would imply that the image, which is recognized by the quantum associative net, becomes the "object of our conscious experience". The image-bearing eigen-waves, which interfere in the quantum associative net, were proposed to be infomax-produced, quantum-rooted Gabor wavelets.

I thus suggest that the neocortex uses three types of image representations: the Gabor coefficients as sparse neuronal codes for automatic processing, the dendritically-implemented Gabor wavelets as spectral-PDP codes for associative visual cognition, and the V2-reconstructed spatial image used in our "direct" conscious experience.

Because the perceptual image seems to match precisely the original object in its external location, holographic back-projections by phase-conjugation have been argued to be necessary. Since neural or dendritic nets alone cannot realize such out-to-space projections, the only medium which is common to holography and to brain networks was declared ultimately responsible for conscious image processing — the quantum system.

The common content of this book are systematic studies of analogies in holographic, associative artificial-neural-net, spin-system and quantum-interference processes which could be harnessed for parallel-distributed information processing. Possible (bio)implementations of these processes were indicated.

Computer simulations of quantum associative networks and applications for image recognition and object recognition were presented. For the first time, quantum-(like) interference of neurally-shaped quantum-implementable Gabor wavelets was used for (quantum) associative memory. This could be experimentally realized as a sort of quantum holography. Because of astronomically many "pixels", image-bearing eigen-vectors are almost orthogonal, therefore our quantum Hopfield-like net is able to provide much better recognition results than a classical Hopfield neural net -- either because of orthonormality OR because of the use of Gabor wavelets.

Mathematical details of an original comparative neuro–quantum study, and original derivation of the model of quantum associative net from the simulated neural-net formalism are presented in Auxil. Chapters. Some novel suggestions for flexible image processing (e.g., "fuzzification") are also described. (An own computational example of recognition of protein-patterns with a Hopfield-like ANN, which is the predecessor of the quantum associative net, is presented briefly.) The Hopfield neural-net input-to-output transformations are directly translated into isomorphic quantum-implementable wave-dynamics. Thereby, all net's information-processing dynamics can, at least in principle, be implemented in a quantum medium also. A direct relation of the quantum associative net and the commercially-successful Holographic Neural Technology (HNeT) was found, and thus the quantum implementation of HNeT is proposed here. Therefore, I can assume with much optimism that the quantum associative net, if really quantum implemented (as also, in a way, probably in brain), would realize efficient image recognition and related associative processing. In cooperation with other brain structures, the result of such an image processing would probably be conscious.

ENDNOTE

1 These observations are unavoidably tentative, since many tentative rapidly-developing computational models are being proposed.

Section 2

The main chapters (1-5) have explicitly presented the central issues of the book in direct relation with the Holonomic Brain Theory of Karl Pribram, whereas the following Auxiliary chapters provide additional or detailed discussion of some issues which are either needed to understand the main chapters better or to go into depths which might not be interesting to all readers. In both cases, the Auxiliary chapters do NOT constitute the Holonomic Brain Theory. They enable a comparison of the holonomic theory and the broad or mainstream neuro-scientific framework of vision research and some quantum analogies which might in the future turn out to be significant for it.

Chapter 6 Visual Processing as Described by Contemporary Main-Stream Neuroscience

This Auxiliary Chapter complements the main part of the book by presenting the conventional views on vision for those readers who are not familiar with them. It also allows comparisons with the Holonomic Brain Theory and critiques, and also some mutual complementing of views and descriptions, although they are presently not compatible. For instance, the work of large numbers of neuroscientists provides many details and experimental data relevant for modeling purposes.

6.1 RETINA

Eye. The light enters the eye at cornea. The optic flow then proceeds through the *pupil*, a circular hole with muscle-conditioned variable radius, then through the

DOI: 10.4018/978-1-61520-785-5.ch006

lens, which bends (refracts) and focuses (because of ciliary-muscle-conditioned flexibility) the rays, to the eye's back where it impinges on the pigment-layer of *photoreceptor cells*[1] on the surface of the retina. A 2-dimensional projection of the 3-dimensional world emerges on retina.[2]

Retinal image. Stimulus intensities, which affect the activities of retinal cells, are mainly determined by: the light source, the surface reflectance and shape of the viewed object, the viewer's standpoint and angle of view. Each location in the image can be described by two primary characteristics of the wave-modulated stimulus (Berne & Levy, 1993):

1. *Brightness (luminance)*, the intensity of light falling on retina, which determines the photoreceptors' activation-level;
2. Spectral decomposition (frequency or *wavelength*) which determines the differential activation of different cone types, leading to the perceived color.[3]

Derived information. Let us mention here that other characteristics can be in later stages (up to the cortex) derived from the primary retinal features 1 and 2. Such secondary characteristics are (De Yoe & Van Essen, 1988):

3. 2-dimensional velocity of the retinal image;
4. Spatial contrast;
5. 2-dimensional orientation of the stimulus;
6. Binocular disparity (which is also a source of stereopsis, i.e. 3-dimensional vision).[4]

Sensation. Photoreceptor cells are *cones*, which are color-sensitive and used for low-sensitive (for daylight), high-acuity vision, and *rods*, which are high-sensitive (for dark), "color-blind" and provide a rough "sketch" only. In the central part of retina, *vis-a-vis* to the lens-center (on the visual axis), lies *fovea* where visual acuity is the highest, because there is extremely high density of cones. Density of rods is, in contrast, high at all other parts of the retina except at fovea and at the location where the optic nerve starts to extend towards the visual cortex. In response to a flash of light within the receptive field[5], a photoreceptor runs a biochemical cascade process and changes voltage. Over a limited range of stimulus intensity, the change in the light-modulated signal is proportional to the change in the light stimulus (Smith in Arbib, 1995). Retina realizes a sort of scale-invariant gaussian-profile sampling (details in Van Essen & Anderson in Zornetzer, Davis & Lau, 1990, pp. 45-47).

Retinal net processing. Retina is a multi-layer neural network. Details of its non-trivial hierarchical feed-forward[6] processing, in spite of its importance for further visual processing, will not be presented here. However, description of

Figure 1. Circularly-symmetrical receptive fields of typical LGN neurons. Left: A parvocellular neuron is excited (+) in the small centre by red light (R) and inhibited (–) in its surround by green light (G). / Right: A magnocellular neuron is excited by all wavelengths in the centre (R+G+B) and inhibited by all wavelengths in its surround (–R–G–B).[18] (Redrawn from Livingstone & Hubel, 1988, p. 741)

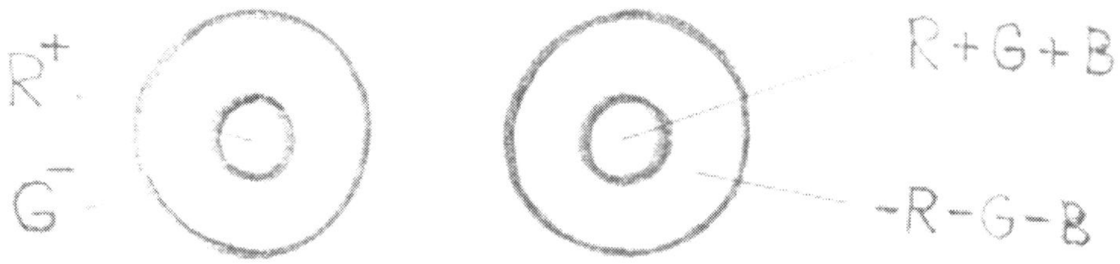

processing has some remarkable similarities to those presented in Auxil. Chapter 11 for competitive cortical circuits even taking into account that the retinal net has only very partial feedback. Retina shares roughly several characteristics with later stages of visual processing (e.g., LGN), like the circular center-surround and similar receptive fields (Figure 1).

Let us just list some basic anatomical and physiological findings. Signal mainly flows directly (but also indirectly over horizontal cells in the 2nd layer) from photoreceptors (1st layer) to bipolar cells in the 3rd layer and proceeds to ganglion cells in the 5th layer. Horizontal cells also mediate feedback to photoreceptors. Amacrine cells in the 4th layer are mediators in both directions between bipolar and ganglion cells (3rd and 5th layer). Bipolar and ganglion cells have *circular center-surround receptive fields* of two types: "on-center" and "off-center".[7] "On-center" means that the center of the receptive field is excitatory (+), but surround is inhibitory (–). "Off-center" receptive field has an inhibitory center and an excitatory surround.[8] Both types lead to *lateral inhibition* between the central and surround neurons — i.e., effectively, the neuronal majority of the surround is an opposition to the dominant neuronal oligarchy of the center. Amacrine cells have, because of inputs from combinations of "on-center" and "off-center" bipolar cells, receptive fields which are a mixture of "on-center" and "off-center" regions. Ganglion cells get inputs from amacrine cells only/mainly, or from bipolar cells only/mainly, or mixed (Van Essen & Anderson in Zornetzer, Davis & Lau, 1990; Berne & Levy, 1993).

Photoreceptors respond with inhibition, i.e. with a graded hyperpolarization (more negative membrane potential), to an increase in light intensity. On the other hand, a half of bipolar cells responds by excitation, i.e. by depolarization (more positive membrane potential) ("on"-bipolars), and a half by hyperpolarization ("off"-bipolars) (Kandel *et al.*, 1991).

6.2 VISUAL PATHWAY(S): L.G.N. AND MAGNO-PARVO-BRANCHES

Note that the contents of the following sections of this chapter, especially its next three sections, are the views of the *main-stream neuroscience*.[9] These views – a mosaic of specialized, isolated studies – are often in more or less *significant discrepancy with the Holonomic Brain Theory* of Pribram (ch. 2). Both paradigms are first faced in Auxil. Sec. 6.12.

Visual pathways. There are three visual pathway families:

1. Retina — lateral geniculate nucleus (LGN) of the thalamus — visual cortex;[10]
2. Retina — (pre)tectum, superior colliculus — visual cortex;
3. Retina — hypothalamus relays — pineal gland (etc.).

Pathways may also exhibit feedback-loops, via pulvinar, etc., and have connections to many other centers (Levine in Kosslyn & Andersen, 1992).

The first (retino-geniculo-cortical) pathway is the main one, especially for primates, while the second family of pathways serves for orientation of the constantly-moving and stimulus-searching eye, control of pupil size, etc. (Berne & Levy, 1993). Only the main (geniculate) pathway is considered in this book. The signals (optic stream) propagate along the optic nerve of each eye to the optic chiasm where the optic fibers exchange sides.[11]

LGN. Neurons of the *lateral geniculate nucleus* (LGN) have similar characteristics to ganglion neurons of the retina. LGN neurons can be, similarly to retinal ganglion cells, classified into X- or Y-cells (there are also W-ganglion-cells), and have similar center-surround receptive fields (excitatory center with inhibitory surround, or vice-versa).[12] After receiving inputs from 90% of all retinal ganglion cells, LGN mostly projects them further (in approx. 1:1 relation) to V1, mainly to its layer 4. This layer probably acts as a switching and regulating network with many outputs. A hypothesis of a mechanism or circuit for dynamic translational shift or re-mapping of receptive fields over remarkable distances is in consideration (Van Essen & Anderson in Zornetzer, Davis & Lau, 1990).

LGN also receives input from the visual cortex (so there is a *geniculo-cortical feed-back loop*!) and from the thalamic *reticular nucleus* and several nuclei of the brainstem *reticular formation*. These inputs very probably influence filtering visual information, suppressing featureless regions and enhancing information-rich parts, and help in establishing selective attention (e.g., Berne & Levy, 1993; Crick, 1984). Substantial feedback from layer 6 of V1 is necessary for synchronization of neuronal activity between the different eye-specific LGN layers (Weliky & Katz, 1999).

Magno/parvo. According to Schiller & Logothetis (1990) and many other sources[13], there are two processing pathway-divisions from retina through LGN. One passes through those part of LGN which incorporates big (magno) cells, therefore it is called *magnocellular* stream. The second pathway goes through those division of LGN which has small (parvo) cells, so it is named *parvocellular* stream. This division is preserved in the visual cortex: a part of it receives parvo-inputs, the other one magno-inputs[14] (Kosslyn & Andersen, 1992). However, these two systems are not entirely segregated — there is overlap, multiplicity of processing and information exchange, especially in the cortex. There is no simple, one-to-one projection (De Yoe & Van Essen, 1988). There is some evidence[15] that:

- *Magno-channel*
  - processes "color-blind"-information about movement and depth, and detects flickering; specialized for (global) spatial relationship, but "overlooks" "color";
  - processes higher temporal frequencies;
  - magno-cells are sensitive to more global contrast because of larger receptive fields.
- *Parvo-channel*
  - executes early color-processing[16] and mainly contributes to shape-processing;
  - processes higher spatial frequencies;
  - parvo-cells are sensitive to local contrast because of smaller receptive fields.

According to Livingstone & Hubel (1988), the two channels are supposed to differ in color selectivity, contrast sensitivity, detection of temporal and spatial properties, acuity and speed of response. The color selectivity of the parvo-system might enable us to see borders using color information alone, including subtle borders that might be "overlooked" by the color-blind magno-system. The parvo-system, which is well developed only in primates, seems to be useful especially for perceiving subtle and detailed properties of shape, color and surface of objects for their identification and association — in contrast to the magno-system which concentrates on basic features needed for pragmatic behavior. For instance, inferring distance and 3-dimensional shape from 2-dimensional image can be realized using luminance information (not color) as the magno-system is able to do. In fact, stereoscopic depth perception, motion perception, perspective, spatial (size) and figure/ground relations, and probably also depth perception using relative motion, are lost at equiluminance (magno-system fails).[17] Both systems seem to participate in processing of some characteristics like orientation and shape. (Livingstone & Hubel, 1988)

Parvo-cells process color information by virtue of the spectral center-surround opponency of their receptive fields (Figure 1). By such an opponency-processing we well perceive relative differences, but perception of absolute rates is much worse.

6.3 TO STRIATE CORTEX AND BEYOND

Continuations of magno/parvo channels. Lateral geniculate nucleus (LGN) has six layers: four are parvocellular, two are magnocellular. Six LGN layers receive input alternately from the left-half of the retina (to three LGN layers) and the right-half of the retina (to another three LGN layers). So, individual LGN neurons can be stimulated from one eye or another eye, but not both eyes. (Cortical neurons are increasingly more synthetic. E.g., V1 neurons usually get input from both eyes.) (Logothetis, 1999) Livingstone & Hubel (1988) reported the anatomical and functional parvo/magno segregation[19] which originates at the retina and is preserved through the entire retino-geniculo-striate pathway up to higher visual cortical areas. The segregation is evident from specific (color versus motion, form or depth/stereo) perceptual impairments following selective lesions, and from anatomical and psycho-physiological research. In the projections of LGN's parvo-channel onto V1, two sub-divisions can be found: the *blob* and the *inter-blob* systems.[20] These two sub-systems, although different, work in a complementary way. According to Livingstone & Hubel (1988), Stillings *et al.* (1995, ch. 12) and other sources, blob cells are explicitly color-coding and non-selective for orientation of the stimulus (e.g., lines or edges). High-acuity interblob cells, oppositely, respond to orientation, and so help in extracting shapes via contours, but mainly ignore color. In V2, blob and interblob sub-dichotomy is[21] preserved with parvo-stripes and parvo-interstripes. Then both parvo sub-pathways unite again in the inferior temporal cortex (ITC) or even earlier in V4.

V1 processing. Like LGN, V1 contains a *retinotopic map*[22], but V1 neurons have more complex receptive fields and include both eyes. However, Aine, Supek & George (1995) and Aine *et al.* (1996) provide evidence that, although there is exact point-to-point projection of the retina on striate and extrastriate areas, its functional organization is not rigidly determined by this retinotopic representation, but is somewhat modifiable.[23] Experiments show that V1-activity is determined not only by inputs from both eyes, but by other contextual information also (Bridgeman in Pribram, 1993).

The classical theory postulates a hierarchy of cells:

1. So-called *simple* cells with on- and off-zones in their receptive fields;
2. So-called *complex* cells, with converging inputs from simple cells, but without distinct on- and off-zones, since they make a synthesis of simple-cell inputs[24];
3. So-called *hypercomplex* cells with special receptive fields[25] where complex-cell inputs converge.

Visual areas V1-V8 and others lie (mostly) in the occipital lobe.

From V1 (which lies at the extreme back of the head) to higher vision centers (V2, V3, which are located concentrically around V1, and so on) neurons have increasingly larger receptive fields. The proportion of active neurons also increases[26] as one goes from V1 to higher visual areas (Logothetis, 1999; Perret & Oram, 1998). This indicates that the information processing becomes increasingly distributed.

There seems to be a paradoxical dichotomy of cortical processing. Namely, beside highly-distributed, densely-interconnected networks, it appears to contain many roughly information-specific pathways (which is being considered first in this chapter). Thus, global, distributed, holography-like processing (focused by the "synthetic" holonomic view presented in ch. 2) seems to be complemented by a sort of "skeletal processing structure" (focused by the analytic mainstream-neuroscience view of this Auxil. Chapter) where specific important aspects are concentrated.

"What/where" pathways. According to Mishkin, Ungerleider & Macko (1983) and many other sources[27], there are two visual pathways from the striate to extrastriate cortex (Figure 6.2). The ventral pathway from the occipital lobe (over V4) to the inferotemporal cortex (ITC) is specialized for *object recognition* (the "*what*" pathway) (Kosslyn & Andersen, 1992). The dorsal pathway from the occipital lobe (over MT) to the posterior parietal cortex is specialized for the perception of *spatial location* (the "*where*" pathway). The object-perception pathway (to ITC) segments the fixated figure under attention from the unattended background, but the spatial and contextual relationships are complemented by the position-perception pathway (to posterior parietal cortex). The latter (dorsal) stream also finally regulates motor actions, so it may be called "how-and-where-to-do-it" pathway (Nelson in Arbib, 1995).

Challenges for simplified classifications. The dichotomy of the "what"-pathway to ITC and the "where"-pathway, or pathway over the middle temporal (MT) area for visual processing of motion, respectively, seem to us on first sight to have some parallels in the dichotomy of the so-called X-cell pathway and the Y-cell pathway. X-cells have (to say briefly) spectral opponent properties; Y-cells, in contrast, respond to luminance and not to wavelength-differences (Berne & Levy, 1993). Cortical X/Y dichotomy is "inherited" from previous (LGN's and retinal) dichotomies. Here, the rough cell-type analogy seems as follows: parvo corresponds to X, magno to Y cell-type.[28] LGN's parvo-cells get exclusive input from retinal X-cells

(De Valois & De Valois, 1990). Nevertheless, all such apparent segregations of pathways and their cells should not be taken *too* seriously[29], since there is enough cross-communication, overlap and "mixing or sharing of tasks" along the overall visual stream (e.g., Vidyasagar, 1999).

Moreover, Pribram, Lassonde & Ptito (1981) have found in their experiments on the cat visual-cortex cells that, regarding their properties, X-cells are equivalent to simple cells and Y-cells to complex cells. This report reveals complications regarding several attempts of classification of cells based on response to *single* stimulus-dimension (feature).[30] Because the cells were found to show *multiple-feature* selectivity, the authors conclude that *it is better to make classifications based on networks of cells, instead of classifying single cells*. Such receptive-field network properties are proposed to be classified into: 1) those reflecting the input, 2) those reflecting the interactive properties of the field, and 3) those reflecting the network as a whole, or the net-output, respectively.

6.4 HIGHER VISUAL AREAS

Perception of movement, which is direction-specific, is realized along a pathway that includes middle temporal (MT) visual area (=V5) (Zeki, 1976), and also V3. Perception of color and wavelength analysis (AuxLit 3, 5) is supposed to be executed in a pathway that includes V4 (Zeki, 1976; De Yoe & Van Essen, 1988; Perret & Oram,

Figure 2. Centers of visual perception and the "what" and "where" pathways. (From Kosslyn, S.M. & R.A. Andersen (Eds.) (1992): Frontiers in Cognitive Neuroscience. Cambridge (MA): MIT Press.)

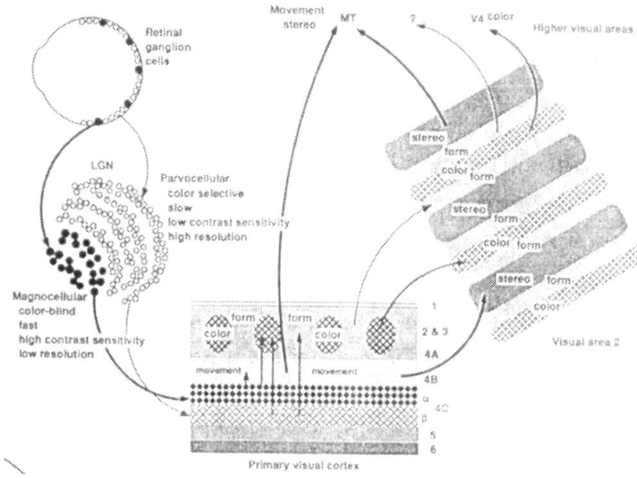

1998), and V8 should also process color vision (Logothetis, 1999). By (primarily Pribram's) lesion studies in primates, followed later by tracing dominant activity using fMRI or PET while subjects watch some patterns, it was found that the *inferior temporal cortex (ITC)* is probably important for perception of form and recognition of objects (like faces)[31], V4 for shape perception (in particular with help of color contrasts), LO for large-scale object recognition (Logothetis, 1999).

Cortical areas are densely bidirectionally and reciprocally connected between themselves as well as with subcortical nuclei, and also between both hemispheres. Without entering numerous details, we can stress that processing is hierarchical and simultaneously parallel-distributed within particular levels or layers. There are numerous inter-level and intra-level feedbacks (Burnod, 1990).

ITC. Inferior temporal cortex seems[32] to encode 2-dimensional appearance of image features like shape, texture, color, intensity gradients and relationships between shapes. It was reported[33] that in posterior ITC there are cells selective for specific shape or for specific relationships between two shapes[34] or two textures[35], or particular cells selective for intersecting edges of specific orientations.[36] There is mainly or probably no (single-cell's) selectivity for object concepts (like a pine-apple), but encoding of simple qualities (like green and star shape). Perret & Oram (1998) further-on review reports on even greater response-specialization: Within anterior ITC, temporal pole, anterior medial and superior temporal sulcus cells have been found, they say citing various sources, which respond selectively to complex patterns like bodies, faces, hands, or to arbitrary patterns (e.g., fractals, wire frame objects) *after extensive training* with such a specific stimulus.

Critique. I should stress again that importance or permanence of such a narrow specificity remains questionable: is it not just temporary or even just by chance? Namely, for instance, Pribram (1991) mentions his finding that a neuron responded most to a *green* hand, and *green* low spatial frequency, after a monkey had been rewarded for responding to a set of *green* stripes. A Fourier-based alternative, presented in ch. 2, seems to be suitable for explanation of specific selectivity (e.g., to a green hand) of individual cardinal neurons (Auxil. Sec. 6.9) or assemblies of neurons, since it is compatible with selectivities distributed among receptive fields (each being selective for, e.g., green in general, finger-like specifically-oriented stripes, etc.).

6.5 GENERAL NEOCORTICAL ARCHITECTURE AND CONNECTIONS

Uniformity. The gross division of the cortex (or 90% of it which is the neocortex, in humans) into areas is genetically determined and is, in adult, almost not modifiable. Ebdon (1993) reviews tentative reports on global uniformity of anatomy and physiology of cerebral neocortex.[37] It should be uniform in the sense that roughly the same circuits and computational mechanisms are used in so-to-say all its parts. In spite of the fact that there is no uniformity of functional properties of cortical neurons, there is *approximate uniformity per unit area over the whole six-layered cortical sheet in the following respects*:

1. The number of neurons;
2. The presence of different types of neurons and the proportion of various neuron types;
3. The laminar distribution of the different types of inputs and outputs;
4. The presence of non-specific fiber systems;
5. The basic pattern of intrinsic circuitry[38] (according to present limited knowledge);
6. The profiles of receptive fields (Pribram, 1971).

Modules. There is nevertheless modularity which is combined with uniformity (in the above-mentioned respects) (Körner *et al.*, 1999). Modules, which are more local than areas (i.e., cortical *minicolumns, columns, hypercolumns, maps*) (Calvin in Arbib, 1995), incorporate approximately the same computational mechanisms, including synaptic plasticity[39], and the same type of input encoding. However, modules differ in information contents which they process. Therefore, uniformity of neural circuitry[40] does not mean uniformity of cognitive structure (Ebdon, 1993). The latter is based on attractor-patterns (Auxil. Sec. 6.8) and their virtual hierarchies (Peruš, 2000a, 2001), and it actually is heterogeneous because of specific information contents which are encoded in modifiable connections. By the way, the cerebellar cortex is even more uniform in the mentioned respects than the cerebral neocortex.

Connectivity. According to contemporary models, the cortical network (Körner *et al.*, 1999) exhibits "connectivity matrices" (Burnod, 1990, e.g. Figure 3-19, 3-20). Hippocampus and cerebellar cortex both have a Hebbian matrix-organization at the neuronal level. In the (neo)cortex, a modular band of columns of the map A and another such band from the map B cross their afferent fibers. The pattern of connections (or junctions of these fibers) forms a "matrix". We note that such a "matrix" is, according to "mathematical structure", roughly analogous to a hologram (ch. 2). Such connectivity matrices emerge on different levels: their elements can be junctions of column-fibers or map-fibers (incorporating inputs of whole bands of columns), etc. Large cortico-cortical connections should be organized in symmetrical, "mirror"-like fashion, where the mirror-axes are located on the gyri and sulci (Burnod, 1990).

Figure 3 shows the sparse cortico-cortical magno-stream connectivity-matrix, and Figure 4 shows the cortico-cortical parvo-stream connectivity-matrix, both necessarily much simplified. The latter figure also illustrates the influence of attention-windows in the parvo-stream. Although there are many physical connections (hatched regions in Figure 4), only some of them (those inside dark-shaded elipses in Figure 4) are effective; the others are presumed to be suppressed by shunting inhibitory top-down influences.

6.6 SELECTIVE VISUAL ATTENTION

Systems involved in attention. Attention helps in object identification by support of V4 and ITC, but these areas are not alone. Mesulam (1981) proposes that the importance of posterior parietal cortex is in mainly subserving sensory, spatially-directed ("where") aspects of attention, the limbic (cingulate) cortex adapts and directs the attentional focus according to subject's interests, and the frontal cortex, together with midbrain's superior colliculus, should take part in motoric support for attentional processes (cf., Pribram, 1991). The motoric support includes eye-

Figure 3. Matrix of connections between cortical areas within the magno-pathway. (Connections are represented by the shaded diagonal stripes.) Information flows "from lines (dendrites) that are parallel to x-axis to lines (axons) that are parallel to y-axis" (i.e., "from dendrites running horizontally over junctions to axons curving around and running vertically") – as the small upper right sub-figure shows. Stippling on the main matrix indicates which connections actually exist. Bottom right sub-figure illustrates how sparsely the network is interconnected. (From Van Essen & Anderson in Zornetzer, Davis & Lau, 1990, p. 64)

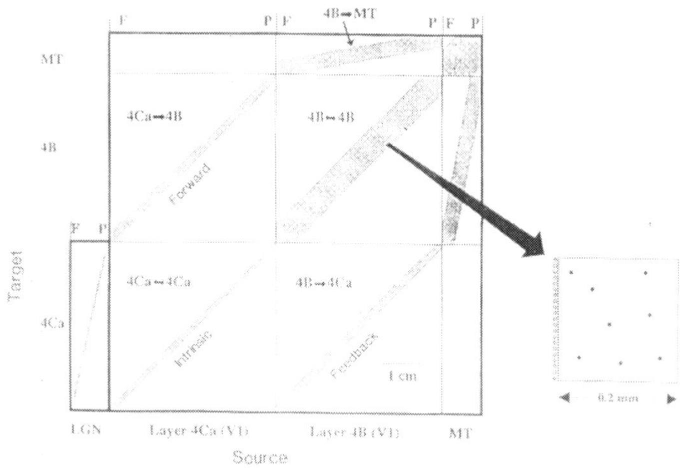

Figure 4. Matrix of connections between cortical areas within the parvo-pathway (to be read as the previous figure). There is V1-to-V2 topographic mapping, followed by V2-to-V4 and then by V4-to-ITC mappings. Hatching shows actually existing connections (for simplicity, only feedforward ones are shown). The dark elipses show the attention-window for V1-(to-V2-to-V4)-to-ITC mapping: Only the "attended" part between a1 and b1 (on the left side of the matrix) is mapped, over V2's a2-b2 and V4's a4-b4 windows (along the regions between fine-dotted lines), to the entire ITC. (From Van Essen & Anderson in Zornetzer, Davis & Lau, 1990, p. 66)

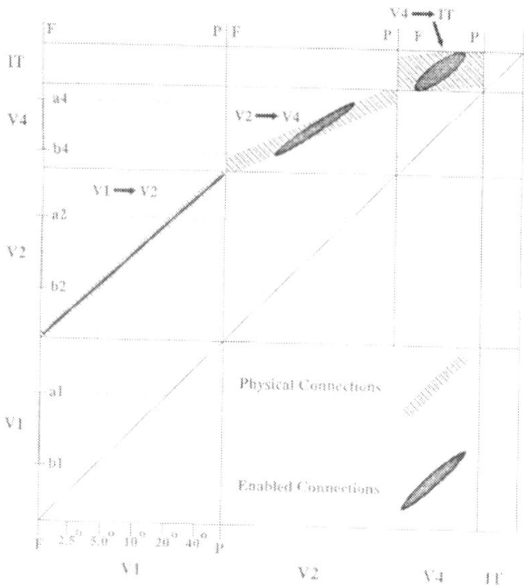

movements, called *saccades*[41], for searching and scanning within extrapersonal visual space; and processes like fixating and reaching are also parallel to attention (Wurtz *et al.*, 1980; Bickle, Worley & Bernstein, 2000). The reticular formation arouses the whole cortex (Luria, 1973; *et al.*).

According to the "searchlight" hypothesis by Crick (1984), selecting and moving-around of the restricted region of the visual field while subject attends is regulated by the reticular nucleus of the thalamus.[42] Vidyasagar (1999) writes that attentional spotlight has origin in the dorsal magno-pathway (toward posterior parietal cortex), but acts in serially searching the visual field for targets in the ventral (or temporal, i.e. toward ITC) parvo-pathway (so, first "where", then "what").

Attention windows. Interesting experiments showed that ITC neurons have very large receptive fields, covering often the most of the visual field, in anaesthetized animals.[43] In alert animals, on the other hand, receptive fields are restricted to the attended window (Moran & Desimone, 1985) which is constantly changing, moving around, adapting to the situation and shifting spatial scale in focus. Moran & Desimone (1985) also showed that when there are two stimuli in the cell's receptive field, those cell (in V4 or ITC) responds much more to the attended stimulus, even if it was ineffective, than to the unattended stimulus, even if that one was effective. Let us note that effectiveness is a property of stimulus, but attention a property of the (conscious) visual system.

So, indeed, attention is a real process, based on *top-down inhibitory influence* which focuses the information flow to important features. For instance, evidence that attention activates the visual nucleus reticularis of the thalamus from V1 using cortico-geniculate top-down pathways is reported by Montero (2000). Bickle *et al.* (1999) simulated effects of these back-projections including the resulting lateral inhibitory influence of the *nucleus reticularis* (NR) onto LGN. Computational simulation of the LGN–V1–NR circuit, using the so-called interactive activation and competition ANN by Grossberg, gave plausible results: Activity in V1-columns with less stimulated neurons decays relative to the most stimulated striate cortical column. Beside voluntary bases, attention has internal automatic bases (Stillings *et al.*, 1995, ch. 12). The visual system somehow "gets attracted" by the stimulus itself.

Attention manifests mainly in V4 cells and more widely in ITC cells, but usually not in V1 cells which do not give such results in the Moran & Desimone (1985) experiment as just described. However, the top-down attentional influence is realized, among others, by feedbacks from ITC and V4 onto V1 [44] which affect changes in position, size or shape, and scale of the window, "trying systematically" to increase the information content. There is one exceptional case where attentional spotlight is effective as early as in V1: when several competing objects (not just one) are presented in the visual field simultaneously (Vidyasagar, 1999).[45]

Figure/ground segmentation. Each perceived pattern usually has the figure-part (with figures, i.e. object-forms) and the background-part. The representation in ITC is probably the figure, extracted by an *image segmentation* process (i.e., extracting the figure from its uninteresting background) which is supported by appropriate attention-window positioning. Such figure/background separation-process exploits differences in luminance, size, color, binocular disparity and motion of figure relative to background (Grossberg in Arbib, 1995). Segmentation of a figure, e.g. a face, is accompanied by object-centered coordinates. Sompolinsky & Tsodyks (1994) invented a network of oscillators which can selectively segment stored objects[46] (see Auxil. Sec. 11.3). Belonging to the same object is encoded by the *coherence of the phases* of individual oscillators. Without the phase information, all objects of the image would be equivalent for the net, and segmentation would not be realizable. So, also selective attention would not operate, because the whole image, encoded by a single "Hebbianly-produced" attractor, would remain undifferentiated. Namely, objects of the image would be undividedly trapped together in the attraction-basin of the common pattern–attractor.[47]

Top-down filtering, or even ignoring the unwanted information, probably has origins in conscious process, or its motivational, volitional and ego-centric components, respectively. For now, let us only remark that the conscious I and volition are supposed to have neural correlates in the prefrontal cortex.

6.7 ELEMENTS OF ACTIVE VISUAL PERCEPTION

Early vision. Poggio, Torre & Koch (1985) list the processes which are usually denoted by the term "early vision" (cf., Poggio, Gamble & Little, 1988):

1. Edge detection;
2. "Guessing" distance, color, reflectance, lightness of perceived objects, or texture of perceived patterns;
3. Spatio-temporal interpolation, approximation and limited extrapolation of feature-data;
4. Computation of optical flow (area-based or contour-based) and its velocity field;
5. Reconstruction of shape-information from shading, contours or texture[48];
6. Reconstruction of object's structure from motion and from binocular stereo-information;
7. Reconstruction of surface, including orientation and color of the surface, etc.

Depth and stereopsis. In visual cortex, inputs from both eyes continuously compete. This *binocular rivalry* has been relatively well experimentally researched (Logothetis, 1999). *Stereopsis* is 3-dimensional vision capability as a consequence of binocular depth perception. It must be cortical, since it depends on inputs from both eyes as Julesz's experiments proved. Beside stereopsis, 3-dimensional forms can also be perceived from detection of motion only, but rarely and less efficiently (DeAngelis, 2000).

Depth, i.e. the distance of an object from the observer, can be primarily "guessed" by the visual system from binocular image-differences. Namely, if this distance is made larger, the images from two eyes overlap more. Primarily, of course, object's retinal image is smaller if its distance is larger, but this might make confusion when context-dependently "guessing" object's size. Impression of depth is got also from shadows, from more diffuse illumination and the resulting perspectives, from compositional variety of environment, and from increasing gradients of texture density. Distant objects are often covered by the nearer ones.

Another depth-information source is object's motion. For instance, if an object moves towards the observer, its image expands. Objects which are more distant appear to move more slowly than objects having the same velocity but smaller distance from us. If we move in a direction and fixate our look to an object, things behind that object will appear to move in the opposite direction comparing to things in front of it. Finally, far-away objects appear to have unclear colors and blurred edges, they look darker (more gray-blue). Distant textures become less distinctive than near ones.

Edges, contours, texture. Images usually have various sorts of boundaries or edges. They are often reflected in sudden intensity changes, but this is not necessary: Remarkable intensity gradients might occur within edge-less surface area also. Surfaces are often not uniformly colored and have various sorts of markings, micro-reliefs or textures. *Texture* is manifested, for instance, by a repeating stripe pattern or by an array of spots or bars which may be of various sorts, sizes, orientations, groupings.

So, to interpret the image, the visual system has several non-trivial problems (which might also mislead it): extraction of edges, of gradual intensity changes, depth cues[49], etc. First, to solve above-mentioned problems, the visual system uses parallel processing. Then it starts to categorize, i.e. to cluster locally-similar items into higher-order items. Such cluster-regions may have boundaries – real or virtual ones, which might both simultaneously be simple or, more usually, complex (e.g., like an interrupted set of edge- or line-segments which nevertheless sketches a more or less quasi-continuous contour).

Von der Heydt *et al.* (1984) reported finding of neurons in V2 that responded to virtual contours[50] as if they were formed by real lines. In contrast, V1-neurons have not yet been able to "see" virtual contours. No wonder, since detecting the gestalt-contour is a second-order process which is realizable after the basic elements of the contour (i.e., individual line-segments) have been detected. V1 is at the bottom of the hierarchical network-processing and incorporates many relatively-*simple* cells, but V2 has more *complex* cells which integrate *simple*-cell-inputs. Increase of complexity of cell-responses is consistent with increase in size and complexity of dendritic receptive fields from lower (V1) to higher visual areas (Ullman & Soloviev, 1999).[51]

Contours are examples of *gestalts*. Emergence of gestalts was successfully explained by *attractors* of network dynamics (Haken, 1991; Luccio, 1993; Peruš, 1995a,b, 2001; AuxLit 9). Namely, patterns act as attractors of system's dynamics.[52]

Object perception versus image processing. When we observe an object from various views, its images are different, but we anyway consider it as a single object (e.g., Riesenhuber & Poggio, 2000). *Image processing* is therefore distinguished from *object perception*.[53] The retinal image of the object (e.g., door) is view-dependent and apparently distorted, depending on the projection: E.g., doors do not appear to have rectangle shape any more, when being opened, but different trapezoid shapes, and finally they "shrink" into a vertical line. Nevertheless, the object is recognized as the same object (e.g., door). *View-dependent images converge into the same object-specific (view-independent) attractor.*

The following example also illustrates the difference between predominantly input-determined image-processing and significantly memory-based and top-down influenced object-perception. As a man approaches, we perceive him coming near, not getting bigger, although his retinal image is magnifying. We implicitly know that his image is getting bigger, but our cognition, using past experience and context, cancels or compensates the immediate image perception. We henceforth consider the man having constant height as coming near.

So-called ambiguous images (like often used by S. Dali) have two possible object-perceptions. In a single image one can see one gestalt or another, i.e. one can construct one attractor or another.

Viewer- and object-centered representations. A subject perceives an object in specific viewing conditions (position, orientation, illumination, perspective or angle of view, partial occlusion, etc.). This is the *viewer-centered* (view-selective) representation. However, as subject moves around the object and many very similar objects of that type are seen, the higher visual system produces an "averaged" representation of an object that is independent of any of the numerous possible viewing circumstances. A car remains a car regardless of the view, and a representation of the car must reflect this fact. The latter – generalized, categorized, prototypic – representation of the object is the *object-centered* (object-selective or object-invariant) representation (Pribram, 1991; Stillings *et al.*, 1995, ch. 12). These two types of representation are not the same[54], and different specific neurons or neuronal assemblies in different areas encode a representation of the first or the second type (Perret & Oram, 1998).

This dichotomy must also be connected with the ventral/dorsal ("what/where") visual-pathway dichotomy. If one has to answer the question "what object is this?", he would surely activate the object-centered representation. On the other hand, if one has to guess, for example, how much an object is rotated from the usual frontal position (with respect to the viewer), the viewer-centered representation will be used.

In the last, i.e. view-dependent and thus relative-location-specific, framework, Pribram (1998a) distinguishes *ego-centered* representation (as in the last case), and *oculo-centered* representation (or *allo-centered* representation, based on "out-projection of the view"[55]) for cases like, for instance, in asking whether an object is left or right with respect to the door. Interestingly, Pribram (1998a) experimentally observed that the *allo*centered representation rules the perceptual space *out*side the reach of subject's hand, but the *ego*centered (which is body-centered) representation controls the space *in*side his reach.

Dissociation between the "where" and the "what" visual pathways is reflected also by the following problem. During rapid eye-movements (saccades) the visual space gets suddenly displaced, but subjects are unaware of this – as if this displacement would not effect the perception at all. Subjects do not confuse locations after saccades; their perception and movement coordination remains stable.[56] So, it appears that spatial information is preserved somewhere outside awareness, but other non-spatial aspects are independently visually processed in another area of the brain where awareness is not excluded (Bridgeman in Arbib, 1995). Of course, the experimental evidence of pathway dichotomy (Auxil. Sections 6.2. and 6.3) is in agreement with the observation that the saccade-based spatial sampling and processing proceeds independently enough from pattern recognition. Perhaps the "what" and "where" processing (the latter includes motion) proceed sequentially in alternation, so they would not disturb each other. In any case, the *ego-centric* information about locations appears to be constantly available and reliable for object perception in spite of changes in target—eye positional relations during saccades. This is proven also by observations that subjects perceptually (consciously) misjudge distances to targets, but in the mean time the motoric (automatic) actions are perfectly successful — as if the parallel (i.e., magno) system would not make such (i.e., parvo-based) misjudgments. So, for instance, grasping an object might be performed directly, without subject's conscious awareness which comes later – with final object recognition (Tittle & Todd; Bridgeman in Arbib, 1995).

Processing of spatial cues and relationships. Information about 3-dimensional shape of an object can be inferred, based on early visual processing (already overviewed), from attributes like:

1. Local orientation and curvature;
2. Relative depth among neighboring surface areas;
3. Silhouettes and occluding contours;
4. Changes in size, length or compression-rate of texture elements[57];
5. Shading and hue-gradients;
6. Motion. (Tittle & Todd in Arbib, 1995; Wallis & Bülthoff, 1999)

The attributes of 1 and 2, used for what I have called the geometry-based object perception, are largely analyzable using the Lie-group model.[58] Similarly, but not equivalently, occlusions introduce gross (inter)geometric relations and point 4 local (inter)geometric relations into perception of (many) objects or their internal patterns.

Visual recognition of object's shape from gradations of illumination of object's shaded surface is an example where a PDP computational model was used – the *back-propagation* network (Lehky & Sejnowski, 1988). This ANN-model (original algorithm see in ch. 8 of vol. 1 of Rumelhart, MacClelland & PDP group, 1986) was used also for "automatic" computing of the change of visual inputs because of eye-movement (Zipser & Andersen, 1988), although the popular error-*back-propagation* algorithm has up to now not been recognized as directly biologically plausible.[59]

Grossberg's network simulation[60] combined into an interplay two systems: a boundary-contour detection and completion system, and second, a featural filling-in system. The second system (possibly based in V4) spreads features, e.g. color, but this spreading over surface is inhibited at boundaries by the first system. Using cooperation of both systems, object is recognized in a higher system where ITC is involved (according to Grossberg in Arbib, 1995).

Object perception and group theory. The theory of mathematical structures called *groups*, especially the Lie groups, has been found useful for modeling object perception in main-stream neuropsychology, but less than in the holonomic theory where its phase-space (Hilbert-space) generalization is used.

Hoffman (1966, 1968, 1970) used the Lie transformation groups for description of *perceptual invariance* (constancies), like contours of objects. Retina and the cortical map of retina are considered as a mathematical manifold with corresponding *response vector-fields* that, for instance, describe neuronal activities specific to orientations of stimuli (e.g., edges) (Figure 5). Neurons in V1 (and beyond) are considered as Lie group germs (Hoffman, 1968), because they cooperate in for-

Figure 5. Contours with the corresponding vector field. An arrow depicts a vector (curve segment, edge) (From Hoffman, 1968)

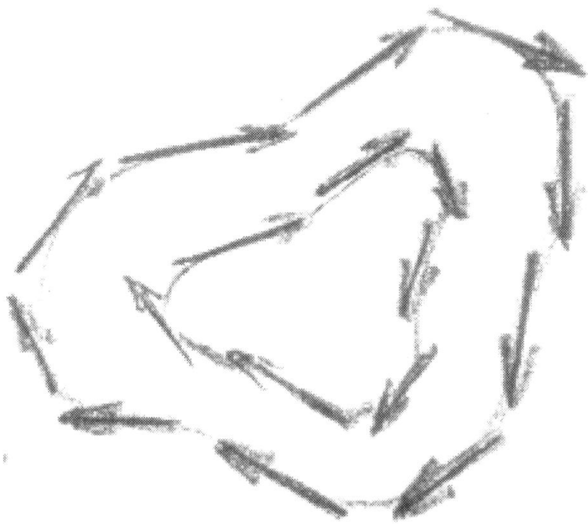

mation of gestalts (or attractors, respectively) in a way that significant perceptual features remain unchanged (invariant) in spite of changes or "flows" in the visual field (e.g., Tsao *et al.*, 1991).

Lie groups are a special sort of mathematical structures, called groups, which satisfy specific conditions.[61] They are used as descriptive categories of various transformations — in our case, transformations in the perceptual field which preserve basic characteristics of the form of a perceived object.[62] In mathematical terms, a contour (depicting the perceived object) is invariant under a given Lie group (Luccio, 1993).

As already mentioned, the image of the object reflects all the transformations performed on it, but is perceived as an object which is invariant in its essence. *Perceived shape and size of the object remain invariant under translations and rotations of the object relative to us.* It is implied here that these translations and rotations are further-on reflected in corresponding *projection-changes of object's retinal image in our eyes.* In the model of geometry-based object perception, after performing certain Lie transformations (listed below), the following object's characteristics remain invariant, if the Lie groups of the model are constructed appropriately:

1. Because location is changed under translation, shape constancy (or size constancy[63]) is ensured by the translation group (or by the dilatation or magnification groups).
2. Because orientation is changed under rotation, orientation constancy (e.g., constancy in relative angles of edge-segments on object's surface) is ensured by the rotation group.
3. Special Lie groups are constraining fusion of binocular stimuli to unified stereoscopic (cyclopean) percepts, and accompanying distortions. There are also motion groups and more abstract ones. (Mathematical details in: Hoffman, 1966, 1968; Dodwell, 1983.)

Various phenomena of gestalt psychology, including perceptual after-effects (like the spiral and movement ones), illusions and hallucinations (like "seeing internal stars, sparks, wheels, disks") were discussed in the Lie-based mathematical framework (Hoffman, 1966; Caelli, 1976).[64]

Ullman & Soloviev (1999) claim that even shift-invariance (which would enable recognizing a translated object as the same one)[65] is not a built-in mechanism, but a learned skill achieved after seeing many multi-view examples.

Binding. An essential property of the higher vision system is *binding* of perceptual parts into a whole, or categories into higher-order categories. According to the main-stream neuroscience, this is realized, for example, by *synchronous oscillatory activities* of neurons or their assemblies which encode specific parts or categorial elements (e.g., Wang, 1999). Binding could also be triggered by visual *attention* which leads to *enhancement of activity-rates* of neurons involved in representation of a consciously observed object (Roelfsema, 1998).

Enhancement- or coherence-based perceptual binding is a multi-aspect and multi-level phenomenon (sec. 4.7). In general, without entering the open debate on neural correlates of binding (does it come from 40-Hz oscillatory neuronal responses or/and quantum process or/and...?), one can say that *units representing items that are bound together fire coherently* (i.e., *in phase*, or at least *with constant* delay or *phase difference*).

Gray *et al.* (1989, 1990) found experimentally that *neurons in spatially separated columns of the visual cortex are able to synchronize their oscillatory spiking-activities*, so that inter-columnar coherent oscillations have *no phase difference*, on average. Because this is dependent on global properties of stimuli, the authors assume that such synchronization might be a mechanism for perceptual binding of features into a whole. This is the binding candidate of the main-stream neuroscience; competing sub-cellular and quantum candidates, favored also by the holonomic theory, are presented in section 4.7. [66]

Lee & Blake (1999a,b) have researched interdependence of detection of temporal correlations (Wang, 1999) and spatial phase relationships during dynamic image processing and binding. Lee & Blake (1999a) provide evidence that spatial visual gestalts can emerge also from exclusively *temporal synchrony*. Other possible spatial cues for gestalt-formation, which are effective in various circumstances (these cues are, for example, luminance, texture density, color, disparity, common motion and contour orientation), were systematically excluded in this specially designed experiment. They studied correlated temporal changes in direction of local motion of "waves" inside Gabor fields. Temporal synchrony in these simulations, as discussed by Lee & Blake (1999a), can be used for figure/ground segmentation and shape perception, but only if computation of higher-order statistics is involved. Importance of departure-times and relative arrival-times of action-potentials (as opposed to their firing rates), temporal coincidences and time-delays has also been emphasized by Hopfield (1995) to be useful for encoding and processing of information (even more than "amplitudes of neuronal activities" and their correlations). On the other hand, experiments of Lee & Blake (1999b) show that relations of spatial phases, including those used for detection of edges (alignments in sine-phases), influence neural registration of temporal structure, and consequently binding.

Imagery. There is (much) evidence that *visual imagery* shares representation and neural correlates with visual perception (e.g., Baars, 1997; Changeux, 1983), and that waking and dreaming visual imagery also share underlying processes. On the other hand, visual imagery might be autonomous with respect to other memory and reasoning processes.[67] It is no wonder that imagery shares perceptual areas since it bases on remembered perceptions. Even visual illusions may appear in imagery (Kosslyn, 1988). Imagery is nevertheless divided into two kinds of processes — those using stored shapes and those using separately stored relations between shapes while arranging an image and its derivatives (Kosslyn, 1988). Tootell *et al.* (1998) ask whether the representation, which is during imagination reconstructed from memory in higher visual areas (especially V2, V3, also V4), is projected back, via existing point-to-point topographic back-projections, to V1 where "it appears on mental screen".

There is enough evidence for existence of *mental images* (Kosslyn, 1994). If subjects were solving a task where they had to identify objects, depicted on one side, with their rotated views, depicted on the other side, they used an amount of time proportional to the angle of difference in views (Shepard's experiment). This shows that they "mentally rotated" the depicted objects to match pictures of both sides. Kosslyn's subjects had to recall a map. For that, they used an amount of time proportional to distances between the locations on the map which they had to recall. Namely, they "mentally travelled" from one to another recalled location, as on the real map before. The mental image is not clear and has somewhat better resolution only under introspective attention. It is stored in an abstract, condensed form. We have impression of a unified image, but this is much misleading. As Dennett said, "we cannot recall how many stripes a tiger had." Eidetic people are a remarkable exception. Subjects are usually able to make a "mental zoom" to trace some details in their mental image.

Perky demanded that subjects visualize a tomato on a screen, but have eyes fixated to a light-point. Then he displayed a tomato-image on the screen. Subjects have still been convinced for a while that the tomato is imagined – they confused their mental image with the real image. Segal and Fusella made subjects to visualize a tree on a screen where a weak light-point with slowly increasing intensity was projected – until they saw the light-point. They needed more time to finally see the light-point than when they had to imagine bell-ringing. In contrast to the bell-sound case, imagining the tree caused a direct competition of the tree-image with the light-image. This is possible only when the *same* perception-center is activated or there is at least much neural relation. So, imaging and perception have to *share* the neural substrate for image processing or be significantly connected. (Changeux, 1983)

6.8 VISUAL MEMORY

Neuronal processes. Let's first recall basics of neural cell physiology. Biological details and mathematical models will not be presented here, since this is done in numerous literatures (e.g., Damask & Swenberg, 1984; Kandel *et al.*, 1991; AuxLit 2; Peruš, 2000a; Berne & Levy, 1993 / AuxLit 10).

Neurons are connected into networks or circuits by their "branches" (dendrites, axons) with synaptic contacts or junctions. An excitatory synapse depolarizes[68] the electric field potential across the membrane of the postsynaptic neuron. An inhibitory synapse hyperpolarizes the cell.[69] A neuron generates a pulse, called *action potential* or *spike*, along its axon towards the postsynaptic neuron (output), *if* the membrane potential exceeds a threshold value. The membrane potential is determined by the sum[70] of pulses from presynaptic (input) neurons, the pulses being enhanced or inhibited to a certain extent at the synaptic cleft.

In inter-neuronal signals, "information is encoded" in the *number of action potentials per second* (called "density [of spikes / pulses]", or "[firing] frequency [of spikes]"), or in *duration* of the "comb-like" firing-sequences (*bursts*) of action potentials, respectively. When a burst is longer or its spike-density is greater, this indicates greater response (more important information). Namely, if the intensity of the stimulus is increased, the heigth (amplitude) of the spike always remains the same, but the frequency of spikes increases, or the length of "spike-train" incerases.

In physiological terms, the postsynaptic response, if triggered by repeated presynaptic stimulation, is determined by stimulus-proportional mean discharge frequency, and also by time of firing, temporal pattern and duration of bursts.

Hippocampus. (Visual) cortex is connected with limbic structures, especially with the hippocampus which is important for *organizing storage of memory traces* into cortical areas and for acquisition and consolidation of memory-items. Although much evidence shows that hippocampus has a structure for sparse Hebbian associative matrix memory, implemented by CA3 and CA1 pyramidal neurons[71], it might (after establishing appropriate associations) rather guide distribution of memory elements into the cerebral cortex for ultimate storage (Rolls in Zornetzer, Davis & Lau, 1990). So, the hippocampus might act as a relatively-temporary buffer which finds economic long-term storage in cortical areas that it cooperates with.

Hebbian memory-storage. The following paragraphs of this section present the favourite model of memory held by main-stream neuroscience. It is not the only possible model.

Long-term potentiation (LTP) is a long-lasting use-dependent increase in the rate of excitatory synaptic transmission, and so subserves several Hebbian-like processes (review in: Gardner, 1993; Brown & Chattarji in Arbib, 1995).[72] It is supposed to be important for memory-storage, like in hippocampus and cerebellum, maybe partially also in the visual cortex.[73] It is a matter of debate whether LTP arises from enhanced release of neurotransmitters, or from increased postsynaptic sensitivity to neurotransmitter, or both, or from increased probability of neurotransmitter release stimulated by retrograde signals from the postsynaptic cell (cf., S. Araujo in Dubois, 2000b) which so enhances neurotransmitter release from the presynaptic cell.[74] Bear (1996), and Tang *et al.* (1999), among others, report that LTP *and* LTD (long-term depression) occur at many synapses in hippocampus and neocortex, thus strengthening the Hebbian paradigm.[75] The resulting much-simplified physiology led to the *Hebb rule*, used for computational models: Synaptic transmission increases (LTP) if the connected neurons are both activated or both inhibited, and decreases (LTD) if one of them is activated and one of them is inhibited. Its mathematical expression states that the synaptic transmission J changes (in time t) proportionally to multiplication (*correlation*) of activities q of both neurons i and j (details see in: Haykin, 1994; Amit, 1989; Peruš, 1995a; AuxLit 10):

$$\Delta J_{ij}(\mathrm{t})=\lambda q_i(t)q_j(t) \quad (\lambda \text{ is a small constant}) \tag{6.1}$$

Memory traces. Memory-storage (or "neuronal learning", colloquially) is based on regulating transmission of chemicals called neurotransmitters, or (more appropriately) neuromodulators, across synaptic clefts. On the level of neuronal networks or circuits, *memory-patterns are parallel-distributively encoded in (large) assemblies of neurons* and, especially, their *synapses*. According to present main-stream models, *patterns of neuronal activities act as attractors* of network dynamics and are thus *neuronal correlates of mental states* (AuxLit 1). A mental state, as a sort of "virtual" "influence of a pattern–attractor", emerges from the situation that the pattern–attractor has the power to "attract" all similar-enough neuronal configurations[76] in order to change them into itself, if/because it is correlated with some meaningful external image, for example. "Patterns" or "matrices" of synaptic transmission-rates represent stored memory-units (patterns of environmental stimuli). (More elaborate and detailed presentation see in: Burnod, 1990; Gardner, 1993; Peruš, 1995a; Peruš & Ečimovič, 1998; AuxLit 7, 10).

Since visual memory has not been shown to be significantly different from other modalities of associative memory, the whole discussion of this section has up to now been valid for brain's associative memory in general.

Locations of memory for vision. Neurons that seem to be involved in visual memory[77] are, according to Fuster & Jervey (1982) and Bear (1996), concentrated in inferotemporal cortical (ITC) region called superior temporal sulcus (in its lower bank). Temporal-cortical (especially ITC) structures, which participate in encoding of information during visual perception, should also participate in storage and recall of that information, at least in short terms (Miyashita & Chang, 1988; Fuster & Jervey, 1982).[78] Electrical stimulation of temporal areas might evoke visual imagery.

"Where" and "what" aspects of visual information, as roughly processed along different pathways, are probably also memorized in different structures: Neurons selective for shape and pictorial information were found in monkey's anterior ventral temporal cortex (Miyashita & Chang, 1988). On the other side, a map of spatial locations of visual stimuli might be, for example, in the dorsolateral prefrontal area.[79] Visual information, like features of objects or faces, is used for cognitive purposes in the so-called working memory which is located in the inferior position of prefrontal cortex.[80] According to Desimone (1996), prefrontal cortex is also responsible for memory-based attentive biasing, i.e. for giving priority to relevant stimuli (as expected by the subject) or to those which "have won the competition", and for enhancing corresponding activity in ITC. Feedbacks from prefrontal to extrastriate and other parts of the sensory cortex execute this.

Critique. Although huge experimental efforts have been invested in studies of LTD and LTP, and their molecular and genetic backgrounds (Goldman-Rakić, 1996; Desimone, 1996; *et al. ibid.*) many problems remained, also because of enormous complexity of the multi-level process. There are many variants of so-called Hebbian processes, each bringing some benefits for biological plausibility of modeling of some concrete (often species-specific) structures, but also making any attempt of global model or view vague and relative (e.g., Gardner, 1993; Abbott & Nelson, 2000).

Furthermore, hippocampus mainly merely organizes memory-storage into cortex for permanent storage there. So, evidence that supports Hebbian processes[81] in hippocampus should not be directly generalized for cortex and be applied for (visual and behavioral) learning.

In ch. 2 the holonomic alternative based on dendritic net processes is presented. It remains roughly partially Hebbian-like, i.e. Hebbian in amplitude-correlations and beyond Hebb in phase-processing. It seems that, going deeper into the tissue, so-called Hebbian processes become more refined ("more quasi-holographic") and the model has to be generalized by incorporating higher-order correlations ("quasi-interferences") of oscillatory processes.

High-level visual and multi-modal processing (AuxLit 3, 5, 7) will not be considered (see: Palmer, 1999; Kosslyn, 1994). For some heterogeneous ideas on global mathematical models of cortical processing see: Körner *et al.* (1999), Clement, Coveney, Jessel & Marcer (1999), Hoffman (1997, and his references within), Ingber (1998), Liljenström (2002), and also Diaz (1997), Kant & Levine (in Wang, 1998). Unresolved questions outnumber comprehensive and reliable studies which are much scattered and so do not provide any substantial overall theory of common agreement.

6.9 ORIGIN OF SELECTIVITY FOR VISUAL FEATURES: A MODEL

Tracing edges. Edges are sudden, i.e. usually non-continuous, changes in brightness of the image, often from one extreme to the other (luminous/dark). In a *basic computational model*, edges are traced by numerous parallelly-performed local comparisons of intensity values *I* in neighboring points (pixels) of the image. Figure 6 illustrates the simplest computational edge-detection. Figure 6(a) quantifies our exemplar "image". Numbers represent neuronal activities which are proportional to intensities of light-stimulus at particular retinal locations. Figure 6(b) quantifies the difference in intensities *I*, as in Figure 6(a), of particular pairs of pixels (i.e., left and right from their common border). This is called the first-order difference-operator and would in the case of "infinitely"-dense array of pixels be called first-order derivative or gradient *dI/dx* (*x* measures distance on horizontal axis). Going from left to right, we get negative values of *dI/dx* (elements of the array of Figure 6(b)) if right is darker than left, positive *dI/dx* if right is brighter, and zero *dI/dx* if there is no change. The *dI/dx*-array of Figure 6(b) is called convolution matrix (each particular element of it is a *convolution*). Namely, a comparison or correlation or, in this case, difference of values at two points is calculated and multiplied by a "weight", or an operator-value, at each location of the array or matrix. This matches

Figure 6. Illustration of computational detection of edges in an "image" (a) by calculating intensity changes (gradients), so getting the convolution matrix (b), and by calculating gradients of gradients, so getting the Laplacian matrix (c)

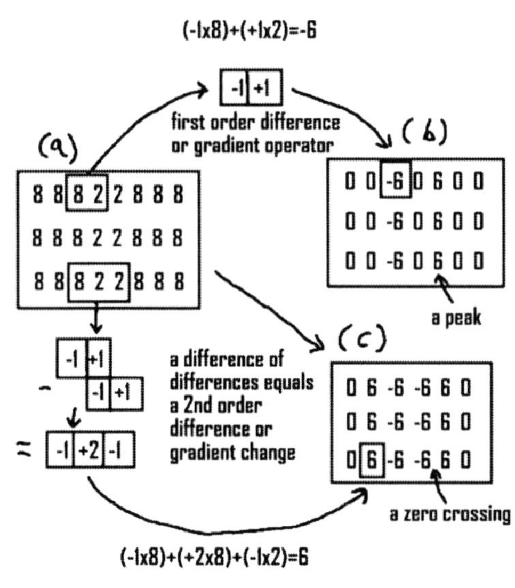

with the definition of convolution. Convolution is defined as an integral (sum) of a value of a function which describes the local state (*I* here) and the "weight" of an operator (difference here). The "weight" is, in this case, the sign of the spatial derivative (gradient) *dI/dx* (+1 or –1) determining increase or decrease in brightness *I*. For example of Figure 6(a)-(b): *dI/dx* [for the third/left and the fourth/right upper pixels] = *8 (-1)* [left *I* times "weight" (–1)] + [the sum!] *2 (+1)* [right *I* times "weight" (+1)] = *–6*.

"Weighting" profiles. Psychophysical experiments, measuring responses to patterns of sinusoidal gratings with different spatial frequencies (i.e., densities of *sine*-like intensity variations) have shown that the visual information is processed along different "channels" having "weighting" profiles of different size (although typically a similar "Mexican hat" form).[82]

For the simple example of Figure 6 (after Stillings *et al.*, 1995, ch. 12, based on Marr), where intensities *I* change only in horizontal direction *x*, gradient *dI/dx* (change of *I* in direction *x*) detects the step-like edges. However, if the changes are gradual, there might be several gradients, and such non-sharp "edges" or a series of slight edges (e.g., "border" of a ball) would not be detectable well enough. Therefore the second-order derivative *d(dI/dx)/dx*, gradient of the gradient (change of *I*-change in direction *x*), is calculated. Maximal and minimal changes (maxima and minima of gradients) manifest as zero second-order gradients. This is described, if we trace an edge of *any* orientation (partially in *x* and partially in *y* direction, if *I* changes horizontally and vertically, or combined: *I=I(x,y)*), by the following expression (Laplacian operator giving zero):

$$d(dI(x,y)\,/\,dx)\,/\,dx + d(dI(x,y)\,/\,dy)\,/\,dy = \frac{d^2I(x,y)}{dx^2} + \frac{d^2I(x,y)}{dy^2} = 0\,. \qquad (6.2)$$

Graphically, this operator corresponds to the profile of intensity increases/decreases as in Figure 6(a), which is generalized (by a "rotation around the centre") from the one-directional *I*-change profile of Figure 6(b), the latter illustrating the difference-"weight" of Figure 6(c). It turns out that even more effective "weighting" profile is the "Mexican hat" of Figure 6(c) which looks the same as the receptive-field profile found in retinal and LGN cells. The Gabor-profile of V1-cells is roughly similar, but actually is a much more specified and selective "weight function" for more sophisticated visual processing. The "Mexican-hat"-profile is obtained by using Laplacian operator on the Gauss function or gaussian (the bell-shaped curve of statistics), i.e. by calculating a second-order derivative of the gaussian data-distribution. The "Mexican hat" is a result of connections, with excitatory and inhibitory influences, between retinal neurons (as described in chapters 6.1 and 6.2).

Convergence of information to a cardinal cell. Figure 8 illustrates the neurocybernetic origin of information condensation and specialization of a "feature-selective" *simple* or *cardinal* cell, in the model. The *simple* cell makes a synthesis of local information, which has been gathered from lower-level cells, in order to transform it into a more global, comprehensive code. First-level cells "observe" the oriented line or edge locally only, so they can just "detect" segments of the edge and contribute information about its location. The second-level cell (called the *simple* cell) automatically extracts common aspects (i.e., correlations) of all local-information contributions from the first-level cells. In the case of a line-edge, this common feature is the same orientation of all edge-segments. Thus, (x,y)-coordinates (i.e., the first-level code) are effectively re-coded into polar coordinates (r,φ), i.e. the distance from an origin of the coordinate system (r) and the orientation-angle (φ) (i.e., the second-level code). Instead of many coordinate-pairs (x,y), the new, global code includes only one (polar) coordinate-pair (r,φ). Replacement of many cells with a single cell as the carrier of information means that condensation of information was achieved. The coding role of a cell is, of course, determined by its specific position in the network and its specific connectivity.

Such a process of information convergence is universal for all so-called *perceptron* tree-like net-architectures. Instead of edge-selective response, other features could be selected if other patterns would be presented into the receptive field of the perceptron-cells. A detailed, illustrative presentation of emergence of dominant, "feature-representative" second-level (or "hidden layer") cells, called *cardinal* neurons, is available in Peruš (1995a,b), therefore it will be omitted here.

Ullman & Soloviev (1999) write that complex shapes are detected by conjoint activation of *simple* cells which detect fragments like edges. However, Pribram (1991, p. 81) emphasizes that this contradicts evidence[83]:

Figure 7. Bird-view profile (a) used for edge detection in our model, and side-view profile (b) which graphically represents the convolution "weight" used of finding intensity differences. A more sophisticated and experimentally supported side-view profile is the "Mexican hat" (c)

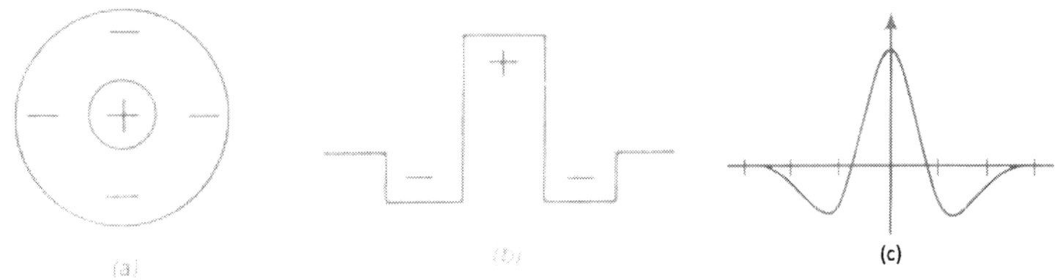

Figure 8. Edge-"selectivity" as a local-to-global information transformation — manifested also as condensation of information, edge-encoding, specialization, and realized through cartesian-to-polar coordinate transformation. See text for details

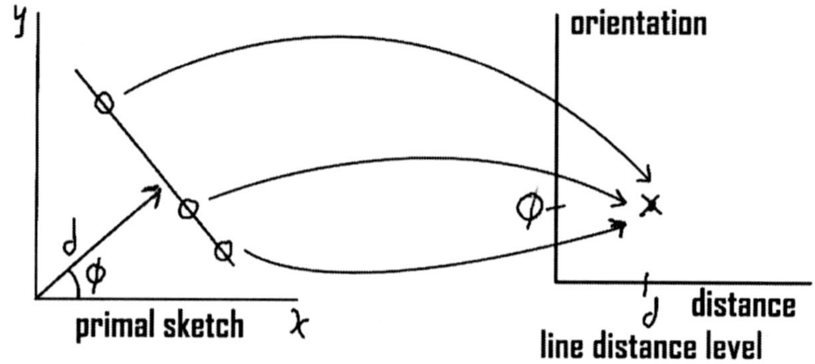

If the feature detecting cell, rather than the feature per se, were the basic representational unit to be classified, a cell might, for example, be complex or hypercomplex, but not both. The fact that a cell can simultaneously be both, and in addition be color- and direction-specific, as well as velocity- and luminance-sensitive, indicates that the representations of these properties, features, are already conjoined within the receptive field of the cell.

In Pribram & Carlton (1986, sec. 1) it is concluded:

Sensory perception results not from processes which conjoin properties, but which simultaneously select these properties from a pool in which they are already to some extent haphazardly conjoined.

My critique. Feature-selection models, based on cardinal cells (called simple cells in early vision), work very well in, usually simplified, computer simulations, but not necessarily in brain — due to its dynamic complexity and richness of perceived scenes and their contexts. Although feature-selective (including edge-selective) neurons are often reported by experimentalists, such cardinal cells might not retain their reported speciality forever, but rather temporarily. Moreover, their specialization is fuzzy, dependent on circumstances (e.g., connectivity, input signals) in PDP systems dynamics. So, if we speak about a "feature detector", we actually mean a neuron or neuronal group[84] which responds more to those feature than to other possible features, and more than other individual neurons or their assemblies do. Thus, a feature-selective neural unit (a cell, whole assembly of cells, a local circuit connected into a cortical column, etc.) should have a relatively exclusive response to the feature, and the response must also be stronger than all responses of other units. The "definition" is very relative, like the responses of neurons are.

6.10 GLOBAL RETINOTOPIC MAPPING

Maps. Up to now, the focus has been on essential physiology rather than on its *global* anatomic manifestation like the structure of cortical maps. Let us see now what knowledge about maps and their formation can help us in modeling image processing.

Each V1-neuron receives its inputs usually from a small region of the retina. This region is called receptive field. Only stimuli that are presented to the center of this area trigger a measurable response of the neuron. The assembly of V1-neurons constitutes the first map in a chain of topographic maps of the visual cortex.

Ocular dominance stripes. In V1 there are regions, "mixed, interspersed" one with another, of response triggered mainly by the left or by the right retina. Although their function is unclear, maybe there is even no function[85], many modelers relatively successfully tried to recreate them in learning-based "evolutionary" computer simulations. Anatomically evident stochastic stripes manifest alternating regions of dominant left-eye influence and regions with predominantly right-eye influence. These stripes belong to assemblies of *ocular-dominance columns*. They are functionally related to cortical columns with preferred response to a particular orientation of stimuli like edges, called *orientation-preference columns*, and to global retinotopic mapping. Ocular-dominance stripes and orientation-preference stripes even anatomically overlap, although they are not identical. (Such neuronal and columnar shared multi-tasking is no surprise in bionetworks.) Swindale (1996: list in Table 2 on p. 181) presented numerous computational models for developing regions of ocular-dominance columns and orientation-preference columns. They were mainly based on Hebbian synaptic plasticity and intracortical competitive interactions (after Kohonen, 1982, 1995, or Kohonen, 1984). No phase-Hebbian (HNeT- or ICA-like) model has been tried, but results were nevertheless roughly successful, i.e. ocular-dominance maps and orientation-preference maps occurred in simulations.

Models usually incorporate V1-neurons (in the layer 4 – input layer to V1) each of which gets inputs from the left eye over the left LGN as well as from the right eye over the right LGN, and also from other cortical neurons over lateral cortico-cortical connections. A computational Hebb-based model of Miller, Keller & Stryker (1989) reproduced formation of ocular-dominance stripes. Such "stripes of columns" do not develop when neural activity is blocked in the eyes or in the cortex, or when input-stimuli are given to V1 from both eyes synchronously. Ocular-dominance columns develop as a result of asynchronous stimulation and are a consequence of binocular rivalry. The simulated evolution of ocular-dominance stripes went from the initial state of binocularity across the whole V1 (because of equilibrated inputs from both eyes) gradually toward segregation of V1 into patches with predominantly right-eye influence and patches with predominantly left-eye influence (these are thus the precursors of ocular-dominance columns). The emergence of these columns is accompanied by emergence of monocular cells (i.e., cells dominated by one eye only, the other-eye's influence is suppressed) and by topographic refinement of receptive fields. Simulations show that the widths of patches depend on intracortical connections, and the degree of monocular specialization depends on inter-LGN (anti)correlations.[86]

Orientation-selective columns. Orientation-selectivity of specialized columns is much more important for image processing, therefore we will list some models of these orientation-preferent columns. (Some of them also model ocular-dominance columns.). Swindale (1996) reviews that competitive Hebbian models (based on Kononen, 1984) were somewhat better than usual correlational (non-competitive) Hebbian models.[87] Among the first ones, the following self-organizing models with laterally inhibitory and excitatory interactions[88] were emphasized: the von der Malsburg model, the Swindale model, the Kohonen Self-Organizing Maps model (details in Appendix B), the Linsker model (sec. 3.1), and the so-called elastic-net model or simulated annealing model (presented in Mezard *et al.*, 1987).

Mapping onto V1: global versus local. Obermayer, Blasdel & Schulten (1992) show, by a mathematical and computational analysis using statistical physics based on Kohonen's Self-Organizing Maps, that ocular-dominance columns and orientation-preference columns emerge in V1-model-net as a stable column system only if certain order-parameters, as they define them, exceed a well-defined critical value. Below this threshold, there are no correlations between V1-regions incorporating ocular-dominance columns, orientation-preference columns and retinotopic mappings; above the threshold, however, they are strongly correlated. Specifically, iso-orientation slabs usually criss-cross the ocular-dominance bands approximately-perpendicularly.

Early-vision *mapping*, simulated by above-mentioned models like Self-Organizing Maps, has to be distinguished from early-vision *processing* which is simulated by information-maximization models (ch. 3) compatible with the holonomic theory. The former is a global physiological "envelope" of the latter — i.e., retinotopic mapping is an "external", even anatomical, relatively-macroscopic manifestation of "internal", relatively-microscopic image-processing.

Let us also mention here that the models like Self-Organizing Maps might be good for modeling early-vision mapping, but not for object perception. For instance, Wang *et al.* (1999) note that "winner-takes-all"-networks, like Kohonen's Self-Organizing

Mapping net, do not encode (global) spatial relations (except as much as they are an essential ingredient of an image itself). Such networks therefore cannot perform perceptual processing of inputs where we need information about spatial relations between relatively autonomous parts of an image, i.e. objects or their distinguished parts. Wang *et al.* (1999) therefore propose a new "Locally Excitatory Globally Inhibitory Oscillator Network (LEGION)" where each object is represented by a synchronized group of oscillators corresponding to the object, and different objects of an image are represented by different oscillator-groups which are desynchronized from each other. (Cf., Auxil. Ch. 11.3)

Cortical cascade of maps. Tootell *et al.* (1998) write that visual information arrives from retina to V1 relatively undifferentiated. However, the retinotopic mapping involves spatial rearrangement of the visual field into brain-map locations which are its "opposites"[89] (e.g., right lower inner-most part of the retina is represented in the left upper outer-most part of V1). Geometric variations, as described by locally different cortical magnification factors, are significantly related to corresponding differences in receptive-field size, and also to corresponding spatial resolution or visual acuity, at different distances from the center of gaze. In the visual cortex, there are maps of preferred response to specific spatial frequency which are remarkably similar to organization of areas of retinotopic eccentricity.[90] From the complete retinotopic map of V1, images are furthermore roughly-isomorphically (i.e., point-to-point) projected into each of the later topographic areas V2, V3, V4 (Tootell *et al.*, 1998; Supek *et al.*, 1999). However, they get subdivided into quarter-field (or, in some areas, into half-field) representations, as indicated by selective loss of vision for a quarter-field (or half-field). Beside that of V1, there are at least 20 complete or partial representations with a similar topographic order in extrastriate areas. (Kandel *et al.*, 1991; Kirvelis, 2000).

Visual "geometry". Images of the left and the right retina get processed along the left and the right visual pathway including the "interchange" of pathways at the optic chiasm. The encoded images then "merge" not earlier than in V1, leading to so-called *cyclopean perception*. This enables stereopsis (as proved by the Julesz stereograms) and near-field depth-perception – probably in the V1-to-V2 connection-array. Far-field depth-perception at more than 30 m, where both images are approximately the same, is possible by a single eye by using previous experience on size-relationships (distant objects appear smaller), shadows, occlusions (nearer objects hide distant ones), motion-parallax[91] and knowledge on perspectives (Kandel *et al.*, 1991).

DeAngelis (2000), reviewing evidence on (binocular) stereopsis, reports that disparity-selective simple-cells were found (in cats etc.), organized in topographic maps, in V1 and also in V2, V3, MT and in ITC, but not in LGN, of course. Binocular disparity could be encoded by positional or phase differences[92] between two eyes, or both. The phase possibility is more likely and more experimentally supported. According to phase-difference model of a binocular simple cell, the first-eye's receptive field is a symmetric Gabor wavelet, but the left-eye's receptive field is an asymmetric ("inverse symmetric") Gabor wavelet shifted by a distance *d* relative to the first-eye's receptive field.

Detailed physiological computational mechanisms of stereopsis are largely unknown, and are computationally modeled in functionalist fashion (DeAngelis, 2000).

6.11 THE ROLE OF COLOR-INFORMATION IN EXTRACTION OF EDGES AND CONTOURS

Since the topic of this book is image processing, we will not discuss the whole range of color-vision phenomena (see R.L. & K.K. DeValois, 1990). In spite of different processing pathways for color- and form-perception, there are some aspects, to be discussed now, where color-information helps in image processing.

Basic characteristics of color. Subject's impression of color consists of experience of *brightness* (intensity of the effect by light imposed on cones), *hue* (the "folk-color" – proportion of activities of the "red", "blue" and "green" cones), and *saturation* (how "rich, clear" is a hue, how much it is unspoiled by grayness[93]). Saturation is determined by the degree to which all three cones are stimulated to the same degree by the object and its background.[94]

Color phenomena can be divided into those dependent on (Trstenjak, 1996):

1. Coloring matter or dye-stuff;
2. Illumination and its context or circumstances;
3. Psychical integration or completion with other perceptions (the so-called subjectivity).

Three-cones system. The wavelength of light only determines the *probability* that the photon will be absorbed during an all-or-none conformational change of the pigment in the outer segment of a cone-photoreceptor, but the electrical response of the cone's inner segment is not affected by the light directly. In spite of the fact that a cone reacts (if it does) always with the same electrical signal regardless of the wavelength, the number of photons absorbed by the cone does vary with wavelength. A cone responds equally to a light of any wavelength as long as the light-intensity compensates for the cone's absorption-rate at that wavelength. Cones have different absorption-probabilities, or sensitivity-maximums, respectively. So, in spite of the fact that different cones are maximally sensitive to different light-wavelengths, *responses of the three types of cones need to be combined, in their* *relative strengths, to recognize the color*. Individual cones are not sufficient for detection of an arbitrary light-wavelength (Kandel *et al.*, 1991, ch. 31). Each of the three classes of cones responds maximally to its own wavelength corresponding to phenomenal blue, red and green.

(Double) opponency. Many ganglion neurons of retina, LGN- and cortical neurons are excited by one class of cones and inhibited by another. Such *opponent cells* encode the hue so that they subtract the output of one of the three cone-types from the output of one of the others – in various combinations. These opponent interactions of the red-, green- and blue-preferring cones are responsible for the phenomenon of color-mixing: e.g., an object that is both red and green appears yellow because the red and green neutralize each-other. Apart of the simple *color contrast* between opponent colors, there is a more complicated ("double") opponent phenomenon called *simultaneous color contrast*: e.g., a gray object with a green background gets a tinge of red.

Color is processed in retina and LGN by so-called *color-opponent cells.*[95] Some other cells react to brightness of the center compared to its surround, and do not contribute to color detection. Color information is further processed in cortical blobs by *double-opponent cells*[96], found in V1 (orientation-selective ones) and area 18 (*complex* ones), which combine inputs from *single*-opponent cells.[97] Saturation of a color is encoded neurally by the proportion of the total response by the color-opponent neurons rather than by the non-opponent (achromatic) neurons (De Valois & Jacobs, 1968).

Light-wavelengths which trigger excitation in a (single)-opponent neuron, denoted +Y–B because of having an excitatory (+) center of the receptive field in response to yellow (Y) light and an inhibitory (–) surround for suppressing the "yellow" center "with blue (B) centripetal influence", are those which *appear yellow* to a subject. De Valois & Jacobs (1968) continue that colors[98] correspond to the points of maximal response of the corresponding color-opponent cells rather than to the absorption-peaks of the cone-photopigments.

Color constancy. The brain is able to recognize color in various circumstances like the ambient light, because it compares not only the cones directly affected by the object-stimulus, but also all other cones. This neutralizes to some extent the contextual effect of illumination (which cannot be neglected), therefore the color of an object anyway appears roughly the same in spite of different illuminations of the environment-space. This is called *color constancy*. It is top-down constrained, similarly as object perception is. Like object-constancy is the next phase of perception after image processing, color-constancy follows color-imaging (Pribram, 1991). For the effect of color constancy also, red/green, yellow/blue and black/white double-opponent processes are needed.

Color experience diminishes in Ganzfeld experiments (i.e., using a pattern-less, uniform color-field). This demonstrates the importance of object–background comparison also in color perception. Experiments suggest that object's color is detected, or recognized, respectively, by comparing all the objects in the scene and its background (Kandel *et al.*, 1991, ch. 31).

Trstenjak (1996) observes that color contrast and color constancy are antagonistic effects, although interdependent and finally integrated. Color-contrast mechanism, which is mainly early-physiological (retinal), tends to "pour together" the colors of illumination and of the illuminated object into an unified / uniform visual perception, so that the object loses its intrinsic color. Color-constancy mechanism, which is conditioned mainly by memory and cognition, tends to separate effects of illumination from the intrinsic color of the illuminated object in order to produce perception of the object in its "real", more-or-less "permanent" color, regardless of the illumination and shading conditions.

Color perception of edges. The boundaries of color-fields usually fit the boundaries of object-forms, therefore color helps in edge detection.[99] Within the boundary of an object, colors cancel each-other out by action of color-opponent neurons; across the boundary, however, they support each-other's diversity by action of double-opponent neurons. Apart of the cases of "competing" color-patches with blurring "edges" *within* the form of an object, color perception enhances contrast of the object versus background. Special *double-opponent neurons* are responsible for *contrast-enhancing by color-differences* at the edge of an object. The internal and the external color are compared synergistically, i.e. perceptual responses of contrasting colors enhance each-other across the boundary because of double-opponent neuronal action.

Color helps in contour-integration. Mullen *et al.* (2000) are more optimistic about the role of color in form-perception than Livingstone & Hubel (1988) were (remember sec. 6.2). Color-detecting system as well as luminance-detecting system are both able to trace edges. Color- and luminance-edges are not coincident in shadows, for example, but are mainly coincident at object's boundaries. Mullen *et al.* (2000) concluded that optimal color-based contour-integration is only slightly worse than the luminance-based one. Contour's curvature was first established, in their experiment, to be independent from contrast. They compared red/green, yellow/blue and luminance postreceptoral contour-integration systems while linking simple oriented Gabor wavelets into a contour-gestalt. It was concluded that all three systems use a common contour-integration process which is also sensitive to color-contrast and phase of the stimuli.

6.12 DILEMMAS (MAIN-STREAM MODELS VS. HOLONOMIC THEORY)

My comparative observations. Although cognitive neuroscience provided, with huge efforts, relatively detailed and impressive experimental studies of brain structures, there are several drawbacks which pose dilemmas to modelers:[100]

• Experimental access to brain structures is severly restricted. If anyway succeeded, invasion (e.g., by lesions) and/or measurement may, even to a large extent, disrupt the natural process.

- Non-invasive techniques such as PET, fMRI and (X-ray) computerized tomography[101] are global only (i.e., they reflect activity in relatively large spatial and temporal regions). They are also indirect — some of them manifest biochemical, metabolical processes that accompany neuronal activity, others sketch the "hard tissue" anatomy (Kandel *et al.*, 1991), but attempts of direct access to physiology are always limited and imperfect. EEG and MEG are not much more local, since they measure collective electric or magnetic dynamics in large groups of neurons. EEG primarily reflects synaptic potentials in cortical pyramidal neurons (Kandel *et al.*, 1991). (Micro) electrodes are used for measuring electrical activities of single neurons or small neuronal groups. An EEG-based computer-averaged measurement of sensory activity is scalp-recording of evoked potentials. Experiments using radioactive or chemical tracers may also not have completely innocent effect on the observed process.
- The information most lacking for plausible computational modeling are empirical data on detailed *connection-structures*. Because networks process holistically, these data would have to be quite comprehensive. This will be extremely hard to achieve.
- Because of understandable difficulties with attempts to obtain reliable and detailed anatomical and especially physiological data on the microstructures, local connectivity, detailed informational pathways, feedbacks, interaction junctions etc., and especially their function, it is very hard to build theories and models. Busy with experimental work and patients, many neuroscientists have left them to a large extent to neurophysicists, mathematicians and computer scientists who, on the other hand, lack "biological experience". Current functionalist[102] models are facing, beside the problem of biological implementation, also the problems of complexity and artefact-modularity. They should so be considered as tentative only.
- In brain and mind, like in microscopic physics, process and interaction, adaptation through change, complementarity of competition and cooperation in order to optimize performance and exploitation of resources (material and

informational), etc., are essential. Brain and mind are thus dynamic, synergetic complex systems. However, our models (Haken, 1996; AuxLit 9; Peruš, 1995b) are able to tackle this merely on a general level, without incorporating much detailed empirical data, if available. So, our models basically remain "algorithmic metaphors". Best models[103] are obtained by learning through integration of multiple partial data (views, aspects).

These problems are characteristic for the entire experimental and computational (cognitive) neuroscience and are mainly unavoidable for now. However, several important critics can be posed to its mainstream, and alternatives can be suggested:

- The *Holonomic Brain Theory* seems quite exceptional, since it can manifest complementary experimental and theoretical work by the theory's author himself. With the holonomic theory, in comparison to other contemporary models, we come closer to a desired synthesis, with bilateral constraining, of experiment and model.
- Tracing locations of functions and "feature detectors" of various sorts and levels is remarkable and extremely important for medical and surgical practice, but it does not provide firm understanding of information processing. The holonomic theory goes beyond this limitation of locationalist main-stream cognitive neuroscience, although the latter provides useful basements.
- Information about processing pathways and "seats" for particular functions alone, without a holistic synthesis, enables only relatively "discrete", mechanistic models which can serve better for engineering and artificial intelligence, or machine learning, respectively (Kononenko, 1997), than for life sciences or even consciousness studies.
- The main-stream experimental cognitive neuroscience, and accompanying theories, concentrated on structure at the level of neurons, mainly because it is more accessible and measurable than sub-neuronal structures which remained neglected. But this might turn into actually being a "search of the needle under the light, not where it actually is (in dark)". The holonomic theory took into account the sub-neuronal structure much earlier than the recent increasing move of attention to that level. Even more, the holonomic theory does not exclude a serious possibility (still largely ignored) that quantum processes are essential.

The situation is, as has always been in science (and elsewhere), a mixture of huge success and crisis, which always means great perspectives. The Holonomic Brain Theory is a relatively perspective alternative (I believe) for a *large-scale* theory of the new century.

ENDNOTES

1. The details of the photo-reception chemistry in the receptor's outer segment, based on photoeffect-triggered transformations of the pigment-molecules like rhodopsin, can be found in Hucho (1986) and Kandel *et al.* (1991), among others. In the inner (neural) segment of a photoreceptor cell, chemical changes trigger ion flows through membrane as in usual neurons.
2. Gibson and Pribram disagree with this statement – see ch. 2.
3. Frequency f and wavelength λ are connected by the wave-velocity c in the formula $f\lambda = c$. So, higher the frequency – lower the wavelength, and vice-versa. Color is, roughly, a phenomenal perceptual aspect of the electro-magnetic wave characteristics (wavelengths or frequencies, respectively) of the stimulus (in the so-called visible-light part of the electro-magnetic spectrum, i.e. λ between about 400 and 700 nm.). The light is those "ingredient" of electro-magnetic waves which can trigger phenomenal visual perception in organisms. Light is thus the phenomenal aspect of electro-magnetic waves as they interact with the visual cortex and consciousness, respectively.
4. (Binocular) disparity is disparity between perceptual features as detected by two eyes.
5. Photoreceptor's receptive field is the region of outer space and time over which it responds to light-stimuli.
6. Feed-forward processing: information flows forward (from layer to layer, in one direction) only; there are no feed-backs. Feed-back processing is two-directional.
7. Let us recall that a receptive field of neuron A is the region where all those neurons lie which send signals to the target-neuron A. Excitatory (inhibitory) part of the receptive field means that the synaptic connections between the target-neuron A and those neurons within A's receptive field are excitatory, i.e. stimulating or "positive-valued" (inhibitory, i.e. antagonistic or "negative-valued"). That is, the excitatory (inhibitory) part of the receptive field of A includes those neurons which excite (inhibit) the target-neuron A.
8. Let us mention here, by the way, that there are no nerve impulses till the ganglion-cell layer (Pribram, pers. commun.), but just slow-wave electrical field potentials. This increases significance of the holonomic theory.
9. The most comprehensive review of vision science is Palmer (1999). Many data are in Buser & Imbert (1992).
10. Thalamus acts, in brain's center, as a relay-gateway to the cerebral cortex (Mumford in Arbib, 1995).
11. The brain's right (left) hemisphere receives inputs from both eyes, but only from those regions subserving the left (right) part of the visual field. This two-

eye-based left/right segregation is to a large extent preserved up to cortical ocular-dominance stripes (e.g., Van Essen & Anderson in Zornetzer, Davis & Lau, 1990).

12 Pribram, Lassonde & Ptito (1981) provide comprehensive and much-less simplified data to be discussed later.

13 E.g., Van Essen & Anderson in Zornetzer *et al.* (1990); Livingstone & Hubel (1988); Stillings *et al.* (1995).

14 Parvo/magno dichotomy roughly corresponds to X/Y-cell dichotomy.

15 See all the references cited in this paragraph.

16 Of course, color is finally constructed in cortical areas like V4 – from the parvo "color" (pre-color) processing.

17 Two equiluminant colors are perceived to be the same shade of gray on a black-white photo. Neuroanatomical parvo/magno distinctions were thus supported by psychological observations that the parvo-system distinguishes equiluminant red and green, for example, and detects their border, but the magno-system does not (Kandel *et al.*, 1991, after experiments by Ramachandran and Gregory).

18 Minus signs denote hyperpolarizing processes; plus signs denote depolarizing processes. A part of bipolar and ganglion cells of the retina has the same type of receptive field.

19 Metaphorically, there is roughly a perceptual dichotomy like in painting (parvo) versus drawing (magno).

20 A blob is a "cylinder", made of neurons, inside a cortical column (or parallel to it, respectively). An interblob is a region, populated by neurons, around the blob or between blobs.

21 Very probably, but it has not yet been finally confirmed (Pribram, personal commun.).

22 Retinotopic map is a one-to-one map or "spatial" code of retinal activity-patterns in a subsequent brain area.

23 They remind that, up to now, retinotopic organization of human striate or extrastriate areas has not been unequivocally demonstrated. The visual cortex responds differently according to stimulus characteristics and context.

24 Complex cells also have shorter latencies (response times) than simple cells (Pribram, personal commun.).

25 E.g., elongated with inhibitory both ends (producing "end-stopping") and flanks (Pribram, pers. com.).

26 Although the level of activity of each of those neurons, in response to a stimulus, is smaller.

27 E.g., Van Essen & Anderson in Zornetzer, Davis & Lau (1990); Kosslyn (1988).

28 However, see De Valois & De Valois, 1990, p. 91-3, 102-3, 110, for details.

29 Pribram agrees with that. He says (pers. commun.) that there are special layers with cells having concentric receptive fields in LGN and cortex, and there are layers of mixed simple and complex cells, but simple and complex cells are not segregated into *own* extra assemblies or layers. In ch. 2 it is presented how alternative evidence (Pribram, 1991, pp. 75-77) transcends the problems of single-cell-based classifications discussed here.

30 Complex details on visual cell-types see in Lund, Wu & Levitt in Arbib (1995; with figs. on p. 1017). They write that neurons of V1's layer 4C seem to compute response from a simple ratio of magno versus parvo inputs.

31 This is a rough statement. Pribram provided the first lesion-based evidence (microelectrode-works followed) on importance of ITC for visual categorization, or (as he says) comprehension. He distinguishes the choice of visual category from object-perception. The latter he localizes to areas V2—V4 and V5, but the former to ITC.

32 ITC appearances seem to be 2-dimensional to main-stream neuroscientists, whereas the "holographic view is 3-dimensional". In Pribram (1991, p. 194, after Gross *et al.*) 3-dimensionality and spatial frequency (and color) are added to the list of ITC-cell's selectivity, allowing to describe a half of ITC receptive fields as *Fourier descriptors* of tangential lines which collectively encode the *overall* shape of a stimulus (object-form). Such a Fourier-description of a "gestalt" is *independent on position and size* of the stimulus (more in ch. 2).

33 In Perret & Oram (1998) after K. Tanaka *et al.* (1991): *J. Neurophysiol.* **66**, 170-189.

34 E.g., "an opaque circle above a clear circle; a light rounded area at the base of a dark egg shape; a dotted brown disk attached to a narrow bar." (However, one should ask whether such reported specificity has sufficient evidence and whether it is enough stable or permanent at all.)

35 E.g., "a horizontal striped area above a separate horizontal striped area."

36 E.g., "T-junctions but not Y-junctions nor the right angled components of T-shapes." (Such junctions should, of instance, be important for recognition of box-like 3-dimensional objects.)

37 Cortex is relatively homogeneous only in eucortex (eu ~ normal), not in allocortex (allo ~ else, (an)other), such as the hippocampus. Further difference is that the sensory *eu*cortex has *hypergranular* 4th layer, but the motor *eu*cortex is without the 4th layer. (Pribram, personal commun.).

38 Ebdon (1993, p. 380) summarizes findings on the *V1 intrinsic circuitry*, together with its looped connections with other structures, which is supposed to be replicated also in other cortical areas:

"The two main sources of input are feedforward connections (from the thalamus, the claustrum, and from other cortical areas lower in the hierarchy [all these three parts also get V1's feedbacks, as Ebdon says later (note by M.P.)] which terminate in the middle layers, and feedback connections (from the limbic system, from areas higher in the hierarchy, and from certain thalamic sources) which terminate in layer 1 and layer 6. Information is processed in layers 2-4 by circuitry consisting of excitatory (collaterals of *piramidal* and spiny stellate neurons) and inhibitory (collaterals of other nonpyramidal cell types) components. [...]"

39 Plasticity refers to flexibility, modifiability and adaptability of the tissue (e.g., synapses) in response to stimuli.

40 Its details are modeled in Körner *et al.* (1999) and Burnod (1990).

41 Saccades may be voluntary or a part of several reflexes (Berne & Levy, 1993).

42 Pulvinar should be responsible for engaging attention, but parietal cortex for disengaging it, after Olshausen & Koch in Arbib (1995) who cite a work by M.I. Posner & S.E. Petersen (1990): *Annu. Rev. Neurosci.* **13**, 25-42.

43 R. Desimone & C.G. Gross (1979): *Brain Res.* 178, 363-380.

44 Lower-order cortical areas may be modulated by higher order-areas. So, V1 neurons are indirectly influenced by attention. V2, getting feedbacks from higher visual areas, acts with strong feedback further on V1 and even LGN (Aine, Supek & George, 1995 – after: M. Mignard & J.G. Malpeli (1991): *Science***251**, 1249-1250).

45 This is reported also in: S. Luck *et al.* (1997): *J. Neurophysiol.* **77**, 124-142. In V1, "complex attentional window is consistent with object-based attention, but is inconsistent with a simple attentional spotlight," is concluded after fMRI experiments by: D. Somers *et al.* (1999): *Proc. Natl. Acad. Sci. USA***96**, 1663-1668.

46 Memory-storage is realized by a learning procedure which I call *phase-Hebbian* and introduce in sec. 3.5 and Auxiliary Chapters. It combines the Hebb correlation, or convolution, with phase-coupling for associative memory.

47 A distributed pattern of activities of an assembly of neural units acts as an *attractor* if it represents an image with a meaning. Attractor means here that other neuronal configurations, which are within the *basin of attraction* (of the pattern acting as attractor), transform into the attracting pattern. See Auxil. Sec. 6.8 and Peruš (1995a) for details.

48 This is well summarized in the abstract of: A. Li & Q. Zaidi (2000): *Vision Res.***40**, 217-242:

"By measuring observers' abilities to estimate the relative 3-dim[ensional] curvature along a textured surface from 2-dim. projected images, and concurrently examining the local spectral changes occurring in the projected image for various texture patterns, we have found that correlated changes in oriented energy along lines corresponding to the lines of maximum and minimum curvature of the surface are crucial for conveying 3-dim. shape of the surface. [...] Our results show that: (1) It is neither necessary nor sufficient to identify individual texture elements or texture gradients in order to extract the shape of the surface. (2) 1-dim. frequency modulation is insufficient for conveying complex 3-dim. shape. (3) Veridical ordinal depth is seen only when the projected pattern contains changes in oriented energy along lines corresponding to projected lines of maximum curvature of the surface. (4) For a surface corrugated in depth about the horizontal axis, this pattern of oriented energy arises from energy along the vertical direction in the global Fourier transform in the pre-corrugated pattern. (5) Local orientation changes along lines of minimum curvature can also be critical for conveying shape. (6) These correlated orientation changes along lines of maximum and minimum curvature are entirely lost in parallel projection. Hence, texture can be a useful cue for shape if the image is a perspective projection. (7) Only some natural textures will provide sufficient monocular cues to support veridical shape inferences, and this can be predicted from their global Fourier transforms."

49 Even line-drawings appear to incorporate depth information, although this is a result of cognitive "top-down" influence based on memory. This is evident especially in ambiguous drawings like the Necker cube where our visual apparatus oscillates between two 3-dimensional interpretations of a purely 2-dimensional drawing (Any 2-dimensional projection of a 3-dimensional object can be written as a linear combination of 2-dimensional views, after Ullman and Basri.) (Wallis & Bülthoff, 1999).

50 Some "optical illusions" arise because of gestalt formation (e.g., a slightly interrupted line is sometimes perceived as continuous; elements get grouped together into a non-existent virtual "object").

51 This is also the conclusion from: G.N. Elston & M.G. Rosa (1998): *Cereb. Cortex***8**, 3278-3294. Increase in number and density of dendritic spines also increases response-complexity, due to this morphological study.

52 I.e., a pattern of neural-units' activities, which represents an environmental image, is an attractor. See Auxil. Sec. 6.8.

53 This distinction has long been emphasized by Pribram, but is now shared by the main-stream neuropsychology. The same happened with the view- and object-centered representations (see next).

54 The holonomic and main-stream theories nowadays all consider these various relative representations (Kosslyn & Andersen, 1992).

55 Allo- refers to outer. Originally, the Greek word "allos" means: else, (an)other, different, outside our set.

56 Detailed correction mechanism, called "corollary discharge", is described by Jeannerod in Arbib (1995). The complicated location-difference-driven genera-

tion of eye saccades by brainstem and superior colliculus is described also in Van Gisbergen & Van Opstal in Arbib (1995). Additionally to eye movements, the *head* movements also disrupt the gaze. The vestibulo-ocular reflex, among others, preserves stable retinal image during rapid head rotations by producing compensatory eye movements in the opposite direction and as much as the head has moved (Anastasio; Mazzoni & Zipser in Arbib, 1995).

57 Texture provides information about 3-dimensional shape, because the projected size and distances between texture-elements increase when observer approaches, and because the projected shape of texture-elements gets "deformed" or more compressed at the nearer-to-us side if they are viewed less frontally, i.e. from an increasing angle.

58 This model, presented a few paragraphs later, is compatible with the holonomic theory.

59 As we move the eye, the location of target-object's image on retina changes while the target's absolute location outside our body remains the same. In order to reach the target by hand, for instance, the visual locations (after the eye-move) should be re-coded by the brain. Such transformation of the retinal coordinates of visual targets to target's coordinates with respect to the body (which are necessary for steering the hand toward the target) was studied by a backpropagation-simulation. The backpropagation-net automatically "computed" the coordinate-transformation by effectively multiplying the response profile of the retinal receptive field by the function describing eye-positions. These outputs of artificial neurons were compared with activities of neurons in the area 7a of posterior parietal cortex which should manifest such coordinate-transformations in human vision.

The conclusion of Zipser & Andersen (1988, p. 683), which is important also for present reproductions of V1 receptive fields with the *ICA-algorithm (which lacks direct biological plausiblity as much as backpropagation),* is the following: "The striking similarity between model and experimental data certainly support the conjecture that the cortex and the network generated by back-propagation compute in similar ways. As back-propagation generates optimal solutions which produce the least error, these results also suggest that the brain chooses optimal solutions with respect to error. [...] Presently the back-propagation paradigm is structured at a level higher than implementation, and obviously cannot be applied literally to the brain, because the information does not flow backwards rapidly through axons. That the back-propagation method appears to discover the same algorithm that is used by the brain in no way implies that back-propagation is actually used by the brain. One approach to understanding the physiological significance of these results is to generate models incorporating features found in the brain, such as Hebbian-like learning at synapses and reciprocal back-projection pathways for propagating error, and determine whether these models generate similar results. In this regard it is interesting to note that all cortico-cortical and thalamocortical connections have reciprocal feedback pathways." In ten years, back-propagation has not got much more biological support in the sense of direct neural implementation, although models of the so-called *antidromic* and the *retrograde chemical-synapse-mediated* (Gardner, 1993, pp. 37-43 & 58-68) *back-signaling*, and the model of *nanoscopic back-propagation of error along microtubules* (ch. 12 by Dayhoff *et al.* in Pribram, 1993) have been proposed. Nevertheless, global processing algorithms like back-propagation and ICA remain, due to astronomical complexity of connection-structures, important in computational models of large-scale cortical processing and even get serious experimental support (as in the last example-case).

60 S. Grossberg & E. Mingolla (1985): *Percept.&Psychophys.* **38**, 141-171; summarized in Stillings *et al.* (1995).

61 Lie transformations *A,B,C*, forming a group (e.g., rotation group), satisfy the following rules: There is a transformation among them which does not change anything (zero rotation); each transformation can be inversed (rotation by the same angle in the opposite direction); a succession of two (or more) transformations is equivalent to a third transformation of the *same* group; and: $A(BC)=(AB)C$. (An introductory presentation to Lie groups for modeling geometry-based object perception see in: Dodwell, 1983).

62 Pribram (1991, Lecture 5) incorporated Lie groups into the holonomic theory of object perception (which is treated in this book in sec. 2.4 only). He did it by extending the group theory (of the Hoffman or Palmer type) from the space-time domain to the spectral domain (i.e., to the phase-space, or Hilbert-space, respectively). It is remarkable that Lie groups are very important also in quantum theory, and there seems to be at least an indirect connection to our problem (for details see AuxLit 16). In particular, Lie groups and related mathematical structures are used for quantum physics of spins and analogous quantities of subatomic particles (e.g., angular momentum, isospin).

63 Size-constancy is related to fractal-like scale-invariance which is a characteristic of such Lie groups.

64 Detailed presentation of the mathematical theory of Lie groups as applied in perception, and of Lie directional derivatives, which quantify the slopes of edge-lines, can be found in: Hoffman, 1966, 1968, 1970; Caelli, 1976; Dodwell, 1983; Luccio, 1993; Tsao *et al.*, 1991.

65 Not to mention invariances in more complex transformations.

66 Doubt that stimulus-related oscillations in range of 30-70 Hz serve a function in monkey, as opposed to cat, is expressed in experimental work by: M. Young, K. Tanaka & S. Yamane (1992): *J. Neurophysiol.* **67**, 1464-1474.

67 This is the result by: M.J. Farah (1984): *Cognition* **18**, 245-272. It is interesting to add that, according to blood flow experiments, thinking as purely-abstract mental process never activated association areas involved in perceptual processes of various types. Only other different task-dependent cortical areas, e.g. superior prefrontal cortex, were used – as reported in: P.E. Roland & L. Friberg (1985): *J. Neurophysiol.* **53**, 1219-1243.

68 Depolarization/hyperpolarization happens when the potential difference (voltage) across the neuron's membrane changes from its resting potential (-60 to -70 mV) to *less/more* negative values.

69 Up to now it was observed that a single synapse can be either excitatory or inhibitory, but not both. Some recent indications for the cases of a single synapse being both excitatory and inhibitory, depending on its postsynaptic receptors or the amount of neurotransmitter released, were reported in: Y.-W. Jo & R. Schlichter (1999): *Nature Neurosci.* **2**, 241-245. This would support the Hebbian paradigm and Hopfield-based ANN.

70 Summation is usually assumed, but recently multiplication of signals is also taken into account.

71 See, for example: Brown & Chattarji (especially p. 456) and Fregnac (especially p. 460) in Arbib (1995).

72 Pribram emphasizes that the "learning" paradigm which is today widely known as "Hebbian" has several predecessors, e.g. Freud, Pavlov, Lashley, and that Hebb merely expressed their observations in such a way that they became a cornerstone of memory-storage models (in ANN-language: "learning models"). Although they are unjustified, I cannot afford replacing deeply-rooted terms like "Hebbian" or "Hebb rule" with more appropriate ones which would hardly avoid new confusions. Abbot & Nelson (2000, p. 1178) write: "Over time Hebbian plasticity has come to mean any long-lasting form of synaptic modification (strengthening or weakening) that is synapse specific and depends on correlations between pre- and postsynaptic firing."

73 Hebbian LTP is experimentally well supported for hippocampus and cerebellum (Gardner, 1993, espec. p. 86; *et al.*), but for neocortex very poorly. Some evidence for Hebbian learning and memory (or LTP, respectively), or for their generalizations or analogs, in the visual cortex has been reported in: M.J. Friedlander *et al.* (1993): *Progress Brain Res.* 95, 207-223 (Hebbian); A. Artola *et al.* (1990): *Nature* 347, 69-72 & *Euro. J. Neurosci.* 2, 254-269; J.M. Greuel *et al.* (1988): *Science* 242, 74-77 (Hebbian-like, with thresholds).

74 The last option is the most recent one and experimentally supported, from hippocampus, by: J.M. Bekkers & C.F. Stevens (1990): *Nature* 346, 724-729. (It is also advocated in: Gardner, 1993, p. 54, and in: Artun, Shouval & Cooper, 1998.) For complicated details on mechanisms of synaptic modifiability see Abbot & Nelson (2000).

75 Experimental support for a generalization of the Hebb rule which incorporates LTD in the form of covariance was proposed by G.S. Stent (1973): *Proc. Natl. Acad. Sci. USA* 70, 997-1001, and by P.K. Stanton & T.J. Sejnowski (1989): *Nature* 339, 531-533, and in Zornetzer, Davis & Lau (1990, ch. 19), for hippocampus. For cortico-striatal synapses, LTD was indicated by: D. Centonze *et al.* (1999): *Brain Res. Rev.* 31, 1-5. See also Gardner (1993, p. 94) and Koch & Segev (2000).

76 Such configurations, which are within so-called basin of attraction, form a *potential well* having the *pattern–attractor at its* (well's) *minumum*. Thus, an attractor has an explicit manifestation (the special configuration of neuronal activities) and a virtual manifestation, i.e. the "region of its contextual influence" – the mental state.

77 It has been believed for a long time that, roughly, in general, the location of *memory* of (visual) perceptual representations coincides with the location of (visual) perceptual *processing*, i.e. the (visual) neocortex, at least at the first, uni-modal level (before numerous associations with other modalities are made) (Bear, 1996).

78 This is similar to statements by: R. Roland & B. Gulyas (1995): *Cereb. Cortex* **5**, 179-193. They found that recognition and learning of the same patterns activated identical visual areas, and also recall from memory activated a subset of the same areas. However, these tasks activated different *extra*visual regions afterwards.

79 Experiment by: S. Funashi, C.J. Bruce & P.S. Goldman-Rakić (1989): *J. Neurophysiol.* **61**, 331-349. [See also Riesenhuber & Poggio (2000, p. 1202). Area TE of ITC is suggested for visual memory by: M. Mishkin (1982): *Phil. Transac. Royal Soc. London B* **298**, 85-95. As mentioned in Desimone (1996), complex behavioral events, episodes (and also associated scenes and imagined maps) are usually (visually) memorized in MT lobe or with help of hippocampus.]

80 According to: P.S. Goldman-Rakić (1996): *Proc. Natl. Acad. Sci. USA* **93**, 13473-13480.

81 We can neglect other mentioned Hebbian examples, like cerebellum, which are irrelevant for visual memory.

82 See Stillings *et al.* (1995, p. 473) where one can find the summary of experiments on contrast selectivity, from: C. Blakemore & F. Campbell (1969): *J. Physiol.* **203**, 237-260. See also R.L. & K.K. De Valois (1990).

83 Experimental evidence on limitations of feature-detector analysis is extensively reviewed in: Pribram (1971, pp. 132-138) and Pribram (1991, pp. 79-81). Experiments showing multiple rather than specialized propensities of neurons in V1 are reported in Pribram, Lassonde & Ptito (1981).

84 Edge-sharpness is not represented by response amplitude of individual neurons, but by the spatial distribution of excitatory peaks across the multi-neuronal representation of the cortical retinotopic map, is concluded, for example, in experimental study by: J. Siguenza, W. Heide & O. Creutzfeldt (1987): *Vision Res*. **27**, 701-717.

85 Swindale (1996, p. 168) writes in his comprehensive review of computational models of ocular-dominance columns in V1: "The possibility that they [such columns] may not have a function at all should perhaps be taken seriously. Modelling results [...] show that mechanisms, whose main purpose is to ensure corresponding topographic mappings from two retinas onto a single sheet of cells, usually produce ocular dominance stripes."

86 For instance, addition of opposite-eye anticorrelations increases the monocularity, addition of anticorrelations within each eye decreases it. Details in Miller, Keller & Stryker (1989), and also on p. 486 of Arbib (1995).

87 Authors based their arguments for biological plausibility of Hebbian learning, which they used for simulation of orientation-preference columns (and ocular-dominance columns), on these experimental works on visual cortex: M. Constantine-Paton, H. Cline & E. Debsky (1990): *Annu. Rev. Neurosci.* 13, 129-154; J. Rauschecker (1991): *Physiol. Rev*. 71, 587-615; M. Friedlander, Y. Fregnac & J. Burke (1993): *Progress in Brain Res*. 95, 207-223; H. Reiter & M. Stryker (1988): *Proc. Natl. Acad. Sci. USA* 85, 3623-3627; J. Gruel, H. Luhmann, W. Singer (1988): *Science* 242, 74-77; Y. Fregnac *et al.*: *Physiol. Rev*. 64 (1984) 325-434, *Nature* 333 (1988) 367-370, *J. Neurophysiol.* 71 (1994) 1403-1421. Which of the numerous variants of Hebbian-like plasticity should be relevant for visual learning and topography formation was rarely specified in these studies, listed in (Swindale, 1996) and in T. Brown, Kairiss & C. Keenan (1990): *Annu. Rev. Neurosci.* 13, 475-511.

88 It is mentioned by Artun, Shouval & Cooper (1998) after Ferster *et al.* that even non-interacting neurons are as orientation-selective (because of internal mechanisms?) as interacting ones.

89 Specifically: left–right, lower–upper, inner-most – outer-most.

90 Geometric details on the retina-to-V1 *log(arithm)-polar* transformation of representation are given in Buser & Imbert (1992), in Davis, Zornetzer & Lau (1990; pp. 54-56) and in MacLennan (in Pribram, 1993, ch. 7, sec. 2.2).

91 I.e., apparent change of location of an observed object because of observer's movement.

92 In the latter case, angle-differences between preferred stimulus-orientation in the left and the right eye are meant. There are indications that absolute binocular disparity is manifested in V1, but relative disparities between two or more features or targets are typically processed in higher visual areas.

93 Similarly, DeValois & Jacobs (1968) define saturation of a light as the measure how much the light is chromatic in contrast to achromatic one.

94 These "definitions" were based mainly on color-correlates in the activity of retinal *cone*-cells. We can later trace deeper physiological correlates of color in LGN- and cortical (assemblies of) cells.

95 E.g., response of such a neuron (in V1), determined actually by its receptive field (!), is increased to red and decreased to green. The other pair of complementary response is yellow/blue. Color-opponent cells are thus of the following types: 1. red excitatory & green inhibitory (+R–G); 2. green excitatory & red inhibitory (+G–R); 3. yellow excitatory & blue inhibitory (+Y–B); 4. Blue excitatory & yellow inhibitory (+B–Y). (Cf., Figure 6.4.)

96 The receptive field of such a neuron (in V4 or sometimes V1) has a central excitatory region with inhibitory flanks responding in opposite manner to a color and to its opponent color (*ibid.*).

97 See Figures 31-5, 31-6 & 31-8 in Kandel (1991) for illustrations of various concentric receptive fields where their internal circles excite / inhibit stimuli-components of a specific basis-color (e.g., red), but their surrounds (vice versa!) inhibit / excite stimuli of the color which is opponent (complementary) to the first one (e.g., green is opponent to red, and yellow to blue).

98 I.e., the color-experience – which is, however, not confined to any of the mentioned cell-groups, I would say.

99 Pribram (1991, p. 104) notes: "For changes [in color] to be perceived, the color must remain invariant within boundaries [between figure and ground]. But color constancy is dependent on the entire extent of the perceived space. There are parallels in the changes of organization of this processing space for color and for shape from retina to striate-peristriate and prestriate systems." (Cf., Grossberg)

100 The following statements are my personal view.

101 PET is positron emission tomography; fMRI is functional magnetic resonance imaging; EEG is electroencephalography; MEG is magnetoencephalography. In our context, brain-imaging techniques based on quantum holography, as in Schempp (1997), are especially interesting.

102 Functionalism, often present in cognitive science of last decades, was satisfied with computer-supported models of information processing independently of implementation in brain tissue. We should now try to go beyond this early approximation.

103 What the best model is, of course, very relative (to other available models) and tentative (a new model can defeat it). Even best models are never perfect. They can never be equivalent to reality. But they are useful.

Chapter 7
Comparison of the Mathematical Formalism of Associative ANN and Quantum Theory

Important note. From now on, in all the following Auxiliary Chapters and Appendices, all terms will be used exclusively in the context of ANN theory, i.e. the computational model, not in the context of neuroscience. Terms like "cell" or "neuron" could rather be understood formally (as in Perus, 2000a), i.e. as "formal neurons" or as processing "units" of connectionists, or could be identified (e.g., cardinal units) with whole cortical columns (For the purpose of model simplification).

7.1 MAIN NEURO-QUANTUM ANALOGIES AND THEIR INFORMATIONAL SIGNIFICANCE

It is best to introduce quantum associative nets by first presenting parallels between quantum processes and neural-net processes. Many mathematical analogies of the

DOI: 10.4018/978-1-61520-785-5.ch007

theory of Hopfield- like associative ANN and the quantum theory can be found. Because we know that our

ANN simulations perform well (e.g., Perus, 2002, and refs. within), we can infer, using the correspondence list below, about similar effective information-processing capabilities on the quantum level also. Here is an overview of neuro-quantum analogies:

1. Neuronal State-Vector ↔ Quantum Wave-Function:

In artificial neural network (ANN) theory, the state of the system of neurons[1] is described by $q(\vec{r},t)$ which denotes the activity of an individual neuron (located at $\vec{r}$) at time t. Neuronal patterns are special neuronal configurations $\vec{q}$ which represent some meaningful information. In quantum theory, the state of the quantum system at location r and time t is described by the wave-function $\Psi(\vec{r},t)$ (Bohm, 1954; Messiah, 1965). They both represent a state vector describing a parallel-distributed configuration of a complex system.

2. Neuronal State is a Superposition of Neuronal Patterns ↔ Quantum Wave-Function is a Superposition of Quantum Eigen-Wave-Functions:

A neuronal configuration q may be described as a linear combination of neuronal patterns $\vec{v}_k$ ($k = 1,..., P$). P is the number of patterns represented simultaneously in the combination. Similarly, a wave-function Ψ can be described as a linear combination of eigen-states Ψ_k ("quantum patterns"):

$$q(\vec{r},t) = \sum_{k=1}^{P} c_k(t) v_k(\vec{r}) \tag{7.1}$$

$$\Psi(\vec{r},t) = \sum_{k=1}^{P} C_k(t) \psi_k(\vec{r}) \tag{7.2}$$

Neuronal patterns and eigen-wave-functions can represent some object of conscious experience on different levels. So, as opposed to other configurations, they represent informational states which have a meaning, because they are correlated with some environmental objects.

3. Both sets of pattern-vectors, $\vec{v}_k$ and Ψ_k, usually have the properties of Orthogonality (mutual scalar product is zero) and Normality (their "length"

is 1). The first property means that patterns do not have"anything in common".

If this is satisfied, it ensures error-free recall of patterns, because it excludes overlap and mixing between patterns or corruption of a pattern because of this. So, the interference between patterns must be destroyed during recall in order to get one of them out properly, and this is realizable if patterns are orthogonal. However, interference must be eliminated during pattern-recall only; in associative memory, interference between patterns is essential. If eigenstates $\vec{v}_k$ and Ψ_k are not orthogonal completely, recall is not so pure, but associations between patterns may become more effective (cf., ch. 8.5).

(Quasi) orthogonality is a comparative property of biologically-processed images as far as their representations have high dimensionality (a lot of encoding units), or if their coding is sparse (Baird, 1990). This is often the case, unless very similar images are observed, which get anyway "melted together into" the same attractor.

4. Coefficients of the Series: Synergetic Order Parameters ↔ Quantum Probability Coefficients:

C_k are the quantum probability coefficients, and c_k are the neural or synergetic order parameters. They are sometimes represented by cardinal neurons or oligarchies of neurons with selective response. In the linear combination, each pattern with index k is represented by a corresponding coefficient. The coefficients c_k / C_k describe how much a specific pattern is represented in the system's actual state $\vec{q}$ / Ψ. Namely, c_k / C_k is scalar product, denoted by angle-brackets in equations (7.3 /7. 4), of $\vec{v}^k$ / ψ_k and $\vec{q}$ / Ψ. In another interpretation, c_k is the probability for the corresponding pattern $\vec{v}^k$ to be recalled (reconstructed, or explicated from the superposition of many patterns). Thus, the time-dependent coefficients encode quantitatively the "meaning" (associative influence)of their patterns. They describe how strong role a given pattern has in contextual system-dynamics (Haken, 1991).

$$c_k(t) = \langle \vec{v}_k, \vec{q} \rangle = \int \int v_k(\vec{r})\, {}^*q(\vec{r}, t) d\vec{r} dt = \tag{7.3}$$

$$C_k(t) = \langle \psi_k, \Psi \rangle = \int \int \psi_k(\vec{r}) * \Psi(\vec{r}, t) d\vec{r} dt \tag{7.4}$$

Asterisk denotes the operation of complex conjugation (i.e., imaginary component for the complex-valued variable is replaced by its negative). If variables $\vec{v}_k$ or

ψ_k are real, we may delete the asterisk. Mathematically, coefficients c_k and C_k describe projections (in terms of scalar products $\langle.,.\rangle$, in Hilbert space, for example).

5. Spatio-Temporal Integration of Neuronal Signals ↔ Feynman's Version of the Schrodinger Equation:

Equation (7.5) describes activity change of a neuron. It is effected by a spatio-temporal integration (summation) of signals from all other neurons which are connected to that neuron. Namely, it is argued by Barlow (1996) that neurons integrate along spatio-temporal trajectories, not separately in space and time. So, the state of a neuron at position $\vec{r}_2$ and time t_2 is given by J -weighted summation of all signals and the whole history of signals:[2]

$$q(\vec{r}_2,t_2)=\int\int J(\vec{r}_1,t_1,\vec{r}_2,t_2)q(\vec{r}_1,t_1)d\vec{r}_1dt_1 \tag{7.5}$$

where $J(\vec{r}_1,t_1,\vec{r}_2,t_2)$ is the "strength" of an individual synaptic connection. Each J is transmission-rate of a real synaptic connection between two neurons (spatially separated, but realized at the same time) or correlation between states of different neurons at different times, represented at least virtually.

In non-relativistic quantum mechanics, the dynamical equation for the quantum state Ψ is the Schrodinger Equation (7.6). Here it is written, analogously to the neural Equation (7.5), in Feynman's form (Hiley & Peat, 1987):

$$\Psi(\vec{r}_2,t_2)=\int\int G(\vec{r}_1,t_1,\vec{r}_2,t_2)\Psi(\vec{r}_1,t_1)d\vec{r}_1dt_1 \tag{7.6}$$

$G(\vec{r}_1,t_1,\vec{r}_2,t_2)$ constitutes the Green function or an element of the quantum propagator (Bjorken & Drell, 1964, ch. 6). The propagator **G** is a matrix which describes a parallel-distributed transformation of the whole system from an initial state $\Psi(\vec{r}_1,t_1)$ to the final state $\Psi(\vec{r}_2,t_2)$. It essentially manifests a sort of Huygens' principle (Clement et al., 1999). The system transforms itself into a new state by exhibiting numerous internal interactions between its constitutive "quantum points" (some mathematical "basic elements" of the system). Informationally, transformation (7.5) or (7.6) is association. One pattern is transformed into another pattern by a collective informational interaction of input data (encoded in the state vector, i.e. Ψ) with memory data (encoded in the system of connections or interactions described by the Green matrix G) (Slechta, 1993).

6. Memory in Auto-Correlations of Patterns: Synaptic Strengths Obeying the Hebb Rule ↔ Green Functions:

The so-called kernels ("weights"), *J* or *G*, of dynamic Equations (7.5) and (7.6) are given as a sum of auto-correlation functions. The transmission-rate of an individual synaptic connection *J* is determined by the Hebb learning rule (Hopfield, 1982) as a sum of auto-correlations between its two neurons participating in various patterns $\vec{v}_k$:

$$J(\vec{r}_1,t_1,\vec{r}_2,t_2)=\sum_{k=1}^{P} v_k(\vec{r}_1,t_1)v_k(\vec{r}_2,t_2) \text{ or } J(\vec{r}_1,\vec{r}_2)=\sum_{k=1}^{P} v_k(\vec{r}_1)v_k(\vec{r}_2) \tag{7.7}$$

Similarly, the so-called quantum propagator *G*, which is related to the usually-used *Green function G* (Bjorken & Drell, 1964, ch. 6) by G = - $\tilde{G}$, is given as a sum of auto-correlations of individual quantum eigenstates ψ_k:

$$G(\vec{r}_1,t_1,\vec{r}_2,t_2)=\sum_{k=1}^{P} \psi_k(\vec{r}_1,t_1)^* \psi_k(\vec{r}_2,t_2) \text{ or } G(\vec{r}_1,\vec{r}_2)=\sum_{k=1}^{P} \psi_k(\vec{r}_1)^* \psi_k(\vec{r}_2) \tag{7.8}$$

These two expressions determine a (neural or quantum) memory element. Both matrices made of such elements, i.e. **J** or **G**, may represent content-addressable associative memory if they are, of course, coupled with Equations (7.5) or (7.6). The analogy above ensures that quantum associative processing can be at least as effective as those by associative ANN. Processing is equivalent, although implemented on different physical levels, if quantum eigenwave-functions would be real-valued only. However, they are usually complex-valued. The wave-function Ψ is a superposition of *P* eigen-wave-functions ψ_k which can represent plane waves $\psi_k(\vec{r},t)=A_k(\vec{r},t)\exp(\frac{i}{h}\varphi_k(\vec{r},t))$ or wavelets (Lee, 1996; Schempp, 1995).[3] A is the amplitude, and φ is the phase at specific location $\vec{r}$ in specific time t.

Because phase-information was found essential in quantum computing[4] in the newest ANN models (e.g., ICA) and in neuroscience (Pribram, 1991; Jibu et al., 1995, 1996, 1997), we want to use quantum phase differences for an associative memory. This significantly improves conventional Hebbian performance.[5]

To express phases φ and amplitudes A explicitly, we insert $A_k(\vec{r},t)e^{i\varphi_k(\vec{r},t)}$ instead of $\psi_k(\vec{r},t)$ into Equation (7.8). We obtain

$$G(\vec{r}_1,t_1,\vec{r}_2,t_2)=\sum_{k=1}^{P}A_k(\vec{r}_1,t_1)A_k(\vec{r}_2,t_2)e^{i(\varphi_k(\vec{r}_2,t_2)-\varphi_k(\vec{r}_1,t_1))} \tag{7.9}$$

This describes the quantum memory encoding which is two-fold: it is in amplitude-correlations $\sum_{k=1}^{P}A_k(\vec{r}_1,t_1)A_k(\vec{r}_2,t_2)$ (Hebb rule) and in phase-differences $\delta\varphi_k=\varphi_k(\vec{r}_2,t_2)-\varphi_k(\vec{r}_1,t_1)$. With exploitation of quantum phases, the quantum associative nets go beyond the Hopfield-like ANN models (Perus, 2000b).

7. Adaptive Memory in Cross-Correlations: Synaptic Strengths Obeying the Generalized Hebb Rule or Delta Rule ↔ Density Matrices:

The kernels J and G of dynamic Equations (7.5) and (7.6) can be written also as sums of *cross-correlation* functions. The generalized Hebb learning rule using cross-correlations of neuronal patterns is given as follows.

$$J(\vec{r}_1,t_1,\vec{r}_2,t_2)=\sum_{k=1}^{P}\sum_{h=1}^{Q}\lambda_{kh}v_k(\vec{r}_1,t_1)v_h(\vec{r}_2,t_2) \text{ or}$$

$$J(\vec{r}_1,\vec{r}_2)=\sum_{k=1}^{P}\sum_{h=1}^{Q}\lambda_{kh}v_k(\vec{r}_1)v_h(\vec{r}_2) \tag{7.10}$$

In the previous analogy-case (no. 6), strengths of individual synaptic connections were determined by a sum of auto-correlations (Hebb rule) of patterns, i.e. activity-correlations of connected neurons which have a role in the same patterns. In this case (no. 7), synaptic strengths are given by a sum of cross-correlations (generalized Hebb rule) of patterns, i.e. activity-correlations of neurons which have a role in different patterns. If one of these patterns is a difference-pattern (difference between the first pattern and a reference or average pattern), then this generalized Hebb rule is called delta rule.[6] So, in the first case (no. 6), a synapse represents correlation of neurons constituting the same pattern, therefore we talk about pattern-auto-correlation, i.e. correlation of parts of the same pattern. In the second case (no. 7), a synapse represents correlation of neurons constituting two different patterns, therefore we speak about cross-correlation or about correlation between parts of two different patterns.

ρ denotes a central quantity describing the internal (in Bohm's terms: implicate) state of a quantum system: the quantum probability-density matrix or quantum statistical operator, respectively. Similarly to the present case (no. 7) of neuronal cross-correlation, the density matrix ρ is given as a sum of cross-correlations of individual quantum eigenstates ψ_k and ψ_h:

$$\rho(\vec{r}_1, t_1, \vec{r}_2, t_2) = \sum_{k=1}^{P}\sum_{h=1}^{Q} \rho_{kh} \psi_k(\vec{r}_1, t_1) * \psi_k(\vec{r}_2, t_2) \text{ or}$$

$$\rho(\vec{r}_1, \vec{r}_2) = \sum_{k=1}^{P}\sum_{h=1}^{Q} \rho_{kh} \psi_k(\vec{r}_1) * \psi_k(\vec{r}_2) \tag{7.11}$$

If explicit expression of phase φ is done by inserting $\psi = A^{i\varphi}$, we obtain

$$\rho(\vec{r}_1, t_1, \vec{r}_2, t_2) = \sum_{k=1}^{P} A_k(\vec{r}_1, t_1) A_h(\vec{r}_2, t_2) e^{i(\varphi_k(\vec{r}_2, t_2) - \varphi_k(\vec{r}_1, t_1))} \tag{7.12}$$

However, in principle, the density matrix, as in general(ized) expressions (7.23) and (7.24), is calculated at the same time-instant (i.e., $t_1 = t_2$; in general it may be time-dependent). Namely, it is a description of quantum statistics rather than of quantum dynamics which is described by the Green matrix **G**.

Comparing Equations (7.10) and (7.11), and knowing that the generalized Hebb rule for J realizes efficient associative memory, we may infer about information processing capabilities based on ρ. Although ρ is not directly involved in any dynamical equation in the same form as Equation (7.10) is involved in Equation (7.5), it can be interpreted as a descriptor of a quantum implicit memory (in contrast to G which is closer to a sort of "working memory").[7]

So-called diagonal parts of the density matrix $\rho(\vec{r}, \vec{r}) = \sum_{k=1}^{p} \rho_{kk} |\psi_k(\vec{r})|^2$ give a superposition of real-valued and thus measurable quantum probability densities. ρ_{kh} and λ_{kh} are generalized versions of the so-called learning rate λ_r (as in equation (7.28) and discussed in Haken (1991)).

Interaction scheme no. 6 realizes auto-association: completion of a neural pattern or quantum pattern ψ^k from partial information. On the other hand, interaction web no. 7 realizes hetero-association: association form one pattern $\vec{v}_k$ or ψ^k to another pattern $\vec{v}_k$ or ψ^h. Similar quantum information processing using ρ-decomposition was discussed by Ohya (1989) and Matsuoka (see their works in AuxLit 24).

8. Neuronal Pattern Reconstruction ↔ "Wave-Function Collapse":

The most important neuro-quantum analogy is the following. Pattern-reconstruction in a given ANN

$$q(\vec{r},t) = \sum_{k=1}^{P} c_k(t) v_k(\vec{r}) \Rightarrow q(\vec{r},t_0) = v_{k_0}(\vec{r}) \tag{7.13}$$

is very similar to the "collapse of the wave-function" in a quantum system

$$\Psi(\vec{r},t) = \sum_{k=1}^{P} C_k(t) \psi_k(\vec{r}) \Rightarrow \Psi(\vec{r},t_0) = \psi_{k_0}(\vec{r}) \tag{7.14}$$

Selection of a relevant pattern from a storage of many patterns is a very important feature of memory-based cognitive processes. Processes of Equations (7.13) and (7.14) are both a result of the influence from the system's environment. The environment selects those neural/quantum pattern which is the most similar (or is correlated) to the state of environment.

Neuronal pattern reconstruction and wave-function collapse are results of a transition from the implicate order (Bohm's term for latent, implicit, inactive, potential non-Shannon information) to the explicate order (manifest, active, realized Shannon information) (Bohm & Hiley, 1993). The implicate order represents a combination of very many possible states or processes (left-hand side equation in (7.13) and (7.14)). It is analogous to a set of so-called "parallel worlds" or parallel sub-branches of the general wave-function proposed by Everett (AuxLit 26). Explicate order, on the other hand, represents a state or process which is at a moment physically actualized (right-hand side equation in (7.13) and (7.14)). It is "chosen" from a set of potential (implicate) states, or is a result of their optimal "compromise". "Taking choice" may be an "act of conscious process" which takes some information out from memory or subconsciousness (implicate order), and makes it explicit and conscious.

9. Neural Duality and Quantum Duality:

An interesting neural analogy of the Heisenberg uncertainty principle of quantum mechanics is represented by an inability of simultaneous determination of patterns in the system of neurons $\vec{q}$ and of patterns in the system of interactions or connections **J**. We are unable to make a pattern explicit in a system of neurons, and to make a pattern explicit in the system of connections at the same time. Only one pattern, which is at the moment realized in the system of neurons, is made explicit. All the others are present only implicitly in the system of interactions, i.e. in the rates of synaptic coupling. We have a similar situation in cognition: we can be aware of one pattern only which has been extracted from memory. Other memory patterns remain unconscious and implicit.

The so-called position (x-) representation of quantum theory can be approximated by the system of neurons $\vec{q}$. The so-called impulse (p-) representation can, on the other hand, be associated with the system of interactions **J** which regulates all transformations of the network-state. Thus, the quantum duality (particles/waves, or particles/interactions, or fermions/bosons, respectively, has a counterpart in neural duality (neurons/synaptic con- nections, or neurons/interactions).

Sometimes we find that a network where neurons and synapses are exchanged has a similar behaviour as the original one. Thus, in general, there is a symmetry of neurons and synapses (or interactions) which represents invariance under neurons-synapses exchange. If we compare, for example, simulations of associative neural networks by Perus (1995a, 2002) and simulations by Menneer[8], we see that explicit processing, which is realized in the system of neurons in Perus's case, is similar to those explicit dynamics which is realized in the system of connections in some Menneer's cases. Menneer in some cases performs "collapse" (pattern selection and reconstruction) in the systems of connections, but Perus performs it in the system of neurons. Menneer also emphasizes analogies of this collective dynamics with processes during quantum measurement. We cannot extract information in the system of neurons and in the systems of synaptic connections at the same time – not in both systems. This is a neural version of uncertainty case. We have to make a "collapse" with output information represented either in the system of neurons or in the system of synapses. Without a "collapse" the interplay of neurons and connections is always an indivisible holistic process.

Like in classical dynamics, for integrable systems, there is a complete equivalence between the Schrodinger equation for Ψ and the Liouville-von Neumann equation for ρ.[9] Both kinds of description are complementary on quantum as well as classical (e.g., neural) level.

10. Quantum and Neural Wavelet Uncertainty Principles:

Some more formalized uncertainty relations, which are similar to the Heisenberg's quantum uncertainty principle, can be found in some ANN and early vision theories. Gabor proved (MacLennan, 1991) that the well-known Heisenberg uncertainty is a special case of Gabor's general frequency/time uncertainty principle: $\Delta f \Delta t \geq 1$. Error in measuring frequency Δf is inverse-proportional to the error in measuring time Δt. Multiplying both sides of Gabor's inequality by Planck constant h, and knowing $E = hf$, we obtain the well-known Heisenberg's version: $\Delta E \Delta t \geq h$. Gabor uncertainty principle $\Delta f \Delta t \geq 1$ is applicable whenever simultaneous measurements of an observable and of its Fourier transform are made (MacLennan, 1991). Previously mentioned x/p_x and E/t Fourier or uncertainty pairs of conjugate variables are thus similar to the *f/t* pair of Gabor's information theory and x/u_x pair of Daugman's vision theory. In the theory of Daugman (1985), spatial frequency u_x determines the wavelets in the frequency-domain like x determines their location in the spatial domain.

The Gabor inequality changes to equality $\Delta f \Delta t \geq 1$ (minimal uncertainty) if *neural wavelets* are modeled by complex oscillatory functions having Gaussian envelope. These are *Gabor elementary functions or wavelets:*

$$g(w,f;t) = \exp\left(-\frac{\pi(t-w)^2}{\sigma_t^2}\right)\exp(2\pi i f(t-w)) \tag{7.15}$$

They are similar to *quantum coherent states or wave-packets*. These wavelets, originally derived by Gabor to satisfy the minimum-uncertainty requirement, are functions in time-domain. (*w* corresponds to the time of the peak.) However, Gabor's mathematics could be used not only for functions of time *t*, but also spatial functions (MacLennan, 1991). It turned out later that spatial Gabor wavelets, i.e. where time *t* is replaced by space co ordinates x and y, best fit the receptive-field profiles of simple cells in V1. Equation (1.2), with the same notation, was such a 1-dimensional Gabor wavelet for vision modeling. His 2-dimensional relative is given here:

$$g(p,q;u_x,u_y;x,y) = \exp\left(-\frac{\pi(x-p)^2}{\sigma_x^2} + \frac{(y-q)^2}{\sigma_y^2}\right)\exp[2\pi i(u_x(x-p)+u_y(y-q))] \tag{7.16}$$

Such 2-dimensional spatial Gabor wavelets constitute basic information cells represented in space $(x,\ y)$ and spatial-frequency areas $(u_x,\ u_y)$.

An arbitrary function γ can be represented by Gabor wavelets $g(p,q;u_x,u_y;x,y)$ as their $\gamma = \sum_{j=-\infty}^{\infty}\sum_{k=-\infty}^{\infty} C_{jk} g_{jk}$ linear superposition, where $C_{jk} = \langle \gamma, g_{jk} \rangle / \| g_{jk} \|$ (if wavelets are orthogonal), analogously to the ANN-state expansion (1) with coefficients (3). Gabor wavelets are, however, in general not orthogonal.

Daugman (1985) presented psycho-physical analogues of Gabor uncertainty principle for two dimensions: $\Delta x \Delta u_x \geq \frac{1}{4\pi}$ and $\Delta y \Delta u_y \geq \frac{1}{4\pi}$, which give together $\Delta x \Delta y \Delta u_y \Delta u_x \geq \frac{1}{16\pi^2}$. MacLennan (1991) naturally added Gabor uncertainty principle for three dimensions: $\Delta x \Delta u_x \geq \frac{1}{4\pi}$, $\Delta y \Delta u_y \geq \frac{1}{4\pi}$, $\Delta t \Delta f \geq \frac{1}{4\pi}$

which give together $\Delta x \Delta y \Delta t \Delta u_y \Delta u_x \Delta f \geq \frac{1}{64\pi^3}$. In this 3-dimensional case, each information cell, representing a "quantum of information", is a 6-dimensional hypercube (Δx,Δy,Δt,Δu_y,Δu_x,Δf) localized to spatial area $\Delta x \Delta y$ and to temporal interval Δt, and furthermore tuned to spatial frequency bands of width Δu_x and Δu_y, and to (temporal) frequency band of width Δf.

Classical and quantum physics have more in common if the system is complex (i.e., many "units" strongly interact). It seems that presented uncertainty relations of the theory of visual cortical processing are as much as Heisenberg's quantum-mechanical original connected with informatics of parallel-distributed complex systems in general, even classical ones such as associative neural nets. Uncertainty relations arise as a result of observational perturbation of quantum systems, but it may emerge also in the case of classical (neural) complex systems. This observational perturbation has much more evident and exact consequence (such as conjugate pairs of non-commuting observable operators) in quantum complex systems, because of their microscopic scale, than in classical complex systems.

Bohm (1954, 1980) noted that if we are starting to think about our own thoughts, we change them with this self-observation act. This is uncertainty of conscious self-reflection which Bohm has related to quantum uncertainty.

7.2 DISCREPANCIES IN PRESENTED ANALOGIES AND BEYOND

Complex variables. There is an essential difference between neural and quantum equations: the latter ones implicitly include the imaginary unit i. It is hidden in complex-valued wave-functions Ψ, and in variables which incorporate them. It manifests also in operations like complex conjugation (connected with phase conjugation, i.e. process of phase reversal) denoted by asterisk.

This prevents us from claiming that quantum systems are a very complex sort of "neural networks". Quantum systems are much more subtle and flexible, because they do not have well-defined fundamental particles and interactions. So, no well defined basic formal "neurons" or formal "connections" can be found, and no well-defined "hidden variables" of Bohm (1980). Anyway, in spite of this difficulty, the mathematics of the usual Copenhagen interpretation of quantum mechanics is essentially very close to the mathematics of associative neural networks. So, the limits of our analogies are also limits of the orthodox (Copenhagen) interpretation of quantum theory.

The equations presented are typical representatives of physics-based associative neural net models and of quantum theory. The neural ones exploit "amplitude" information only (therefore they are real-valued); the quantum ones exploit phase-information also (therefore they are complex-valued). Future developments are expected to erode this difference and turn it into an artifact of early PDP-models.

Phase conjugation. Replacement of complex-valued quantum equations with a system of real-valued equations (Bohm, 1954; Bohm & Hiley, 1993) *or introduction of complex numbers into the neural network theory*, e.g. *for networks of coupled oscillators* (Auxil. Chapter 11; Baird, 1990; Haken, 1991; Mannion & Taylor, 1992; Sompolinsky & Tsodyks, 1994) and *quantum neural holography* (Schempp, 1993, 1994, 1995; Marcer & Schempp, 1997, 1998), would support the search for further analogies of neural and quantum nets. This would then lead to convergence of "neural" and "quantum" PDP-models.

Complex-number variables, usually suitable for description of wave phenomena, in quantum formalism obviously have a very deep meaning. Ψ^*, as opposed to Ψ, is connected with the *phase-conjugated* or time-reversed wave dynamics (cf., Cramer, 1986). So, in quantum systems we have a superposition of forward and backward time-evolution. In associative ANN, this time- flow-superposition is transformed into superposition of associative transformations in the opposite directions in space, or the superposition is completely collapsed into single directed dynamics (time sequence of patterns in the direction of the arrow of time). Observed quantum systems reflect an intrinsic complementary duality which is evident also in the wave/particle double nature, or in location and momentum representations. Probably this apparent duality, and uncertainty principles, are a consequence of an active observer's participation in the observed quantum process.

In spite of the fact that quantum systems are described by a dual complex-valued (real plus imaginary) description, but the usual ANN only in a real-valued description so far, we are able to argue that there are very interesting functional and mathematical analogies between quantum and neural levels, although they are not perfect. If, however, we have an associative network of coupled *neurons with oscillatory activities* (Baird, 1990; Haken, 1991; Mannion & Taylor, 1992), then complex-valued mathematical formalism becomes relevant for ANN also. Our presented neuro-quantum analogies, which in that case contain information about *phase-correlations*, then become more important. Indeed, following the holonomic theory of "infomax"-processing in V1, neural processing becomes even more quantum-like (ch. 3).

7.3 ADDITIONAL ANALOGIES

11. Quantum Neural Holography: Wavelet-Mixing or Convolution:

In associative ANN, memory consists *of (auto)correlations between patterns*. In quantum mechanics *phase differences between different parts* of the wave-function may have a similar role (Bohm, 1954). They control the time evolution of probability distribution involving interference of the contributions of different stationary eigen-wave-functions. In the concept of quantum neural holography by Schempp (1993, 1994, 1995), there are interference patterns of coherent neural wavelets. *Phase shift* $\tilde{\varphi}$ encodes the degree of correlation between components of coherent wavelets. On the other hand, their amplitude encodes the significance of synchronized wavelet-components. Indeed, wavelet mixing $\psi_k(t)dt \otimes \psi_h(t')dt'$, (mapped) spatio-temporal cross-correlation function $\int \exp(2\pi i\nu\tilde{\varphi}t)\psi_k(t+\Delta t)\psi_h(t)dt$ [10] (Marcer & Schempp, 1998), or *convolution*, etc., are examples of bilinear forms similar to the generalized Hebbian (i.e., "phase-Hebbian") learning rule used in ANN. So, quantum holography and theory of associative ANN (in particular if neurons have oscillatory activities) have significant mathematical analogies which are a consequence of similar collective dynamics in both complex systems.

In associative ANN we have "matrices of (synaptic) connections" which encode correlations of two neural patterns. In holography, on the other hand, we have holograms which encode correlations of two quantum-optical signals, i.e. their interference. In both theories we use similar mathematics for similar information processing as long as we can neglect details in the structure of basic elements of the network.

Elements of the system are, of course, different again. This brings differences between quantum holography and associative ANN, but it seems that brain realizes quantum- holography-like and ANN-like processing on different scales. So, in spite of differences in "anatomical" details, we may speak about *fractal-like multi-level synergetic processing* in brain. This is caused by similar "sociology" in networks on various biophysical levels, i.e. *quantum, sub-cellular, neural or virtual networks* or "holograms".

12. Relations of Phases in Quantum Holograms ↔ Strengths of Formal Synapses:

Changing the phase-relations between eigen-wave-functions (e.g., quantum-optical signals) is analogous to the *learning*-process in ANN where new pattern-correlations are added to the synaptic correlation-matrix. As mentioned, conventional ANN do not incorporate neurons with oscillating activities, although this is observed in brain. Only some more advanced neural-net models introduce oscillating neuronal activities and put relevance to phase-relations or phase-correlations between neurons with oscillating firing rates. Phase- correlations are an essential ingredient of (quantum) holography, therefore associative net-works with oscillating neuronal activities are certainly more similar to holographic devices than conventional associative ANN. If we study networks with oscillating neuronal activities, we find similar associative memory capabilities as in conventional associative ANN (Haken, 1991) and similar phase-coupling or frequency-coupling (coherent behaviour) as with wavelets in holography.

13. "Collapsed" Quantum Superposition ↔ Formal Neural Net-Work:

Quantum holography, however, introduces something more. It gives opportunity for introduction of complex conjugate form of a wavelet. Complex conjugate wavelet ψ^* is a time-mirrored version of the original wavelet ψ.

The well-known quantum bilinear form $\psi\psi^*$ represents a combination of a state vector (wave function) ψ with its phase-conjugated (time-mirrored) form ψ^*. Together they give real-valued probability density $|\psi|^2 = \rho$, in the case of a pure state. For "mixtures" a weighted sum of $\psi_k\psi^*_k$ has to be written (Wheeler & Zurek, 1983); in general this gives expression (7.11). Thus, in quantum world we have superpositions of many possible networks, processes and "arrows of time".

Only one network, process and "arrow of time" is, however, realized explicitly during measurement process, and this makes situation similar to the case of ANN. One could state roughly that a generalized (mathematical) ANN emerges as a result of "collapse" of a quantum superposition of implicit "networks". To say as simple and short as possible, a mathematical ANN (network of mathematical points or formal "neurons" which interact intensively) is a "collapsed" quantum system. With the "collapse", complex-valued (quantum) variables turn into real-valued (classical) ones (AuxLit 17). Brain's neural network evolves as a macroscopic replica of these fundamental dynamical principles, it appears.

Fourier-like correspondence rules between the above equations can be shown first for the neural case (7.17), and then for the quantum case (7.18), using the following calculation.

$$q(\vec{r}_2, t_2) = \int\int J(\vec{r}_1, t_1, \vec{r}_2, t_2) q(\vec{r}_1, t_1) d\vec{r}_1 dt_1 =$$

$$= \int\int [\sum_{k=1}^{P} \upsilon_k(\vec{r}_1, t_1) \upsilon_k(\vec{r}_2, t_2)] q(\vec{r}_1, t_1) d\vec{r}_1 dt_1 =$$

$$= \sum_{k=1}^{P} \upsilon_k(\vec{r}_2, t_2) [\int\int \upsilon_k(\vec{r}_1, t_1) q(\vec{r}_1, t_1) d\vec{r}_1 dt_1]$$

$$\Rightarrow q(\vec{r}, t) = \sum_{k=1}^{P} c_k(t) \upsilon_k(\vec{r}, t) \text{ or } q(\vec{r}, t) = \sum_{k=1}^{P} c_k(t) \upsilon_k(\vec{r}) \tag{7.17}$$

In the last step we have used definition (3) for time-dependent patterns. For the stationary case we delete the time integral.

$$\Psi(\vec{r}_2,t_2) = \int\int G(\vec{r}_1,t_1,\vec{r}_2,t_2)\Psi(\vec{r}_1,t_1)d\vec{r}_1 dt_1 =$$
$$= \int\int [\sum_{k=1}^{P} \psi_k(\vec{r}_1,t_1) * \psi_k(\vec{r}_2,t_2)]\Psi(\vec{r}_1,t_1)d\vec{r}_1 dt_1 =$$
$$= \sum_{k=1}^{P} \psi_k(\vec{r}_2,t_2)[\int\int \psi_k(\vec{r}_1,t_1) * \Psi(\vec{r}_1,t_1)d\vec{r}_1 dt_1]$$
$$\Rightarrow \Psi(\vec{r},t) = \sum_{k=1}^{P} C_k(t)\psi_k(\vec{r},t) \text{ or } \Psi(\vec{r},t) = \sum_{k=1}^{P} C_k(t)\psi_k(\vec{r}) \quad (7.18)$$

The following version of dynamic equations can be derived from neuro-synergetic Equations (7.5) and (7.7) which are here somewhat generalized. The so-called *attention parameter* is added after Haken (1991). Attention parameter λ_k of the pattern $\vec{v}_k$ is an eigenvalue of the matrix **J** with eigenvectors $\vec{v}_k : \mathbf{J}\vec{v}_k = \lambda_k\vec{v}_k$.(In quantum case, analogously, ψ_k are eigenstates of the density operator ρ with eigenvalues ρ_k).[11] We insert a generalized version of the Hebb rule (7.7) into the time-dependent Equation (7.5). In the second step we use definition (7.3) for time-dependent patterns. For stationary case we can delete the time integral.

$$\dot{q}(\vec{r}_2,t_2) = \int\int [\sum_{k=1}^{P} \lambda_k v_k(\vec{r}_1,t_1)v_k(\vec{r}_2,t_2)]q(\vec{r}_1,t_1)d\vec{r}_1 dt_1 =$$
$$= \sum_{k=1}^{P} \lambda_k v_k(\vec{r}_2,t_2)[\int\int v_k(\vec{r}_1,t_1)q(\vec{r}_1,t_1)d\vec{r}_1 dt_1] =$$
$$\Rightarrow \dot{q}(\vec{r},t) = \sum_{k=1}^{P} \lambda_k c_k(t)v_k(\vec{r},t) \text{ or } \dot{q}(\vec{r},t) = \sum_{k=1}^{P} \lambda_k c_k(t)v_k(\vec{r}) \quad (7.19)$$

In quantum case we can make the following calculation starting from the time-dependent Schrodinger equation with the Hamiltonian operator $\hat{H}$. In the last step we use the stationary Schrodinger equation $\hat{H}\psi_k(\vec{r}) = E_k\psi_k(\vec{r})$ where E_k is the energy-eigenvalue.

$$i\hbar\dot{\Psi}(\vec{r},t) = \hat{H}\Psi(\vec{r},t) = \hat{H}[\sum_k C_k\psi_k(\vec{r})] =$$
$$= \sum_k C_k\hat{H}\psi_k(\vec{r}) = \sum_k C_k E_k\psi_k(\vec{r}) \quad (7.20)$$

We thus obtain

$$\dot{\Psi}(\vec{r},t) = -\frac{i}{\hbar}\sum_{k=1}^{P} E_k C_k \psi_k(\vec{r}) \text{ or } \dot{\Psi}(\vec{r},t) = -\frac{i}{\hbar}\sum_{k=1}^{P} E_k C_k \psi_k(\vec{r},t) \tag{7.21}$$

Here E_k (an eigenvalue which represents result of a measurement of an observable quantity, i.e., energy) has a similar role as λ_k, the neuro-synergetic *attention* parameter. Note the interdependence of the process of attention and observation which is here, probably not completely by chance, replicated in the mathematical formalism.

The analogy no. 6 can be extended to relativistic domain also. In the relativistic case, where ψ_k are 4-dimensional vectors, the role of the Green-function propagator G is realized by the *S-matrix* (scattering matrix) (Bjorken & Drell, 1964/65). Its elements are:[12]

$$S(\vec{r}_1,t_1,\vec{r}_2,t_2) = -(i)\sum_{k=1}^{P}\sum_{j=1}^{2} \psi_k^j(\vec{r}_1,t_1) * \psi_k^j(\vec{r}_2,t_2) \text{ if } t_2>t_1 \tag{7.22}$$

$$S(\vec{r}_1,t_1,\vec{r}_2,t_2) = (i)\sum_{k=1}^{P}\sum_{j=3}^{4} \psi_k^j(\vec{r}_1,t_1) * \psi_k^j(\vec{r}_2,t_2) \text{ if } t_2<t_1 \tag{7.23}$$

The same Green function which propagates a solution of the Schrodinger equation forward in time, propagates its complex conjugate backward in time. It contains *all* the solutions of the Schrodinger equation, including the bound states as required in the completeness theorem $\sum_{k=1}^{P} \psi_k(\vec{r}_1,t)\psi_k(\vec{r}_2,t)* = \delta^3(\vec{r}_1 - \vec{r}_2)$, with equal weight. In the relativistic S-matrix one half of components (j = 1, 2) corresponds to the positive-energy waves and the second half (j = 3, 4) to the negative energy-waves (antiparticles, i.e. particles propagating into the past: $t_2 < t_1$) which were not present in the non-relativistic case (Bjorken & Drell, 1964/65).

Let us only mention two additional expressions which might act as "learning matrices", similar to those of analogy no. 6, if coupled by an appropriate dynamical equation, like Equation (7.6), for instance. Feynman introduced[13] the quantum propagator

$$G(\vec{r}_1,t_1,\vec{r}_2,t_2) = \sum_k \psi_k(\vec{r}_1) * \psi_k(\vec{r}_2) \exp(-iE_k(t_1 - t_2)) \quad \text{for } t_2>t_1$$

which is inserted into (7.6)-like dynamical equation. This version of the Green function explicitly encodes the "phase"-differences in the exponents.

A similar expression from statistical physics is the Green function of the Fokker-Planck equation (AuxLit 9) [14]

$$G(\vec{r}_1,\vec{r}_2,t) = f(\tilde{r}_1,\tilde{r}_2)\sum_{h=0}^{\infty}\sum_{k=-\infty}^{\infty} \psi_{kh}(\vec{r}_1)\psi_{kh}(\vec{r}_2)\exp(ik(\varphi_1-\varphi_2)-\lambda_{kh}\tilde{t})$$

which is inserted into (7.22)-like equation for statistical distribution. f $(\tilde{r}_1,\tilde{r}_2)$ is an exponential function of new variables incorporating r_1 or r_2.

These short examples are only to remind that there are numerous cases in physics where "neural-net"-like collective processes take place which could be given informational interpretation. Indeed they could, at least in principle, realize information processing similar (or more advanced) than those described in the first part.

Gould (in Pylkkanen & Pylkko, 1995) presented *analogies of the holonomic theory* (Pribram, 1991, Appendices with Jibu and Yasue) and *the ontological interpretation of quantum theory* (Bohm & Hiley, 1993). Gould emphasized the importance of non-local processing in two different synergetic levels, which these two theories describe, using quantum and neural versions of Hamilton-Jacobi equation, based on a general theory of Gould (1989).

Furthermore, he showed that the non-local Schrodinger equation.

$$i\hbar\frac{\partial\Psi}{\partial t} = -\frac{\hbar^2}{2m}\nabla^2\Psi + \int V(\vec{r}_1,\vec{r}_2,t)\Psi(\vec{r}_2,t)d\vec{r}_2 \tag{7.24}$$

is quite similar to the non-local neural wave-equation, derived after Pribram (1991, Appendices with Jibu and Yasue),

$$i\nu\frac{\partial\Psi}{\partial t} = -\frac{\nu^2}{2m}\nabla^2\Psi + \int | U(\vec{r}_1,\vec{r}_2,t)\Psi(\vec{r}_2,t)d\vec{r}_2 \tag{7.25}$$

Integrals in both integro-differential equations take the whole system's spatial volume. The potential per unit volume is denoted by $V(\vec{r}_1,\vec{r}_2,t)$.[15] $U(\vec{r}_1,\vec{r}_2,t)$ is the potential per unit volume at the classical biomolecular level resulting from its many-body interactions. ν has the dimensions of *(length)²/(time)*, because it is a constant of proportionality between the flow-velocity of the ionic "bioplasma" and the gradient of phase. It relates the flow-velocity of "bioplasma" to the density of isophase-contours.[16]

7.4 CONSEQUENCES OF NEURO-QUANTUM PARALLELS

Presented analogies (many others could also be found) show that there may be a subtle "division of labor" and an isomorphism-like relation and cooperation between the neural and the quantum levels. These levels may be in a sort of *fractal*-relationship (i.e., multi-level self-similarity in complex-system structure and dynamics; cf. Bianconi & Barabasi, 2001).

Here the question arises, what is the difference between ordinary quantum processes and mental or even conscious information processing. The first main difference between the physical and mental processes is that a *synergetic (quantum or neural)* system itself is not intentional (does not carry any mental information), but *mind-brain is intentional* (carries specific mental contents). Only conscious process is those which gives mental interpretation and meaning to some synergetic processes, and makes them, by this very act, informational in a psychological sense (i.e., they start to represent information for humans).

The second difference is that the *quantum system itself does not have any relatively independent environment, but brain does*. Therefore the brain models its macroscopic environment in a specific and flexible manner by *using bionetworks as a macro-micro-interface and a (subconscious) preprocessor* for an *unified conscious experience which involves neuro-quantum coherence,* I believe.

In sections 6.8 and 2.1-2.3 it was argued that mind-like processing needs neural or/and quantum virtual structures (attractors). Indeed, every collective state of a complex system may constitute a specific gestalt (a specific virtual unity) which cannot be reduced to the state of constitutive elements of the system alone. The problem is the formation of a specific isomorphic (e.g., fractal) multi-level coherence. Our practice with computer simulations of ANN shows (e.g., Peru s, 2002; AuxLit 10) that we can govern the artificial virtual level implicitly by regulating the artificial neuronal level explicitly ("bottom-up influence"). *If our dynamic equations for neurons and synapses regulate the patterns only, the attractors always accompany this dynamics implicitly!* The opposite can also be done: in computer modeling, by regulating system's "harmony" -function (McClelland et al., 1986) or "energy"-function (Hopfield, 1982; Amit, 1989), in brain-mind by "top-down influences" of conscious process.

Neuronal dynamic equations, used in computer programs, are differential equations (with local range of validity), but attractor-structures may be mathematically described by variational calculus (with global range of validity). We do not necessarily need both mathematical descriptions – one is sufficient. Thus, we may organize one level and the others will follow automatically. This is the reason why the reductionist approach usually works for all practical purposes, and this is a chance for advanced system-based artificial intelligence in form of neural, quantum (AuxLit 24, 25;

Chrisley in Pylkkanen & Pylkko, 1995), and *neuro-quantum* (Dawes, MacLennan, Schempp, Werbos in Pribram, 1993; Kak, 1995; Behrman et al., Linkevich, Peru s, Ventura, Vitiello in Wang et al., 1998; Ezhov et al., 2000, 2001; AuxLit 22, 23) or other *hybrid computers*. They may be artificial and/or synergetic (the latter being parallel computers implemented in natural physical, (bio)chemical or other media).

This list of neuro-quantum analogies has general purposes for my topic. It is not restricted to vision, of course, but provides necessary, I believe, framework for *tracing quantum information-processing backgrounds of conscious experience, based on successful computer simulations of associative* ANN. Although in this Auxil. Chapter a Hopfield-like ANN model was translated into quantum formalism, such a correspondence could also be generalized to other models of "formal neurons and connections", e.g. to subcellular ones. Hence, such an extrapolation to synapto-dendritic networks is realized in sec. 4.6.

ENDNOTES

1 In this Auxil. Chapter also, as usual inANN theory, terms like neurons and synapses will denote formal neurons, or other connectionist units, and their interactions. They could, in principle, be roughly implemented also by brain's synapto-dendritic networks.

2 t_2 is always greater than t_1-because of the "arrow of time"

3 From now on, the usual notation/convention $\hbar = 1$ will be used to simplify the expressions. $\hbar$ is the Planck constant (6.610^{-34}*Joule sec.*, characteristic for all quantum systems) divided by 2π.

4 Ahn et al. (2000); Jones et al. (2000); Berman et al. (1998); AuxLit 25, 24; Anandan (1992)

5 In 1998 I found that these predictions, expressed in Perus (1997a), were marvellously realized by HNeT of Sutherland (1990).

6 In special, but important, generalized cases, the memory matrix J can be written as cross-correlation function of a pattern and a difference between two patterns. In neural network theory we usually use cross-correlation of a novel pattern $\vec{v}_n$ with a difference or discrepancy of this novel pattern $\vec{v}_n$ and a desired or average pattern $\vec{v}_d$ (or $\bar{\vec{v}}$) where $\vec{v}_d$ (or $\bar{\vec{v}}$) represents old or reliable knowledge (Kohonen, 1984; Peretto, 1992). A Hebb-like learning rule which is based on cross-correlation of a (novel) pattern $\vec{v}_n$ and a difference $\vec{v}_n$ - $\vec{v}_d$ (delta learning rule) is used for novelty filtering: new information, not represented in the old neural pattern $\vec{v}_d$, is extracted and stored newly. Therefore the network finally adapts, i.e. optimally condenses and combines

components of the purely-new information ($\vec{v}_n$ - $\vec{v}_d$) with the whole knowledge collected up to now. Such generalized Hebbian rules work also in cases of correlation of two difference-patterns, like ($\vec{v}_n$ - $\vec{v}_d$)($\vec{v}_{n'}$ - $\vec{v}_d$) or ($\vec{v}_n$ - $\bar{\vec{v}}$)($\vec{v}_{n'}$ - $\bar{\vec{v}}$) (a case similar to equation (27) from Appendix A where $\vec{v}_{n'}$ = $\vec{v}_n$).

7 With this term I mean that it regulates current associative processing based on inputs (while ρ is used more like an "archive"), but not like the working memory of cognitive psychology.

8 T. Menneer: Quantum-inspired two-layer neural networks; Tech. rep., Univ. of Exeter, UK, Comp. Sci. Dept. & Tuscon II Conf., 1996. Comparisons with works of Ohya (1989) and Matsuoka (see their references in AuxLit 24), concerning the density matrix ρ and its decompositions, can also be made, and one thus gets similar observations about "symmetry of neuronal and synaptic representations" and their "collapses". Such a case see also in Linsker (1988, p. 110)

9 E.g., I. Prigogine (1995): *Int. J. Quantum Chem. 53*, 105-118

10 ν denotes a fixed spectral parameter. Phase-shift $\tilde{\varphi}$ is equivalent to phase-difference φ_k-φ_h, or $\varphi_k(t+\Delta t)$-$\varphi_h(t)$ respectively.

11 T. Matsuoka, M. Ohya (1995): *Reports Math. Phys. 36*, 365-379

12 Imaginary unit *i* is in brackets, because it depends on definitions of S and the dynamical equation whether *i* will appear in both, giving together -1, or it will be cancelled out.

13 See his book with Hibbs ("Quantum Mechanics and Path Integrals", 1965) and historical papers on the theory of positrons: R.P. Feynman (1949): *Phys. Rev. 76*, 749- and 769-.

14 For a detailed context please see sec. 10.3. "Computer Solution of a Fokker-Planck Equation for a Complex Order Parameter" in Haken's book "Advanced Synergetics" (1987). λ_{kh} is an eigenvalue.

15 It is not just potential as in usual Schrodinger equation.

16 Numerous physiological and biophysical details (e.g., concerning the ionic "bioplasma" in surroundings of dendritic membranes) are presented in sec. 4.6 and originally in: Pribram (1991, Appendices); Jibu & Yasue (1995); Jibu, Pribram & Yasue (1996); AuxLit 19.

Chapter 8
Derivation of Quantum Associative Network from Hopfield-Like ANN and HNeT

8.1 CONNECTIONIST SIMULATIONS APPLIED FOR QUANTUM DYNAMICS

In Chapter 7, I have presented in detail how and where the mathematical formalism of associative ANN models by Hopfield and Haken is analogous to the mathematical formalism of quantum theory. In this Auxil. Chapter, I derive from associative ANN in a straight-forward way an original and fundamental quantum information processing model, or "algorithm", called the *Quantum Associative Network*. The initial ANN "algorithm" is presented first (sec. 8.2). It was chosen, because it is simple but nevertheless effective, technically realizable, and roughly (on global scale) biologically plausible, as a convenient one for presenting ANN-like and HNeT-like quantum information dynamics at a *fundamental* level.

DOI: 10.4018/978-1-61520-785-5.ch008

The statement on rough biological plausibility needs justification. The "algorithm" is relatively biologically plausible on the level of uniform, bi-directionally densely-interconnected network of some physiological units or formal neurons. As discussed in sec. 6.5, the neo-cortex globally meets such a description. Formal "neurons" could even be whole cortical (mini)columns. Additionally to such a macroscopic level, the "algorithm" is even much more implementable on the microscopic level of synapto-dendritic nets.

The Hopfield ANN model is mathematically equivalent to the theory of spin-glass systems (i.e., amorphous assemblies of spins or small "magnets") (Mezard, Parisi & Virasoro, 1987; AuxLit 14). Spins are fundamental in (quantum) physics. Magnetic dipoles are considered as "macroscopic" spins. Electric dipoles, which are used in the holonomic theory (Pribram, 1991, Appendices with Jibu & Yasue), exhibit very similar collective dynamics. So, the present starting "algorithm" is potentially plausible for modelling the mentioned levels of the multi-level image processing in V1, but it must be properly constrained as discussed in section 4.6.

By extensive simulations of the quantum-implementable Hopfield-like ANN model and its generalizations using various concrete data sets, we have effectively realized parallel-distributed content-addressable memory, selective associative reconstruction of image-like patterns memorized in a compressed form.[1] We have found that the results depend very much on the correlation structure of a specific set of input patterns. In other words, beside the "hardware" (implementation) and the "software" ("algorithm"), also the "virtual software" (i.e., the input-data correlation structure) is essential. In sec. 8.5, results of my computational "recognition" of protein-patterns are briefly presented to show very successful performance of this "algorithm".

In the next section, I will present conventional associative ANN in a form which will be in sec. 8.4 translated to quantum formalism in a straight-forward way. I will start from a Hopfield-like ANN model in the approximation of linear activation function (central, "non-saturated" section of the tangens-hyperbolicus-like sigmoid function), because this is sufficient for associative processing and memory-storage, and suitable for our aim to implement information dynamics in a natural quantum system. Image processing is the most successful special case of the "algorithm's" application.

8.2 ASSOCIATIVE NEURAL NETWORKS

From now on, the terms like neural, neuron, synaptic connections, memory, association, learning, recognition, etc., will always be used in the formal sense of the present artificial neural net (ANN) theory.

Discrete model. In our starting Hopfield-like neural net model, the equation for change of *neuronal activities* (right in the whole state-vector $\vec{q}$ and weight-matrix **J**, left in their components)

$$q_i(t_2 = t_1 + \delta t) = \sum_{i=1}^{N} J_{ij} q_j(t_1) \text{ or } \vec{q}(t_2) = \mathbf{J}\vec{q}(t_1) \tag{8.1}$$

is coupled with the Hebb equation for synaptic connections (again, left are components of the right expression)

$$J_{ij} = \sum_{k=1}^{P} \upsilon_i^k \upsilon_j^k \text{ or } \mathbf{J} = \sum_{k=1}^{P} \vec{\upsilon}^k (\vec{\upsilon}^k)^T \tag{8.2}$$

q_i is the current activity of the ith neuron. The state of the network with N neurons is described by: $\vec{q} = (q_1, \ldots, q_N)$. υ_i^k is the activity of the neuron i when taking part in encoding the kth memory pattern $\vec{\upsilon}^k$. $()^T$ denotes transposition: in this case, a column-vector is transformed to a row-vector.

Equation (8.1) describes each neuron's summation of signals from all other neurons weighted by the strengths of their synaptic connections J_{ij}. Equation (8.2) prescribes that the strength of each synaptic connection is increasing when its two neurons are both active or both inactive, and decreases if their activity is not correlated. This is the *Hebb learning rule*. The neuronal correlations belonging to each individual pattern, which are simultaneously stored in the network, are then summed over all (P) stored patterns. **J** is quite appropriately called "memory matrix".

Global optimization. Alternatively, one can regulate the Hopfield emergent collective computation by minimization of the Hamiltonian energy function

$$H = -\frac{1}{2}\sum_{i=1}^{N}\sum_{j=1}^{N} J_{ij} q_i q_j = -\frac{1}{2}\sum_{i=1}^{N}\sum_{j=1}^{N}\sum_{k=1}^{P} \upsilon_i^k \upsilon_j^k q_i q_j \tag{8.3}$$

Continuous model. This system of Equations (8.1) & (8.2) is enough for exhibiting PDP of input data. Let us rewrite it into continuous description of activities of neurons and synapses at position $\vec{r}$ and time t:

$$q(\vec{r}_2, t_2 = t_1 + \delta t) = \int J(\vec{r}_1, \vec{r}_2) q(\vec{r}_1, t_1) d\vec{r}_1 \tag{8.1b}$$

$$J(\vec{r}_1,\vec{r}_2)=\sum_{k=1}^{P}\upsilon^k(\vec{r}_1)\upsilon^k(\vec{r}_2) \tag{8.2b}$$

Memory recall. The recall of pattern $\vec{\upsilon}^1$ from memory is done by $\vec{q}_{output}=\mathbf{J}\vec{q}_{input}=\mathbf{J}\vec{q}\,'$. This can be analyzed as follows:

$$\begin{aligned}q(\vec{r}_2,t_2) &= \int J(\vec{r}_1,\vec{r}_2)q'(\vec{r}_1,t_1)d\vec{r}_1=\int[\sum_{k=1}^{P}\upsilon^k(\vec{r}_1)\upsilon^k(\vec{r}_2)]q'(\vec{r}_1,t_1)d\vec{r}_1=\\ &=[\int \upsilon^1(\vec{r}_1)q'(\vec{r}_1,t_1)d\vec{r}_1]\upsilon^1(\vec{r}_2)+[\int \upsilon^2(\vec{r}_1)q'(\vec{r}_1,t_1)d\vec{r}_1]\upsilon^2(\vec{r}_2)+...\\ &...+[\int \upsilon^P(\vec{r}_1)q'(\vec{r}_1,t_1)d\vec{r}_1]\upsilon^P(\vec{r}_2)=\\ &=A\upsilon^1(\vec{r}_2)+B \text{ where } A\doteq 1(\text{‘}signal\text{’}),\ B\doteq 0(\text{‘}noise\text{’})\end{aligned} \tag{8.4}$$

In Equation (8.4) we had to choose such an input $\vec{q}\,'$ that is more similar to $\vec{\upsilon}^1$, for example, than to any other $\vec{\upsilon}^k, k\neq 1$. At the same time, the input $\vec{q}\,'$ should be almost orthogonal to all the other $\vec{\upsilon}^k, k\neq 1$. In that case, $\vec{q}$ converges to the memory pattern-qua-attractor $\vec{\upsilon}^1$, the other terms give a negligible sum B, as it is shown in the last row of Equation (8.4). Thus, $\vec{\upsilon}^1$ is recalled.

One can describe the above recall process in another, equivalent way using the projection-coefficients $C^k(t)=\int \upsilon^k(\vec{r})q(\vec{r},t)d\vec{r}$ to show the similarity to quantum theory and synergetics:

$$\begin{aligned}q(\vec{r},t) &= \sum_{k=1}^{P}C'^k(t)\upsilon^k(\vec{r})=\sum_{k=1}^{P}(\int \upsilon^k(\vec{r})q'(\vec{r},t)d\vec{r})\upsilon^k(\vec{r})=\\ &=(\int \upsilon^1(\vec{r})q'(\vec{r},t)d\vec{r})\upsilon^1(\vec{r})+...+(\int \upsilon^P(\vec{r})q'(\vec{r},t)d\vec{r})\upsilon'^P(\vec{r})=A\upsilon^1(\vec{r})+B\end{aligned} \tag{8.4b}$$

where A ≐ 1 ('signal') and B ≐ 0 ('noise'). (The first row of equalities in Equation (8.4b) follows, among others, from the neurosynergetic model by Haken (1991) where C^k are called order parameters.)

8.3 HOLOGRAPHIC NEURAL TECHNOLOGY (HNeT)

Oscillator net. If we introduce the *oscillatory* time-evolution of neuronal activities, we get more biologically plausible dynamics (Baird, 1990; Haken, 1991; Schempp, 1993, 1994, 1995). Neural-net variables then become complex-valued: $q_j=q_{j0}e^{i\varphi_j};\varphi_j$

is the phase. This is more akin to quantum dynamics, and to holography also. However, in quantum theory, complex-valued formalism is essential, but in the network of coupled oscillatory neurons complex-valued formalism is just mathematically more convenient.

Holographic/Quantum Neural Technology (HNeT) by AND Corporation, Toronto, directed by John Sutherland, developed a holography-like artificial ANN model which also transcends the ANN concept by essential innovations. As evident from demonstrations, their artificial image processing is very efficient in comparison to conventional ANN models.

Sutherland (1990, 1994) and his group, Manger (1997), Soucek's collaborators and others (AuxLit 13) have shown the efficiency of their computational model in numerous commercial data-processing applications. Schempp (1993, 1994) and Perus (in Wang, 1998; 2000c) presented HNeT, which has been inspired by the holonomic theory, in the context of the triple analogy holography—quantum—ANN.[2] Therefore, let us roughly present the basic part of HNeT which is holography-like and does not go beyond our scope. Namely, several HNeT's extensions and developments, including hardware implementations, transcend the basic processing which could be termed simulated holography or quantum dynamics.

HNeT learning. A sequence of k input–output vector pairs is used for learning. Each input and output component oscillates with a specific activity amplitude and a specific phase (in the exponent). Thus, the kth input–output pair is:

the input or stimulus vector $\vec{s}^k = (s_1^k e^{i\theta_1^k}, s_2^k e^{i\theta_2^k}, \ldots, s_N^k e^{i\theta_N^k}),\ i = \sqrt{-1}$ and
the output or response vector $\vec{o}^k = (o_1^k e^{i\varphi_1^k}, o_2^k e^{i\varphi_2^k}, \ldots, o_M^k e^{i\varphi_M^k})$

These names or interpretations of vectors, used in HNeT, are arbitrary. It works also if $\vec{s}$ and $\vec{o}$ are joined together into a common vector $\vec{v}$ as its first and second parts.

To encode patterns, we let them "interfere" in a way that is mathematically equivalent to optical interference of the "object wave" and the "reference wave". In the real (optical) holography, the "object wave" is those which is directly reflected from the object whose 3-dimensional image will be stored in the hologram. "In the object wave," the information about the object-form is encoded or modulated. The "reference wave" is the second wave reflected from the object, but it is indirect (a reflection in a mirror is made, for example). These two waves, carrying the modulated scene -information from two different angles or points of view, are then made to interfere on the hologram-plate. The constructive or deconstructive interference determines the local rates of optical transparency of the hologram-plate, so that it encodes the pattern-correlations (i.e., memory components in our interpretation).

Mathematically, a small portion of the wave is equivalent to activity of a neuron, and wave-interference is equivalent to interaction between assemblies of neurons with oscillating activities. This is our holography—neural-net analogy.

Using a Hebb-like expression, analogous to Equation (8.2), we calculate the "connection strength between two neurons h and j "

$$J_{hj} = \sum_{k=1}^{P} s_h^k o_j^k e^{-i(\theta_h^k - \varphi_j^k)} \text{ or } \mathbf{J} = \sum_{k=1}^{P} \vec{o}^k ((\vec{s}^k)^*)^T \tag{8.5}$$

$\vec{s}\,*$ means complex conjugation of $\vec{s}$, i.e. $s_h e^{i\theta_h}$ is transformed to its complex conjugate $s_h e^{-i\theta_h}$. $()^T$ denotes transposition. Every J_{hj} encodes the superposition of local input–output amplitude correlations and the phase-angle differences in the exponent. Thus, the matrix of all amplitude-correlations multiplied with corresponding phase-differences represents a "hologram," i.e. the memory. If we interfere pattern-vectors v with themselves sequentially, we obtain $J_{hj} = \sum_k v_h^k v_j^k e^{-i(\theta_h^k - \varphi_j^k)}$, i.e. the original auto-correlation form of the Hebb learning rule "plus" the intra-pattern phase-difference in the exponents. So, the Hebb and the phase parts together constitute the phase-Hebb learning rule. Several optical implementations of such a phase-Hebb associative memory, among others, were realized (Anderson in Zornetzer, Davis & Lau, 1990, ch. 18; Hariharan, 1996).

Actually, Sutherland interprets phases as the "real" information and the corresponding amplitudes as their confidential levels. He reports significant improvement of performance in comparison to conventional ANN because of this (Sutherland, 1990, 1994).

HNeT memory recall. In the real holography, an image $\vec{v}$ is reconstructed from the hologram by sending a new reference wave $\vec{v}\,'$ (it may be carrying partial information only– that's the so-called key $\vec{s}\,'$) through the hologram plate. This is mathematically similar to the ANN recall of Equation (8.4). In Sutherland's simulated "neural" holography, a stored pattern (say, k = 1) can be reconstructed, analogously to Equation (8.4), from the "neural hologram" by "interacting" the key-input with the memory: $\vec{o}\,' = \mathbf{J}\vec{s}\,'$. This is in components (for each component, here for the jth one):

$$\sum_{h=1}^{N} J_{hj} s'_h e^{i\theta'_h} = \sum_{h=1}^{N} \sum_{k=1}^{P} s_h^k o_j^k e^{-i(\theta_h^k - \varphi_j^k)} s'_h e^{i\theta'_h} = o_j^1 e^{i\varphi_j^1} \tag{8.6}$$

We had to choose an input s'_h that is similar to s^1_h, and θ'_h that is similar to θ^1_h. In such a case, $s'_h\, s^1_h \doteq 1$ (if s are normalized) and $e^{-i\theta^1_h} e^{i\theta'_h} \doteq 1$. Thus, for the "holographic" recall, similar processing and conditions have to be realized as in ANN of sec. 8.2, even with the addition of phase-information.

In HNeT recall, a normalization constant $\frac{1}{K}$ is multiplied with the output-result of the reconstruction-process (8.6). $K = \sum_{j=1}^{N} s'_j$.

Advanced HNeT. HNeT uses data pre-processing to transform the distribution of real-valued input-data into a symmetrical (i.e., uniform) distribution for the described "holo- graphic" processing. The reason is to use optimally and proportionally all the data and the "oscillatory network", especially its phases (hidden in the exponent of the polar representation of complex-valued "field activities") which can have angle-values from 0 to 2π. The preprocessing function so transforms a set of scalar input-values, which are unbound (i.e., may extend up to infinite values), into a bound "circular" set of complex values of phases all-around the circle-diagram.

In general, data-preprocessing techniques, by reforming the data-distribution or data- structure (e.g., by symmetrization, i.e. "giving all units equal amount of work"), optimize network's resources by focusing net-processing to important data-features.

In the case of gaussian input-distribution, HNeT uses the sigmoid "activation function" $Sigmoid(x) = \frac{1}{1+\exp(-x)}$ to preprocess each phase θ (executed on "cell's inputs") as well as or rather than to preprocess "amplitudes" (done at the "axonal output of cell's soma") in the form $\theta^k_{new} = 2\pi\ Sigmoid\left(\frac{\theta^k - \overline{\theta}}{\sigma}\right)$ ($\overline{\theta}$ is the mean phase, and σ the variance).

For the general case of non-gaussian data, Manger et al. (in AuxLit 13: Plantamura et al., ch. 4, pp. 81-82) derived the following preprocessing transformation of phase θ in the input to the "holographic" net: $\theta^k_{new} = 2\pi \int_0^{\theta} f(\tilde{\theta}^k) d\tilde{\theta}^k$. Thus, the new phase is to the cumulative or integral distribution of the input-phase θ^k (f is the input distribution k function of phases $\tilde{\theta}^k$, or θ^k, respectively.).

The advanced HNeT feature, which ultimately transcends the usual two-wave interference holography, is higher-order HNeT processing (Sutherland in Dubois, 2000a, pp. 325-329). To extend the information-set for learning, vector-components of $\vec{s}^k$ are: $\Pi_{w=1}^{w\max} s^k_{h,w} e^{i\theta^k_{h,w}}$; h = 1,..., N. Hence, in the case of fourth order

($wmax = 4$), vector-components of $\vec{s}^k$ are: $s_1^k s_2^k s_3^k s_4^k$. These are higher-order correlations. The higher-order "hologram" is then:

$$J_{hj} = \sum_{k=1}^{P} o^k e^{i\varphi^k} \prod_{w=1}^{w\max} s_{h,w}^k e^{i\theta_{h,w}^k} \tag{8.7}$$

Such a learning rule is called "sigma-pi" rule. It is a generalized phase-Hebbian rule. It might turn out to be useful for quantum informatics, because the Slater determinants, which describe many-particle quantum systems, have similar mathematical structure.

Vector-components of the output $\vec{o}^k$, being reconstructed from the "higher-order holo-gram" of Equation (8.7) using new input $\vec{s}\,'$, are: $\sum_{h=1}^{N} J_{hj} \prod_{w=1}^{w\max} s'_{h,w} e^{i\theta'_{h,w}}$.

Another advanced HNeT feature is *enhanced "delta" learning* where stimuli $\vec{s}$ are cor-related with the difference vector $\vec{o} - \vec{o}_{desired}$ ("delta" i.e., discrepancy from the desired output). Since $\vec{o} = \mathbf{J}\vec{s}$, the delta-based "hologram" is given by the matrix: [3]

$$\mathbf{J} = (\vec{o}_{desired} - \mathbf{J}_{previous}\vec{s})(\vec{s}^*)^T \tag{8.8}$$

Sutherland notes that this novelty-filtering rule (cf., simulations of Perus, 1995a, sec. 13.5) is efficient in figure/ground segmentation.

Advanced HNeT and biology. All the described advanced HNeT features and non-described (pre)processing techniques, like vector expansion (Sutherland, 1990, 1994), are not necessary for associative memory, but increase its capacity, resolution and generalization capability (very) much. Biological plausibility of the described techniques is under question, but might turn out to be positive. Does a dendrite sum synaptic signals sigmoidally (like soma)? Sigma-pi learning might in principle be implemented neurally and maybe also by special quantum fields like those given by multi-particle Slater determinants. The "delta" learning has, due to Sutherland (in Dubois, 2000a, p. 330), the following cortical counterpart: A pyramidal cell summes outputs of several stellate cells, and/or subtracts the sum from the desired response. Biological plausibility of non-described HNeT techniques is much more questionable (e.g., vector expansion is artificial).

Higher-order statistics, as needed for infomax-processing, are processed in the higher- order HNeT simulations, of course. This brings some similarities of HNeT to ICA, since both use preprocessing of data by producing superpositions of higher-order correlations. For face recognition, for instance, HNeT uses two artificial "holographic cells" for, first, efficient segmentation of heads (first HNeT cell) and, second, for classification of faces (second HNeT cell). This is a similar method to that of the holonomic theory which proposes two networks, the first one (V1) to perform image processing, and the second one (V2 and beyond) to perform object recognition. Sutherland (unpubl. presentation) wrote:

The HNeT cell may be tuned to recognize human heads only within a range of positional variations, or over all possible variations that may be encountered in the real world. For this application the cell was trained to recognize only frontal shots within a range of angular variation. A second HNeT cell is used to classify between the different individuals.

A single HNeT cell, which resembles the Purkinje neuron of the cerebellum used for motorics, is estimated to be able to process, in the case of 100.000 synaptic inputs, $P = 2.24\ 10^9$ elements / bytes (e.g., "multi-color pixels"). This is far beyond the capacity of all usual ANN models — not only because HNeT's dendrites are summators, too, in contrast to ANN where neuronal somas are summators only.

Holographic memories are universal. Analogous holographic associative memories can be implemented optically (Psaltis et al., 1990), acoustically, quantum-electronically, quantum-biologically (Nobili, 1985; Pribram, 1971, 1991; Schempp, 1993, 1994; Marcer & Schempp, 1997, 1998). Various kinds of holography are a bridge between ANN and quantum information systems (AuxLit 24), similarly to those exploiting quantum coherent states (Vitiello, 1996, 1992; van der Wal et al., 2000), quantum information channels (Alicki 1997; Ohya, 1989), quantum dots (Behrman in Wang et al., 1998; Snider et al., 1999), and the like (AuxLit 22, 23).

8.4 QUANTUM ASSOCIATIVE NETWORK

From ANN to quantum associative net. Before HNeT, we have discussed a Hopfield- like neural-net or spin-glass model in a specific form which is quantum implementable. Now I will explicitly present its quantum relative, with phase implicitly incorporated like in HNeT. The ANN equations will just be translated into quantum formalism in such a way that the information-processing capabilities will be preserved. From the very beginning, let us put attention to the following correspondence scheme (discussed in Auxil. Chapter 7) between the neural (left) and quantum variables (right) ("⇔" means "corresponds to"):

$q \Leftrightarrow \Psi$, $q' \Leftrightarrow \Psi'$, $v^k \Leftrightarrow \psi^k$, $J \Leftrightarrow G$ or $\mathbf{J} \Leftrightarrow \mathbf{G}$

equation(8.1) ⇔*eq*.(8.9);*eq*(8.2) ⇔*eqs*(8.10*b*);(8.4) ⇔(8.11);(8.4*b*) ⇔ (8.11b)

The phase φ is hidden in the exponent of the wave-function Ψ which describes the state of the quantum system: $\Psi(\vec{r},t) = A(\vec{r},t)\exp(\frac{i}{h}\varphi(\vec{r},t))$; A is the amplitude.[4]

The equations in pairs are *mathematically equivalent*, because the collective dynamics in neural and quantum complex systems are similar, in spite of different nature of neurons and their connections on one hand, and quantum "points" $\Psi(\vec{r})$ and their "interactions" described by $G(\vec{r}_1,\vec{r}_1)$ on the other. Variable with a prime means that its quantitative value is close to the variable without prime.

Taking into account this correspondence list, all the descriptions of neural and holographic memory (re)construction is valid for the quantum case also. First, we encode patterns into specific quantum states. Then, interference of these quantum states is produced, thus exhibiting a sort of quantum memory. At the end, in order to reconstruct a "quantum pattern" ψ^k, the quantum state Ψ has to be "collapsed" as it will be explained later.

Quantum "algorithm". Now I will describe the mathematical formalism of this procedure which follows from the fundamentals of quantum theory as described in any textbook on quantum mechanics (Messiah, 1965; Bohm, 1954; AuxLit 16). My ANN-like and HNeT-like quantum associative net combines the dynamical equation for the quantum state[5]

$$\Psi(\vec{r}_2,t_2) = \int\int G(\vec{r}_1,t_1,\vec{r}_2,t_2)\Psi(\vec{r}_1,t_1)d\vec{r}_1 dt_1 \text{ or } \Psi(t_2)=G\Psi(t_1) \qquad (8.9)$$

and the expression for the parallel-distributed interactive transformation of the quantum system

$$G(\vec{r},t_1,\vec{r}_2,t_2) = \sum_{k=1}^{P}\psi^k(\vec{r}_1,t_1) * \psi^k(\vec{r}_2,t_2) \text{ or } G(\vec{r}_1,\vec{r}_2) = \sum_{k=1}^{P}\psi^k(\vec{r}_1) * \psi^k(\vec{r}_2) \qquad (8.10)$$

ψ^k are eigen-wave-functions, energetically stable quantum states. Let us assume that we have succeeded to *encode some information* in them. Eigen-wave-functions have then become "quantum patterns". In such a case, the interaction matrix **G** (so-called propagator) constitutes the *quantum memory*.

Quantum phases. If we, in Equation (8.10) which looks Hebbian, expose the phases φ explicitly, using $\Psi = A\exp(i\varphi)$ we get an expression which is the **quantum phase**-Hebb learning rule:

$$G(\vec{r},t_1,\vec{r}_2,t_2)=\sum_{k=1}^{P}A^k(\vec{r}_1,t_1)*A^k(\vec{r}_2,t_2)e^{-i(\varphi(\vec{r}_2,t_2)-\varphi(\vec{r}_1,t_1))} \quad (8.10b)$$

Like in ANN of sec. 8.2, the process of learning is a result of, metaphorically speaking, a web of quantum interactions "between formal quantum neurons or qubits", described by $\Psi(\vec{r},t)$. Interactions are described by weights $G(\vec{r},t_1,\vec{r}_2,t_2)$ which are elements of the phase-Hebb-like quantum memory matrix **G**.

The difference between the rule (8.10b) and a non-quantum phase-Hebb rule is that in Equation (8.10b) phases φ are quantum phases which are implicitly related to the quantum-specific Planck constant hidden in the exponent (but the usual notation $\hbar = 1$ is used).

"Collapse recall". The quantum memory recall is realized by the *wave-function* "collapse":

$$\Psi(\vec{r}_2,t_2=t_1+\delta t)=\int G(\vec{r}_1,\vec{r}_2)\Psi'(\vec{r}_1,t_1)d\vec{r}_1=\int[\sum_{k=1}^{P}\psi^k(\vec{r}_1)*\Psi^k(\vec{r}_1,t_1)d\vec{r}_1=$$
$$=[\int\psi^1(\vec{r}_1)*\Psi'(\vec{r}_1,t_1)d\vec{r}_1]\psi^1(\vec{r}_2)+[\int\psi^2(\vec{r}_1)*\Psi'(\vec{r}_1,t_1)d\vec{r}_1]\psi^2(\vec{r}_2)+\ldots$$
$$\ldots+[\int\psi^P(\vec{r}_1)*\Psi'(\vec{r}_1,t_1)d\vec{r}_1]\psi^P(\vec{r}_2)=$$

$$=A\psi^1(\vec{r}_2)+B \text{ where } A\doteq 1 \text{ ('}\textit{signal}\text{'), } B\doteq 0 \text{ ('}\textit{noise}\text{')} \quad (8.11)$$

In Equation (8.11) we had to choose such an "input" Ψ' that is more similar to ψ^1, for example, than to any other ψ^k, $k\neq 1$. At the same time, the "input" Ψ' should be almost orthogonal to all the other $\psi^k, k\neq 1$. In that case, Ψ converges to the quantum "pattern-qua-attractor" ψ^1, and the other terms give a negligible sum B, as it is shown in the last row of Equation (8.11). Thus, the memory pattern ψ^1 is recalled (measured). If the condition, well known from our Hopfield-model simulations (see sec. 8.5), that "input" must be similar to one stored pattern (at least more than to other stored patterns) is not satisfied, then there is no single-pattern recall.

It is remarkable to compare the process of Equation (8.11) with Vapnik (1998).

The above process can be described in another, equivalent way:

$$\Psi(\vec{r},t)=\sum_{k=1}^{P}c'^k(t)\psi^k(\vec{r})=\sum_{k=1}^{P}(\int\psi^k(\vec{r})*\Psi'(\vec{r},t)d\vec{r})\psi^k(\vec{r})=$$
$$=(\int\psi^1(\vec{r})*\Psi'(\vec{r},t)d\vec{r})\psi^1(\vec{r})+\ldots+(\int\psi^P(\vec{r})*\Psi'(\vec{r},t)d\vec{r})\psi^P(\vec{r})=A\psi^1(\vec{r})+B$$
(8.11b)

where A ≐ 1 ('signal') and B ≐ 0 ('noise'). (The first row of equalities in Equation (8.11b) takes into account the quantum superposition principle.)

Let us describe the memory recall procedure again:

In the quantum as well as neural case, the new input was let to interact with memory. Mathematically, the input vector was multiplied (convoluted) [6] with the memory matrix (**G** or **J**). Or locally, as in Equations (8.11) or (8.4), the components describing activities of formal neurons (or qubits), which encode the input, were multiplied (convoluted) with the connection weights, determined by Equations (8.10) or (8.2), which encode the memory (correlations of all possible learned patterns). Because the (phase)-Hebb memory matrix is a superposition of single-pattern auto-correlations, we can expand the above multiplication into a series. Each term of this series corresponds to a stored pattern "interacting" with the input. If we choose the input similar to one stored pattern, the corresponding term is amplified ("resonance"), and simultaneously, the others are suppressed. The "resonant" term gives the output; other terms represent the "noise".

However, in contrast to ANN and HNeT, a non-artificial quantum system obeys the closure relation $\sum_k \psi_k(\vec{r}_1, t) * \psi_k(\vec{r}_2, t) = \delta(\vec{r}_1 - \vec{r}_2)$ (Messiah, 1965; AuxLit 16) which is probably not too restrictive under certain conditions related to (quasi)orthogonality (discussed in Perus & Dey, 2000).

If the condition of similarity would not be satisfied, the output would be a superposition of all stored patterns and the input, i.e. an un-intelligible "mixture" of patterns. On the other hand, if the input is identical to a part of a stored pattern, we get pattern recognition, i.e. the stored pattern is exactly recalled from memory as triggered by the input.

8.5 EXAMPLE OF SIMULATED PATTERN RECOGNITION

To demonstrate characteristic performance of the ANN "algorithm" of sec. 8.2, computational recognition of encoded protein patterns (henceforth "proteins") is presented as an example (Tables 1, 2). All detailed results and their interpretations are, of course, not important for this book and beyond its scope, but they can be seen in Perus (2002). In chapter 7, see proof that these computer simulations are also exact mathematical and informational evidence for the possibility of quantum-wave image recognition.

For our purposes, it is sufficient to note low values of the three measures of recognition-error: K (class error), H (number of wrong coding "neurons"), RMS (coordinate distance or discrepancy), and much negative values of "energy". "Energy" refers to the Hopfield energy-function which has a more negative value if

Table 1. Reconstruction of a memorized pattern

"protein"	K	H	energy	RMS	category	Commentary
bgd	37	35	-583	10.66	E	Short 'protein'
Hlm	30	40	-285	11.74	E	Short 'protein'
Gdk	29	29	-3334	11.67	G	-
Lh4	29	32	-3334	12.83	G	-
Emy	12	16	-6896	8.54	Y	-
Mbc	35	33	-7609	9.75	Y	-
Mlf	0	0	-9785	0	R	-
Mll	0	0	-9778	0	R	-
Mlq	1	1	-9760	2.59	R	-
Moa	0	0	-9778	0	R	-
Mtj	0	0	-9760	0	R	-
Yma	14	15	-7060	7.04	Y	-
Spc	0	0	-7659	0	Y	SP = min. 0.9
Vxd	2	3	-7737	2.32	Y	-
2mgb	0	0	-9747	0	R	-
2mgh	0	0	-9717	0	R	-
2spn	1	1	-9766	2.59	R	-
Mlh	0	0	-9785	0	R	-
Bmn	37	28	-347	5.81	E	Short 'protein'
Ilk	32	32	-772	7.9	E	Short 'protein'

there is less reconstruction-failure. Several protein categories were found in the Brookhaven protein data-bank. The ANN's attractors were shown to correspond to categories with codes G, Y, R. E denotes all extra (usually short) protein-patterns not belonging to any category / attractor.

The prototype R was almost orthogonal (i.e., not similar) to other prototypes, therefore reconstruction of a R-type "protein" (its 3-dimensional structure) from partial input-data (2-dim. And 1-dim. structure, i.e. aminoacid sequences) was almost 100% perfect (H, K, RMS zero or very small). Almost 60% of all the "proteins" used in the computational experiment were reconstructed with such a high success — they were mainly members of the strongest R-category. The prototypes G and Y were similar (as measured by the scalar product SP: minimum 0, maximum 1) to each-other, but not to R. Therefore, there were memory-disturbances ("interferences") between Y and G, which resulted in unclear ("mixed") responses, indicated by higher value of H, K and RMS, for Y and G. Some Y-type "proteins" converged to the R-attractor, because of having some similarity (denoted by "SP = min[imally]

Table 2. New response based on similar memories

"protein"	K	H	energy	RMS	category	commentary
bgd	37	35	-354	10.7	E	short "protein"
acp	34	41	-818	23	E	short "protein"
hlm	30	40	-60	11.8	E	short "protein"
gdi	29	29	-3117	11.7	G	-
gdj	28	29	-3067	10.7	G	-
gdk	29	29	-3116	11.7	G	-
gdl	31	26	-3142	10.8	G	-
gdm	31	30	-3154	10.8	G	-
lh4	29	32	-3118	12.8	G	-
2gdm	31	31	-3154	10.8	G	into R-category
2lh3	30	33	-3116	12.5	G	-
2mgg	0	0	-9519	0	R	-
bvc	3	5	-7526	4.57	Y	-
emy	12	16	-6884	8.54	Y	-
fcs	0	0	-9390	0	R	-
fip	38	52	-852	12.4	E	short "protein"
hrm	43	40	-6898	15.1	Y	into R-category
hsy	12	14	-7048	7.33	Y	-
lhs	43	46	-5611	13.3	Y	-
lht	2	2	-5764	1.87	Y	-
mbc	35	33	-7414	9.75	Y	-
mbd	39	42	-7433	14.8	Y	-
mbo	40	41	-7434	14.8	Y	-
mbs	37	34	-6324	9.62	Y	into R-category
mcy	0	0	-9428	0	R	-
mgn	0	0	-9510	0	R	-
mlf	0	0	-9579	0	R	-
mlj	1	3	-9571	0	R	-
mlk	0	0	-9579	0	R	-
mll	0	0	-9572	0	R	-
mlm	1	1	-9557	2.59	R	-
mln	0	0	-9579	0	R	-
mlq	1	1	-9550	2.59	R	-
mlr	0	0	-9552	0	R	-
mls	1	1	-9759	2.59	R	-
mlu	35	32	-9278	10.4	R	SP = max.0.8

continued on following page

Table 2. continued

"protein"	K	H	energy	RMS	category	commentary
moa	0	0	-9572	0	R	-
mob	34	34	-9272	10.1	R	SP = max.0.8
moc	0	0	-9523	0	R	-
mod	0	0	-9523	0	R	-
moh	41	55	-1481	14.8	E	-
mtj	0	0	-9552	0	R	-
mtk	0	0	-9579	0	R	-
mym	0	0	-9559	0	R	-
spc	0	0	-7447	0	Y	sp = min. 0.9
swn	1	2	-7526	2.32	Y	sp = min. 0.9
tcs	0	0	-9272	0	R	-
vxc	1	2	-7526	2.32	Y	sp = min. 0.9
vxc	37	41	-7343	13.6	Y	-
vxf	39	42	-7433	14.8	Y	-
vxg	39	42	-7433	14.8	Y	-
vxh	39	40	-7380	13.8	Y	-
yma	14	15	-6849	7.04	Y	-
ymb	43	46	-7009	12	Y	-
ymb	43	39	-6858	13.3	Y	-
2hif	35	34	-382	11	E	short "protein"
2mb5	40	40	-7367	15.6	Y	-
2mga	33	33	-9283	10.5	R	SP = max.0.8
2mgb	0	0	-9540	0	R	-
2mgc	1	1	-9489	2.59	R	-
2mgd	0	0	-9490	0	R	-
2mge	0	0	-9490	0	R	-
2mgf	0	0	-9523	0	R	-
2mgh	0	0	-9510	0	R	-
2mgi	0	0	-9523	0	R	-
2mgk	0	0	-9579	0	R	-
2mgl	0	0	-9579	0	R	-
2mgm	1	1	-9570	2.59	R	-
2mya	2	3	-7528	2.32	Y	SP = min. 0.3
2myc	39	42	-7433	14.8	Y	-
2spl	0	0	-9579	0	R	-

continued on following page

Table 2. continued

"protein"	K	H	energy	RMS	category	commentary
2spm	1	1	-9557	2.59	R	-
2spn	1	1	-9557	2.59	R	-
2spo	1	2	-9559	2.33	R	-
bvd	6	7	-7487	2.17	Y	SP = min. 0.3
ilk	32	32	-544	7.9	E	short "protein"
irc	0	0	-7386	0	Y	SP = min. 0.3
bmn	37	28	-104	5.81	E	-
mbw	6	8	-9473	3.06	R	-
mlg	0	0	-9577	0	R	-
mlh	0	0	-9577	0	R	-
mlk	1	1	-9555	2.59	R	-
mti	5	7	-9502	2.17	R	-
vxa	1	3	-7219	0	Y	SP = min. 0.3
vxb	1	13	-7125	4.32	Y	SP = min. 0.3
vxd	2	3	-7526	2.32	Y	SP = min. 0.3
2cmm	2	4	-7524	3.37	Y	SP = min. 0.3
2mgj	0	0	-9508	0	R	-
2mm1	37	41	-6865	14.2	Y	-
2myb	38	42	-7397	14.7	Y	-
2myd	1	2	-7526	2.32	Y	SP = min. 0.3
2myc	1	2	-7526	2.32	Y	SP = min. 0.3
4mbn	39	40	-7380	13.7	Y	-
5mbn	2	3	-7528	2.32	Y	SP = min. 0.3

0.3" in Table 2) with the R-prototype which constituted "far the strongest lobby". Some other Y-type "proteins" were similar to each other (see "SP = min[imally] 0.9"), so they constituted an independent (small) Y-attractor. G-type "proteins" were, because of being too few, always attracted into "foreign" attractors.

There were only few R-type "proteins" being less similar to the R-prototype (see "SP=max[imally] 0.8"), so their reconstruction was not successful. These, and other "proteins" with H- and K-errors about 30 to 40, converged to a wrong attractor, called "spurious attractor" in ANN theory, which would correspond to a "mistake" in psychology. This was unavoidable in this case, because of an "ambivalent" key shared by more than one attractor.

The results demonstrate dependency of pattern recognition on categorization, or memory-based prototypes (attractors), respectively.

Table 1 presents evaluations of computational reconstructions of individual encoded protein patterns, named with standard codes in the left-side box "proteins". First, all known data on all mentioned proteins were used in learning (i.e., insertion into the Hebb memory-matrix). After learning, each protein individually was tried to be reconstructed based on the partial "recall-key" (20% of totally 593 "neurons"). Reconstruction means the global graphic presentation of a decoded 3-dimensional "protein", made from a "key" ("seed") of aminoacid sequences and of data on local meso-structure such as α-helices and ß-sheets.

Table 2 shows evaluations of responses which the ANN makes after learning of all "proteins" except the one given in the left-side box, with the "recall-key" incorporating partial data on this given "protein". The other, unknown data on this "protein" were then interpolated ("predicted") by my ANN. The second experiment thus studied responses in somewhat new circumstances from those learned. A response was based on a new key interacting with memories of other "proteins". The network automatically selected the most relevant (similar) examples from memory to produce an optimal response-pattern.

Note added in proof. My Quantum Associative Net model has recently been computationally verified also by successful simulations of recognizing high-resolution images (Loo, Peru s & Bischof). Specifically, that is: selective reconstruction of an image, triggered by an occluded or noisy "key"-image, using a memory-matrix with many stored images.

ENDNOTES

1 I.e., Haykin (1994), Arbib (1995), Perus (tech. reports and conf. presentations, 1993, 1995, 1996, 1998).

2 As a consequence of such mathematical comparisons, HNeT was recently renamed to Holo-graphic/Quantum Neural Technology. Note: I will just somewhat modify the original basic HNeT model to make it mathematically equivalent to quantum processing. The modified model ($J = \sum_k \vec{o}^k (\vec{s}^k *)^T$) is a special case of the original sequential model ($\vec{J} = S\vec{o}$ for different time-instants t instead of k), and is guaranteed to retain all its processing powers.

3 This is again a bit modified, but equivalent, version of the original HNeT expressions – to show analogies and be consistent with other parts of this book. (Cf., paragraph no. 7 in Auxil. Chapter 7.)

4 Notation/convention $\hbar = 1$ will always be used. $\hbar$ is the Planck constant divided by 2π

5 See R. Feynman & A. Hibbs (1965): Quantum Mechanics and Path Integrals. New York: McGraw-Hill. $G(\vec{r}_1, t_1, \vec{r}_2, t_2)$ are the components of the Green matrix G.

6 Because all the pro ducts are summed up by the integral, convolution is the term that describes the process more directly than the term multiplication.

Chapter 9
Quantum Neural Information Processing

Neural vs. quantum nets – discussion. This Chapter is devoted to discussion of neural-to-quantum (or holography-to-quantum, respectively) transforms of connectionist associative processing of Chapter 8. Equations (8.9), (8.10) and (8.11) come from basic quantum physics. Quantum interference is very fundamental and essential. With the Young experiment it was an early experimental predecessor of the Schrodinger quantum wave-mechanics and the Heisenberg quantum operator theory (Messiah, 1965; Bohm, 1954; AuxLit 16).

There is nothing "neural" (in biological sense) in the system of quantum equations (8.9) & (8.10). It is just *similar, according to their mathematical structure and coupling*, to the system of ANN equations (8.1) & (8.2). Because we are certain that the ANN system (8.1) & (8.2), and HNeT's formulas (8.5) & (8.6) as well, realize efficient information processing, I have taken the similar system of equations

DOI: 10.4018/978-1-61520-785-5.ch009

from the quantum-physics formalism in order to discover quantum ANN-like, or better HNeT-like, information dynamics. Using computer simulation experience with the Hopfield model and HNeT's successes, I have merely given a precise information-processing interpretation to a selected and deliberately connected set of quantum equations which are, each one alone, known to all physicists and well experimentally verified.

Since the proposed "algorithm" is realizable in any complex system obeying similar collective dynamics, information processing of the same type could in principle be realized in many such (classical) systems, if we neglect some specific features arising from the nature of *individual* elements of the system. The "algorithm" is thus relatively universal, with exception of some specific characteristics to be discussed now.

There is a difference between the neural "algorithm" (8.1)-(8.2)-(8.4) and the quantum "algorithm" (8.9)-(8.10)-(8.11): Neural variables in equations (8.1)-(8.2)-(8.4) are real-valued, but quantum variables are complex-valued, because they implicitly incorporate the phase. Thus, a quantum quantity or expression usually incorporates, at least implicitly, the imaginary unit i, the Planck constant $\hbar$, and/or the asterisk denoting complex conjugation. Complex numbers with phases as imaginary components accompany oscillatory dynamics, absent in the presented ANN, but present in holography and its simulation – HNeT. (An example of spectral, i.e. wave-dynamic, Hopfield-based ANN see: Spencer, 2001a,b.) Chapter 7 demonstrates explicitly in which case real-value processing is equivalent to oscillatory processing – the latter is usable for quantum-wave image recognition.

Quantum informatics. The quantum memory recall procedure is a special case of usual quantum measurements (Wheeler & Zurek, 1983). The difference is that usual physical processes in our case have an informational interpretation, i.e. our eigen-wave-functions have a specific prescribed *meaning*. Let us describe the reconstruction of the selected quantum eigen-wave-function from the informational or cognitive point of view.

In quantum language, the quantum system was perturbed by the experimenter in a specific way. Informationally, the new input was let to interact with quantum memory. Mathematically, the input vector was multiplied (convoluted) with the memory matrix (the Green-function propagator). Or, the input-wave-components (as shown in (8.11), similarly to (8.4)) were multiplied with the propagator's components (correlations of all possible learned quantum patterns) which encode the memory, and summed together. Because the propagator is a superposition of self-interferences of each eigen-wave-function, we can expand the above multiplication into a series. Each term of this series corresponds to an implicitly-present eigen-wave-function "interacting" with the input, i.e. the perturbation. If we choose the perturbation-pattern similar to one stored quantum pattern, the corresponding term is amplified ("resonance"), and simultaneously, the others are suppressed. The "resonant" term gives the output, i.e. the finally measured quantity corresponding to a selected eigen-wave-function, other terms represent uninteresting implicit states.

Like in ANN and HNeT, if the condition of similarity (or a high rate of orthogonality, respectively) would not be satisfied, the output would be a superposition of all stored quantum patterns and the perturbation (which is our input). This means, the output would be an un-intelligible interference of eigen-wave-functions. On the other hand, if similarity turns to identity (the input is identical to a stored quantum pattern), we get quantum collapse to the nearest stored quantum pattern. The quantum-pattern recall may work well also if the measurement-perturbation is just partially correlated with a stored eigen-wave-function. Quantum holography, like any other kind of holography, realizes the effects described in the last paragraphs, and they are often used in experimental practice (Psaltis et al., 1990; Nobili, 1985).

Neuro-quantum interaction. It seems that, in the brain, neural networks trigger the collapse of quantum wave-function and thus transform the quantum complex-valued, probabilistic dynamics into the neural (classical) real-valued, deterministic dynamics. So, neural nets serve as an interface between classical environment of the organism and its quantum basis (Pribram, 1993; Perus, 1996; Bob & Faber, 1999). The transformation of the complex-valued quantum dynamics into the real-valued classical dynamics during the collapse of the wave-function (a quantum sort of pattern recognition – remember eq. (8.11)) is very probably triggered by the system's interaction with environment (Wheeler & Zurek, 1983; AuxLit 17).

If we try to implement Shannon-type information processing in quantum systems, some essential quantum features, manifested in the complex-valued formalism, are unavoidably partly eliminated (AuxLit 18, 26; Peru s, 1996). However, nature has obviously always performed a sort of non-Shannon-type implicit-information processing (Bohm uses term "active information": Bohm & Hiley, 1993; Hiley & Peat, 1987) where the "users of information" are not human observers, but natural systems (e.g., immune systems) themselves.

Quantum entanglement. Holographic and other oscillatory neural networks (Baird, 1990; Haken, 1991; AuxLit 11; Auxil. Chapter 11) base on complex-valued formalism like quantum theory, but they do not realize the deep essence of quantum dynamics, i.e. the quantum holism — indivisibility and omnipresent continuous non-local interconnection or entanglement (Bohm, 1980; Bohm & Hiley, 1993; Majewski, 1991; Kwiat et al., 2000). However, these essentially-quantum features are broken and dynamics are thus discretized during quantum measurements: Instead of complete wholeness and unity of the so-called quantum vacuum (as named by contemporary physics) or holomovement (as named by Bohm, 1980), networks made of interacting relatively-discrete "connectionist units" — particles or waves, or spins or dipoles, or (on a larger scale) neurons, etc. — emerge. The result of this so-called symmetry breaking is a *fractal-like* (i.e., multi-level, self-similar) hierarchy of "networks" or "dynamic holograms" — from the One to the mechanistic circuits.

The implicit and multiple quantum process enables PDP of extremely high speed and capacity, but it has to be collapsed during the readout (i.e., recall) in order to obtain results of processing (Berman et al., 1998; AuxLit 25). During observation, quantum dynamics become more Hopfield-net-like, e.g. the observable "activities of a network of quantum neurons" are real-valued. Neuro-quantum (classical-quantum) cooperation (Kak, 1995; Nanopoulos, 1995; Nishimori & Nonomura, 1996; AuxLit 20-23), as manifested in the collapse-readout (Hamero & Penrose, 1995, 1996), is thus essential for the brain which combines Shannon-type cognitive information (AuxLit 29) and non-Shannon-type conscious phenomenalism (AuxLit 30-33).

Quantum neural nets. In parallel and independently from the Quantum Associative Network (Peru s in Wang et al., 1998), some other approaches to implement the Hopfield model in quantum systems (Bonnell & Papini, 1997; Zak & Williams, 1998) were developed. Although they are more sophisticated [1], the quantum associative net allows more straight- forward implementation in a usual, unconstrained quantum system (i.e., without need of artificial manipulations, apart from perturbational insertions of inputs into the system).

These similar approaches as well as the fast-developing field of quantum computing (Berman et al., 1998; AuxLit 25) raise the possibility for the quantum associative net being physically implemented in some form – in (sub)biological systems (Fedorec & Marcer, 1996) as well as in computers (Jones et al., 2000). Present technology allows shaping (by quantum holography) of quantum wave-functions used for information encoding (Weinacht et al., 1999), storage and retrieval of information to/from quantum phase (Ahn et al., 2000; Denschlag et al., 2000), and (as far as I can see) all the necessary quantum manipulation and control (Rabitz et al., 2000). Anyway, one should mind the differences between my quantum neural-like network, which has a natural computing capability, and quantum computers based on relatively artificial implementation of logic gates in quantum devices. Technological implementation could be used for our biological purposes to argue that the quantum associative PDP is realizable, and to search its analogs in brain-tissue.

Bonnell & Papini (1997) especially studied quantum attractor dynamics in their quantum ANN. They emphasize the similarities in quantum and ANN attractor dynamics, but relate the quantum eigen-patterns to peaks of the probability distribution of quantum states (Bonnell & Papini, 1997, p. 2873-4):

Memory is connected to the 'wavefunction' by the probability distribution $|\Psi|^2$ *which is peaked 2 at the singularity [...] in the Schrodinger equation. In full analogy with the classical system [ANN (note by M.P.)], this singularity occurs at the same location in the activity space as the resting point of the classical dynamical trajectory [...]. [...]*

An additional similarity is that a certain measurement time is required to gain a precise value for the location of the attractor. In the classical system this consists of waiting for the motion of the pseudo "particle" [i.e., ANN state (note by M.P.)] to exponentially slow to the point where it can reasonably be considered to have stopped [i.e., at the bottom of the potential-well corresponding to the attractor (note by M.P.)]. Likewise, the quantum system requires a sufficient amount of time to perform enough measurements to determine the average probability distribution.

Bonnell & Papini (1997) calculated wave-functions Ψ° which encode memory-attractors in their net for the cases of symmetric weights ($J_{ij} = J_{ij}$) and small-magnitude asymmetric weights. The approximate solution of the Schrodinger equation in the first case are Gaussian wave-functions centred on the activity-point (q°_1, q°_2) in a net of two neurons (with activities q_1 and q_2):

$$\Psi^\circ(q_1, q_2) = A \exp\left[-\frac{1}{4\sigma^2}\left((q_1 - q^\circ_1)^2 + (q_2 - q^\circ_2)^2\right)\right] \tag{9.1}$$

All quantum information systems (AuxLit 22-25) promise miniaturization of PDP implementation, and much greater speed, efficiency and computational capacity. The main source of significant improvements in performance of the quantum associative net and similar models (Wang et al., 1998) is, like in HNeT (as shown in Sutherland, 1990, 1994), the use of *quantum phase* φ, hidden in $\Psi(\vec{r},t) = A(\vec{r},t)e^{i\varphi(\vec{r},t)}$, for encoding and processing of information (details in Perus & Dey, 2000).

See Chapter 7 for explicit mathematical presentation of *quantum-holographic* realization of my Quantum Associative Network model for the special case of *image recognition using quantum waves.*

ENDNOTE

1 Both models are called "Quantum Neural Nets" and incorporate the sigmoid activation-function – in contrast to the quantum associative net which executes linear summation of signals, like the original quantum mechanics.

Chapter 10
Quantum Phase–Hebbian Image Processing

In this chapter, some specific characteristics of quantum-implementable phase-Hebbian content-addressable associative memory and pattern recognition are discussed. Quantum formalism, constrained by the closure relation, is generalized into a flexible information-processing system by a suitable (re)interpretation of quantum states involving "fuzzification" of the orthonormality and closure relations.

Core of the quantum associative net. Let us first repeat the core of the Quantum Associative Network model. In Feynman's path-integral formalism, the Schrodinger equation can be rewritten (AuxLit 16) in the form

$$\Psi(\vec{r}_2, t_2) = \int \int G(\vec{r}_1, t_1, \vec{r}_2, t_2)\Psi(\vec{r}_1, t_1)d\vec{r}_1 dt_1 \tag{10.1}$$

DOI: 10.4018/978-1-61520-785-5.ch010

where the kernel G is the propagator or the Green function having form of a projection-operator (Bjorken & Drell, 1964/65):

$$G(\vec{r},t_1,\vec{r}_2,t_2) = \sum_{k=1}^{P} \psi_k(\vec{r}_1,t_1) * \psi_k(\vec{r}_2,t_2) \tag{10.2}$$

The wave-function Ψ is a superposition of P eigen-wave-functions ψ_k which can represent plane-waves

$$\psi_k(\vec{r},t) = A_k(\vec{r},t)e^{i\varphi_k(\vec{r},t)} \tag{10.3}$$

or wavelets (Lee, 1996; Schempp, 1994, 1995).[1] For simplicity we will not use the Gabor wavelets as wave-packets ψ_k in the formalism explicitly, but this neuropsychologically-important option is allowed.

It is assumed that it is possible to encode information into quantum eigenwaves ψ_k: we let an *eigen-wave-function represent an image*. For each possible vector-basis ψ_k (k = 1,..., P) there is an expression of the same type as Equation (10.2) (Messiah, 1965) which "stores" the eigenpatterns and performs projections to eigen-subspaces. Now it will be shown how such a quantum system can be manipulated in order to realize content-addressable memory storage and associative retrieval.

Opening the closure relation. Because propagator (10.2) must reproduce the initial state in dynamical Equation (10.1) when $t_1 = t_2$, the quantum closure relation

$$\sum_{k=1}^{P} \psi_k(\vec{r}_1,t) * \psi_k(\vec{r}_2,t) = \delta(\vec{r}_1 - \vec{r}_2) \text{ or } \sum_{k=1}^{P} \psi_k(\vec{r}_1) * \psi_k(\vec{r}_2) = \delta(\vec{r}_1 - \vec{r}_2) \tag{10.4}$$

must be satisfied (Messiah, 1965).

Closure relation (10.4) implies the postulate of complete and orthonormal set of quantum eigenstates: i.e $\Psi = \sum_k c_k \psi_k$ (completeness), and the scalar product of eigenvectors ψ_k,which have norm 1, is 0 (orthonormality).

Prescription (10.4) ensures reversible and unitary quantum evolution[2] determined by the linear Schrodinger equation (implying complete orthonormal set of eigenwaves in kernel— Equation (10.2)) *if* the system is closed, i.e. if there is no disturbance from environment. On the other hand, the same kernel, (10.2), serves as a projection-operator realizing non-unitary, non-linear and irreversible "collapse of the wave-function" *if* the system is *open*, i.e. if there is a disturbance from environment (Wheeler & Zurek, 1983). If we "perturb" the system, incorporating an "information input" in it, we can use this projection (or "collapse") to an eigen-subspace as associative memory recall *if this disturbance is "intelligible"*. "Intelligibility" arises from the fact that the system encodes information (i.e., our "inputs") into its eigenstates, and we thus have knowledge about system's memory-states. Without having knowledge about the system's eigenstates, there is a very low probability that an experimenter's interaction with the system (which would in this case be "unintelligible") would trigger the same projection (or "collapse") to an eigen-subspace belonging to a simple quantum "pattern" (Perus, 1996).

Phase-Hebbian propagators. Since quantum phase-differences improve associative memory, we express phases and amplitudes explicitly. Thus, by inserting Equation (10.3) into Equation (10.2), we obtain the phase-Hebb rule

$$G(\vec{r},t_1,\vec{r}_2,t_2)=\sum_{k=1}^{P}A_k(\vec{r}_1,t_1)A_k(\vec{r}_2,t_2)e^{i(\varphi_k(\vec{r}_2,t_2)-\varphi_k(\vec{r}_1,t_1))} \tag{10.5}$$

This describes the memory encoding which is two-fold: it is both in amplitude -correlations $\sum_{k=1}^{P}A_k(\vec{r}_1,t_1)A_k(\vec{r}_2,t_2)$ (Hebb rule) and in phase-differences $\delta\varphi_k=\varphi_k(\vec{r}_2,t_2)-\varphi_k(\vec{r}_1,t_1)$ (We use $\hbar = 1$.)

Differentiating the phase φ, which is here equivalent to action, one gets:

$$\varphi=Et\Rightarrow\delta S=\delta\varphi_a+\delta\varphi_b=\delta Et+E\delta t$$

This implies that two complementary variations of G are possible:

$$G_a(\vec{r}_1,\vec{r}_2)=\sum_{k=1}^{P}A_k(\vec{r}_1)A_k(\vec{r}_2)e^{i(E_2-E_1)t} \tag{10.6a}$$

where the phase difference $\delta\varphi_a=(E_2-E_1)t$ results from different energies E at the same time t, and

$$G_b(\vec{r},t_1,\vec{r}_2,t_2)=\sum_{k=1}^{P}A_k(\vec{r}_1,t_1)A_k(\vec{r}_2,t_2)e^{-iE_k(t_2-t_1)} \tag{10.6b}$$

where the phase difference $\delta\varphi_b=E(t_2-t_1)$ results from the change of state in time, but the energy of the system is invariant.

Fuzzification. Quantum theory has been developed prior to computer revolution, therefore its formalism has been limited to the use of complete set of orthonormal eigenvectors (which represent quantum "patterns" in our information-processing application). Using extensive computer simulations[3], I have realized how it is possible to generalize quantum formalism to "almost orthonormal" eigenwaves (i.e., having scalar product more than zero, but close to zero) (Kainen & Kurkova, 1993). MacLennan (in Pribram, 1993, ch. 6, p. 181) calculated that a 5000-dimensional space (e.g., a net with 5000 "neurons" or "units") has 5000 orthogonal vectors (with scalar product zero) and 82 billion "almost orthogonal" vectors (images) with scalar product less than 0.1. Thus, the "fuzzification" increases the representational capacity enormously. (Mathematical details see in: Kainen & Kurkova, 1993; Kainen, 1992; Kurkov´ a & Kainen, 1996.)

With such a "fuzzified" quantum system, one can implement an associative holography-like information-processing procedure which has some interpolation and generalization capabilities. This goes beyond simple reproduction of parallel-distributively stored data which is realizable in an orthonormal system.

Propagator- kernel G is subject to the strict closure relation (10.4) if eigenstates are mutually orthonormal. If they are not orthonormal (e.g.,$\langle\psi_k,\psi_h\rangle=\epsilon$ for each k and h), kernel G is subject to a "fuzzified" closure relation (Perus, 2000b, fig. 1)

$$\sum_{k=1}^{P}\psi_k(\vec{r}_1,t)*\psi_k(\vec{r}_2,t)=\tilde{\delta}(\vec{r}_1-\vec{r}_2)\text{ or }\sum_{k=1}^{P}\psi_k(\vec{r}_1)*\psi_k(\vec{r}_2)=\tilde{\delta}(\vec{r}_1-\vec{r}_2) \qquad (10.7)$$

where $\tilde{\delta}(\vec{r}_1-\vec{r}_2)$ is a distribution similar to the Dirac delta-distribution, but it does not have zero width, it has a width $\epsilon' > 0$. So, making the closure relation (10.4) more "fuzzy" means that we spread the Dirac delta-distribution to a certain extend (Perus & Dey, 2000, fig. 1), or we increase the values of off-diagonal elements of the memory-matrix **G**. This is achieved by using "*almost* orthogonal patterns" as "teaching inputs" into the network. As a beneficial result, we get more adaptive and more flexible (i.e., not just reproductive) associative processing.

Conditions for image retrieval. A juxtaposition of Hopfield-net-based conditions for amplitudes and holography-based conditions for phases gives the conditions for retrieval of images from memory in the quantum associative net (cf., chapter 7). In order to obtain a selected result of quantum image-processing, we have to trigger the wave-function collapse. We will now discuss what phase relationships as well as amplitude relationships are necessary for unambiguous image recognition. Using Equations (10.5) and (10.3), we rewrite Equation (10.1) into a pair of equations where amplitudes A_k and phases φ_k are separated:

$$A(\vec{r}_2,t_2)=\sum_{\vec{r}_1}\sum_{t_1}\left(\sum_{k=1}^{P}A_k(\vec{r}_1,t_1)A_k(\vec{r}_2,t_2)\right)A(\vec{r}_1,t_1) \tag{10.8a}$$

$$e^{i\varphi(\vec{r}_2,t_2)}=\sum_{\vec{r}_1}\sum_{t_1}\left(\sum_{k=1}^{P}e^{i(\varphi_k(\vec{r}_2,t_2)-\varphi_k(\vec{r}_1,t_1))}\right)e^{i\varphi(\vec{r}_1,t_1)} \tag{10.8b}$$

(10.8a) is valid for my ANN of sec. 8.2; (10.8b) resembles phase-couplings in HNeT and in MacLennan's field computing. The role of amplitudes and phases in information processing is independent for these/our purposes. So, these equations are valid separately as just mentioned, and together – as in Equation (10.1) with the kernel (10.2). In this case, they can also be separated – as at least the computational practice requires (I do not provide any mathematical proof here to support this empirical indication.).

In separated system (10.8), the quantum system has been, just for simplicity of discussion, discretized: we get a network of "points" (or formal neurons) instead of a continuous field. Constructive and destructive interference between amplitudes and phases of interacting (eigen) states at particular locations and times occurs.

My computer simulations with concrete data show the following (consider Auxil. Sec. 8.5 together with chapter 7!): This interference-process is extremely complex and every detail is relevant. "Macroscopic" image-formation or image-reconstruction occurs – only in the case of phase-coupling or tuning, i.e. constructive interference. Let us first ideally assume that the memorized images are mutually orthogonal and normalized, i.e. they form an orthonormal vector-basis. If at least a part of the "input" is equal (or, alternatively, similar) to a chosen memorized image having index $k = k_0$, then the "output" (i.e., the result of the "collapse") is also equal (or similar) to the chosen memorized image with $k = k_0$.

Specifically, in (10.8a), if $A(\vec{r}_1,t_1) \doteq A_{k0}(\vec{r}_1,t_1)$, *then* $A(\vec{r}_2,t_2) \doteq A_{k0}(\vec{r}_2,t_2)$, and in (10.8b), if $\varphi(\vec{r}_1,t_1) \doteq \varphi_{k0}(\vec{r}_1,t_1)$ then $\varphi(\vec{r}_2,t_2) \doteq \varphi_{k0}(\vec{r}_2,t_2)$ [4]

That is because in (10.8a) [5], all terms $A_k(\vec{r}_1,t_1)A(\vec{r}_1,t_1)$, where $k \neq k_0$, are 0 (or close to 0), and $A_k(\vec{r}_1,t_1)A(\vec{r}_1,t_1)$ where $k = k_0$, is 1 (or close to 1), and because, in (10.8b), all terms $e^{-i\varphi_k(\vec{r}_1,t_1)}e^{i\varphi_k(\vec{r}_1,t_1)}$, where $k \neq k_0$, in average cancel out (or their sum remains small, but essential), and $e^{-i\varphi_k(\vec{r}_1,t_1)}e^{i\varphi(\vec{r}_1,t_1)}$, where $k = k_0$, is 1 (or close to 1).

If our previous assumption of mutual orthogonality would not be completely valid, but just roughly (i.e., scalar products between pattern-eigenvectors would be just close to 1), then we have a highly non-trivial situation of subtle "interferences" or "associations" (mutual "enhancements" or "inhibitions") among images. They cannot be discussed analytically in short, but their emergent global result can be observed in computer simulations. Not just the "resonant" terms with $k = k_0$ give contributions to the total-sum-output, but also other terms give small contributions. The latter ones together give a non-negligible output which is responsible for interpolation and generalization. Of course, this interference process is strongly dependent on the correlation structure of data (eigenfunctions ψ_k). All similarities (i.e., correlations) between images are optimally taken into account, but their importance is relative. If the aforementioned conditions are not satisfied, then the output is an "un-intelligible" superposition of possible images, i.e. a network state which is considered meaningless by humans.

Cortical Schrodinger-wave holography. Nobili (1985) presented a model of cortical holography, similar to the quantum associative net, based on Schrodinger (quantum) waves rather than on the usual (D'Alambert) waves. His reason is that the Schrodinger equation does not have any biologically-unjustified double time-derivative, which indicates a sort of "acceleration", of the wave-function as the D'Alambert wave-equation has. ("Accelerations" are "tissue-unfriendly".) Nobili (1985) thus writes the Schrodinger equation in Feynman's form, i.e. (10.1), with a newly- calculated Gaussian-like Green-function propagator:

$$G(\vec{r}_1, t_1, \vec{r}_2, t_2) = [2\pi(\epsilon + i\sigma)(t_2 - t_1)^2]^{-d/2} \exp\left(\frac{-(\vec{r}_2 - \vec{r}_1)^2}{4(\epsilon + i\sigma)(t_2 - t_1)}\right) \tag{10.9}$$

where d is the dimension of the medium, assumed to be made of cortical glia. By using special plane-waves $\psi_k(\vec{r}, t) = A(\vec{k})\exp(i(\vec{k}\vec{r}) - (\epsilon + i\sigma)k^2\tau]$, where is the wave-vector $2\pi/\lambda$ (λ is wavelength), Nobili constituted a complete set of orthogonal eigenwaves. These should interfere in order to realize a sort of large-scale holography in the cortex.

ENDNOTES

1 As always, A is the amplitude, φ is the phase at specific location $\vec{r}$ in specific time t, and k is the eigenstate index. (Eigenstate, eigenwave, eigen-wave-function are synonymous here, and so will also be eigenimage or eigenpattern.)

2 In unitary evolution, the eigenwaves ψ_k are just re-arranged in the superposition $\Psi = \sum C_k \psi_k$, with changed "weights" C_k. Nothing nonlinear (i.e., "nothing extraordinary") and "nothing new" happens.

3 I have tested the associative network on the example of recognition of tertiary (3-dimensional) structure of proteins using the Brookhaven protein data bank. Details see in sec. 8.5.

4 These two if-sentences are valid also if "$\doteq$" are replaced by "=".

5 After rearrangement of terms.

Chapter 11 Computational Models Relevant for Visual Cortex

In this chapter, some ANN models, which are potentially relevant for modeling cortical visual processing in the holonomic context, will be briefly presented. They either produce ICA-like processing or collective oscillatory dynamics with phase-processing. Activity of modules like these could shape the image-bearing wavelets. These models are presented, because they are extraordinarily computationally efficient and because they have some relevant common features in-between as well as with some physiological reports — and such a non-trivial compatibility is remarkable if found in autonomous models.

DOI: 10.4018/978-1-61520-785-5.ch011

ANN WITH INHIBITORY FEEDBACK-LOOP GIVING INFOMAX OUTPUTS

The Harpur & Prager (1996) model is illustrated in Figure 1(a). It performs *infomax* by an unique ICA-*like* process without any explicit phase-processing. It anyway achieves statistical independence by recurrent opposite-weighted connections and by the resulting competition between "neurons". *Output weight-vectors represent Gabor-like wavelets.*

The input vector is $\vec{x} = (x_1, \ldots, x_m)$; the output vector is $\vec{a} = (a_1, \ldots, a_n)$; feed-forward *(j-to-i)* connection-weights are elements in the vector $(\vec{w}_i = w_{i1}, \ldots, w_{im})$; their feedback (*i*-to-*j*) parallels are their negatives: $-w_{ij}$.

The ANN "algorithm", which (iteratively) processes images by minimizing the discrepancy $|\vec{x}\,'|^2$, is as follows:

$\vec{x}\,' = \vec{x} - \sum_{i=1}^{n} a_i \vec{w}_i$ (index *i* belongs to the output "neuron(s)");
$\Delta a_i = \mu \vec{x}\,' \vec{w}_i$, $0<\mu\leq 1$ (μ is adjustable learning rate);
$\Delta \vec{w}_i = \eta a_i \vec{x}\,'$ (η is learning rate).

Comparing the model's connectivity with an experimental report on the rat visual cortex[1], one can see similarities which suggest to relate $\vec{x}\,'$ with V1 and $\vec{a}$ with the extrastriate cortex. Namely, it is reported that feedback connections from the extrastriate cortex "provide input directly to [V1's pyramidal] neurons which make the reciprocal forward connection, and that feedback-recipient forward-projecting neurons are strongly interconnected". However, Harpur & Prager (1996) neglect the *direct* intra-layer connections, so there is just indirect (although essential) mutual influence between "neurons" of the "V1"-layer. *The Baird (1990) model of networks of oscillators* incorporates the reported extrastriate-to-striate "top-down" modulatory connectivity and the reported interconnections of forward-projecting excitatory neurons of V1, possibly.

NETWORK OF UNITS WITH COUPLED OSCILLATORY ACTIVITIES, EMBEDDED IN NEUROPIL

By comparing subfigures of Figure 1, one sees that the Baird (1990) oscillatory ANN model has a loop-structure similar to the Olshausen & Field (1997) net and to the Harpur & Prager (1996) net, and fits the experimentally found striate–extrastriate loop. However, Baird considers the sources of inhibitory feedback as *interneurons* without (significant) connections other than with their excitatory partner in the main

Figure 1. (a) The Harpur & Prager (1996, p. 278) ANN. / (b) The Baird (1990, p. 370) oscillatory ANN. / (c) The Olshausen & Field (1997, p. 3317) net

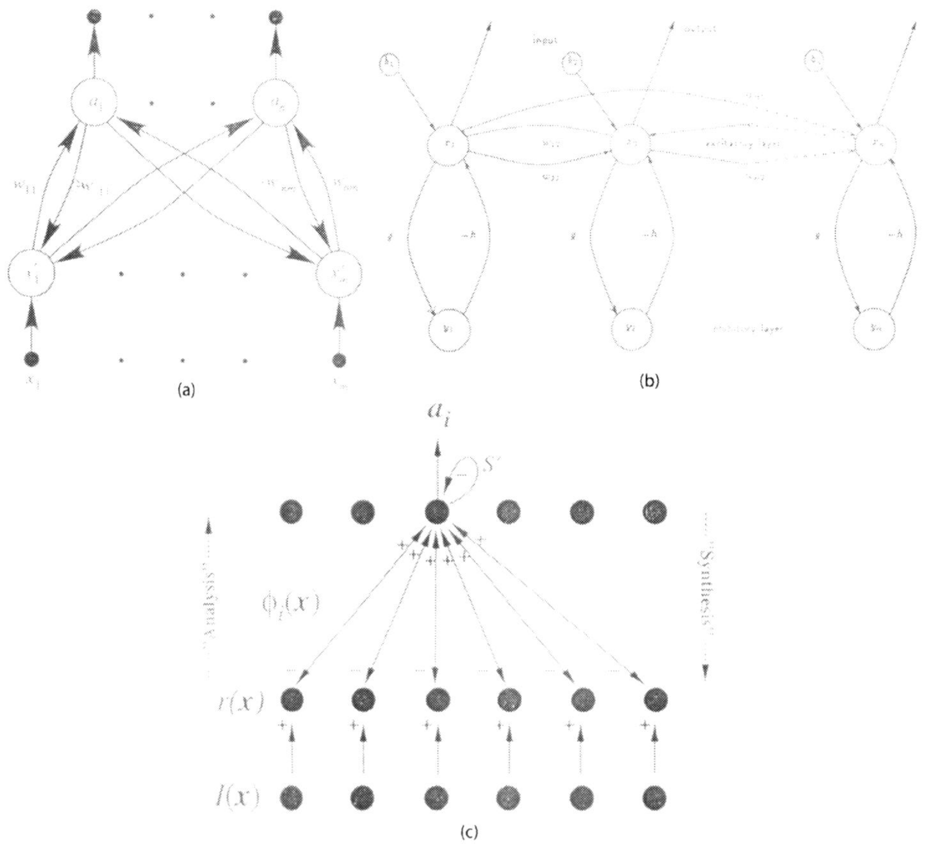

layer. Baird claims that he obeyed biological data and that his net, which can realize oscillatory content-addressable memory, could be relevant for modelling visual cortex (as well as the olfactory bulb). Baird's model might reveal something about (Gabor) wavelet formation, but even more about formation of temporal sequences of images (visual episodes).

Baird's (1990) full connection-matrix is $T = \begin{bmatrix} J & -hI \\ gI & 0 \end{bmatrix}$

b_i is the input; x_i is activity of an excitatory neuron; y_i is activity of an inhibitory neuron –such (inter)neurons act in a local inhibitory feedback-loop, as in Figure 1(b). The coupled dynamics (t is time) of units' activities are described by equations:

$$\frac{\partial x_i}{\partial t} = -x_i - h\sigma(y_i) + \sum_{j=1}^{N} J_{ij}\sigma(x_j) + b_i;$$

$$\frac{\partial y_i}{\partial t} = -y_i - g\sigma(x_i).$$ (σ is a sigmoid function)

The same system of equations, without the $\sum_j J_{ij}\sigma(x_j)$ term, is reported also by Mannion & Taylor (1992) to produce coupled oscillatory activities of neurons.[2] Their equation for temporal changes in phase φ is (in Baird's notation): $\frac{\partial \varphi_i}{\partial t} = b_i + \sum_j J_{ij}H(\varphi_j - \varphi_i)$, where H represents weights of inhibitory connections. Mannion & Taylor (1992, p. 102) describe the circuit's functioning as follows:

The activity of one neuron of the pair excites the other until it is so strongly excited that its feedback inhibition finally turns on the first. This then dies away, so turning off its inhibitory partner, which then releases the first cell from inhibition. This then starts the cycle again.

Baird's (1990) inhibitory feedback "neurons" produce oscillations for the orthonormal real-valued static eigenpatterns—attractors stored in the Hebb matrix **J** with elements $J_{ij} = \sum_k x_i^k x_j^k$ (k is the eigenpattern index). The result are complex-valued eigenpatterns of oscillatory activity-amplitudes of the full matrix T (i.e., J plus negative feedbacks) which is thus *phase-Hebbian* (phases are hidden in the resulting oscillatory activities x_i). Baird then adds a higher-order ("sigma-pi") Hebb matrix $\mathbf{J}^{(2)}$ with elements $J_{ijdl} = \delta_{ij}\delta_{dl} - \sum_k x_i^k x_j^k x_d^k x_l^k$, which ensures that the **T**-eigenpatterns participate in a multiple periodic *2nd* -order attractor— *an attracting periodic trajectory* visiting all the 1st –order fixed-point attractors corresponding to individual images. This is a *temporal sequence* or "episode" made of images (similar models see in AuxLit 11). Such higher-order oscillatory dynamics can turn into standing waves. The quadruple-correlation matrix $\mathbf{J}^{(2)}$ is proposed by Baird to be implemented in (2) the *synapto-axo-dendritic web* called the *neuropil* (Figure 1.2).

FIGURE/GROUND SEGMENTATION BY A NETWORK OF PHASE-COUPLED OSCILLATORS

The Sompolinsky & Tsodyks (1994) network of oscillators realizes segmentation of a visual scene, as introduced in section 6.6, with help of stored scenes where each

Figure 2. Left: Initial state of the net of oscillators (each illustrated as a "magnetic needle" or "spin" which rotates). / Right: Two figures segregated from the common background. (From Sompolinsky & Tsodyks (1994), p. 649)

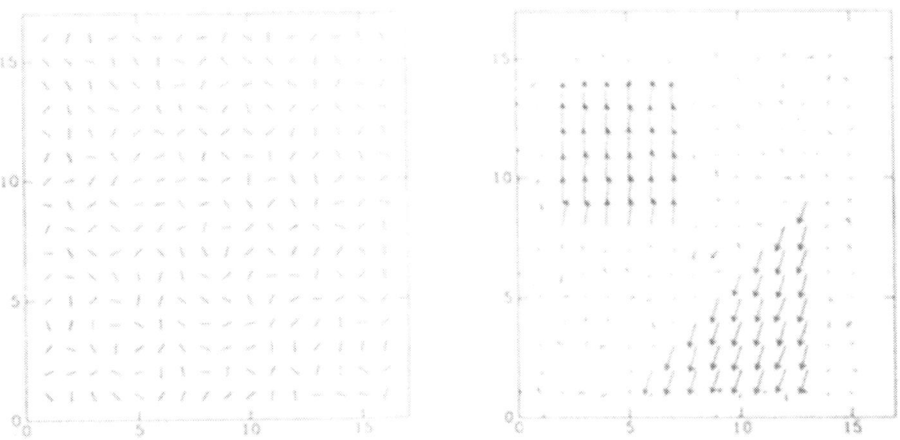

includes similar figures, i.e. object-forms, on a background. The probability that a neuron at position $\vec{r}$ emits a spike at time t is oscillating — this is understood as the oscillatory activity in this model. The coupled oscillatory dynamics lead to *selective phase coherence*: "neurons–oscillators", which collectively co de the figure, synchronize their phases; others, which collectively code the figure's background, keep "oscillating without synchronized phases".

Images with several figures on a background are stored in Hebb-like matrices with elements describing phase-relations. As the phases selectively change, *phases within each individual figure become all the same — this is illustrated by a common orientation of "spins" in a specific direction* (Figure 2). In other words, *couplings between phases, which store the images, are modified in a Hebb-like manner,* so the model is essentially *phase-Hebbian*. The result of phase-changes is labelling each figure of an image by a specific "phase orientation" of all the oscillators that collectively constitute the figure.

To achieve figure-selectivity of each coherence-patch, which each corresponds to a figure, two dynamic equations for phases are used. The first one is for phases $\varphi(\vec{r},t)$ of all image- pixels; the second one is for phases $\phi_F(t)$ of image-pixels belonging to a figure, called cluster-phases (phases specific to the cluster F):

$$\frac{\partial \varphi(\vec{r},t)}{\partial t} = \omega + \eta(\vec{r},t) - \sum_{\vec{r}\,'\neq\vec{r}} J(\vec{r},\vec{r}\,')\sin[\varphi(\vec{r},t) - \varphi(\vec{r}\,',t)]$$
$$\frac{\partial \phi_F(t)}{\partial t} = \eta_F(t) - \sum_{F'\neq F} J(F,F')\sin[\phi_F(t) - \phi_{F'}(t)]$$

ω is the frequency of the neural oscillations and η represents external noise. "Neuron's" index $\vec{r}$ does not only denote its location, but rather its figure-specific preferred local phase θ and the index F of the cluster (of the figure) it participates in.

In both phase-equations above, there are phase-differences (which measure discrepancies among pixels) inside the sin-oscillation-function. The phase-types are related as in equation:

$$\varphi_F(\theta,t)=\omega t+\phi_F(t)$$

The coupling-rates J between phases and those *J* between cluster-phases are both prescribed by a different (phase)-Hebbian rule (they will be omitted for shortness here). Figure-specific local phases θ_F have a specific value ("spin orientation") θ_F^0 if inside the cluster F, and random value if outside it.

This highly non-trivial process of producing figure-selective phase-synchronization or coherence provides a result as in Figure 2-right. Of course, it could in principle be implemented in classical and quantum spin systems (Mezard et al., 1987; AuxLit 14), magnetic or other (bio) physical complex systems.

ENDNOTES

1 R. Johnson & A. Burkhalter (1997): J. Neurosci. 17, 7129-7140

2 Similarly "flip- flop"-connected "neurons" are used also in ANN by Sienko (in Dubois, 2000b). It produces coupled net-oscillations, too, which are then applied for quantum information-transfer and quantum computing

Appendix A: Principal Component Analysis

Principal Component Analysis (PCA) is almost equivalent to Singular Value Decomposition (SVA), or Karhunen-Loeve expansion. It will be presented first as an important computational method for feature extraction from input-data (Ritter, Martinetz & Schulten, 1992; Haken, 1996; AuxLit 10). To perform PCA, input-patterns $\vec{x}^k$ are decomposed into a series, i.e. a linear combination of prototype-patterns $\vec{w}^r$ (r = 1, ..., p'):

$$\vec{x}^k = \vec{w}^0 + \sum_{r=1}^{p'} \vec{w}^r c_r(\vec{x}^k) + \vec{R}(\vec{x}^k) \tag{12.1}$$

DOI: 10.4018/978-1-61520-785-5.ch012

$c_r = \vec{w}^r \vec{x}^k$ are the *principal components* of the input-patterns. Their number is p'. Variable c_r could also be treated as the activity-rate of corresponding dominant or cardinal neuron $\vec{r}$. Index r corresponds to location of a cardinal neuron $\vec{r}$. At the beginning, all principal components c_r are approximately equal and lower than 1. Then we can talk about potentially-cardinal neurons. Later environmental stimulus gives privilege to one pattern and its c_r increases towards 1 (wins). This means that one neuron becomes actually-cardinal, other neurons get subordinated. We are able to store all inputs $\vec{x}^k$ completely (so, there is no need for data compression) if $p = p'$. In this "ideal" case index k and index r are equivalent. If, on the other hand, there is $p > p'$, data-compression causes that a higher number *(p)* of inputs $\vec{x}^k$ is represented by a lower number (p') of cardinal neurons c_r or/and their set of synapses $\vec{w}^r$ (or $\vec{w}^r$, respectively). c_r -representation is a relatively sparse code; $\vec{w}^r$ is a relatively *distributed* code.

$\vec{w}^0$ specifies the center of weight of the input data distribution $P(\vec{x}^k)$ ($\vec{w}^0$ are very often represented by mean values – e.g., as in: Bankman in Pribram, 1993, p. 77). Other p' prototype-vectors $\vec{w}^r$ form a so-called eigen-space basis. These basis-vectors, which are typically *mutually orthogonal, coincide with (relatively pictorial) memory representations*, i.e. so-called *eigen-images*. In fact, these p' vectors are eigen-vectors of the *autocorrelation- matrix* **C** that have *the largest eigen-values* (Ritter et al., 1992):[1]

$$\mathbf{C} = \sum_{k=1}^{P} (\vec{x}^k - \vec{w}^0) \otimes (\vec{x}^k - \vec{w}^0)^T P(\vec{x}^k). \tag{12.2}$$

T denotes the transposed vector (i.e., line vector); ⊗ denotes the outer or tensor product: $(\vec{a} \otimes \vec{b})_{ij} = a_i b_j$. **C** is an (auto) correlation matrix of the difference-vector $\vec{x}^k - \vec{w}^k$, therefore it is a *covariance matrix*. Typically, mutually orthogonal basis-vectors capture the directions of maximal variance (dispersion) of the data.

Equation (12.1) defines a hyper-plane which passes through the center of weight $\vec{w}^0$ and is spanned by principal axes along all $\vec{w}^r$. $\vec{R}(\vec{x}^k)$ is a residual vector which represents a non- vanishing distance from the approximating hyper-plane perpendicular to it. If $\vec{R}(\vec{x}^k)$ would be zero, our approximation with principal eigenvectors $\vec{w}^r$ (prototypes) or, equivalently, with principal components c_r (cardinal neurons corresponding to prototypes) would be perfect.

To summarize, $\vec{w}^r$ are eigenvectors of the covariance-matrix C with the largest eigenvalues λ_r (Ritter et al., 1992; Haken, 1996; AuxLit 10):

$$\sum_j C_{ij} w_j^r = \lambda_r w_i^r \tag{12.3}$$

c_r are *projections of the input-patterns* $\vec{x}^k$ along the *principal axes* defined by the eigenvectors $\vec{w}^r$. In each learning cycle, following each input, one potentially-cardinal neuron wins the competition, others are suppressed. This is equivalent to increase of one principal component c_{r0} to 1, and decay of other projections $c_r (r \neq r_0)$ onto corresponding eigen-patterns $\vec{w}^r$ $(r \neq r_0)$. After the next input, another principal component may get privilege to dominate.

Kohonen's self-organized topology-preserving maps, which are presented in Appendix B, are actually generalizations of the principal-axes-calculations (Ritter et al., 1992).[2] Namely, they use (instead of linear principal axes and linear eigen-spaces) *principal curves* or *principal surfaces* (figure 1).

Figure 1. Approximation of a two-dimensional input-distribution $P(\vec{x}^k)(k = 1, 2)$ with the center of weight $\vec{w}^0$ by a principal axe along $\vec{w}^1$. Left: linear case. Right: non-linear case where a principal curve would fit the data better than a principal axe.

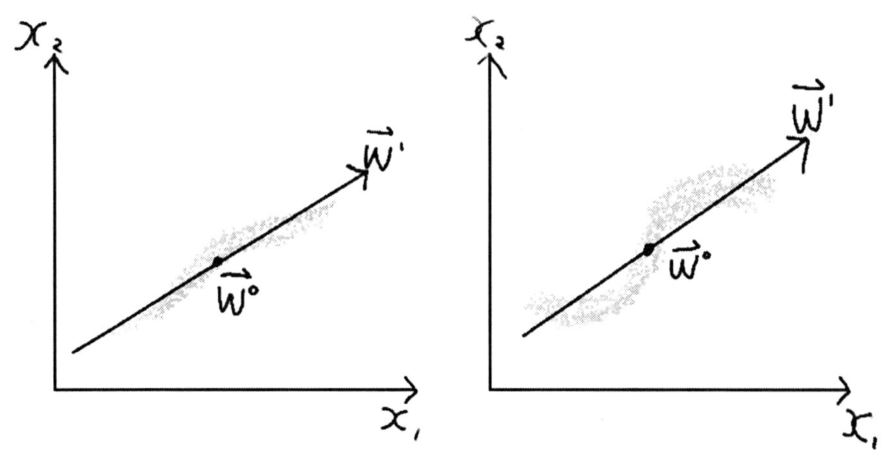

ENDNOTES

1 Definitions of common mathematical terms, like eigenvalues etc., can be found in any textbook on linear algebra, e.g. in D.T. Finkbeiner II (1978): *Introduction to Matrices and Linear Transformations*; San Francisco: W.H. Freeman & Co., 3rd ed.

2 Therefore, the same notation as in Appendix B was used, although a somewhat different PCA could be performed using eigen-patterns $\vec{v}$ (as they will be defined and used) as the basis (Haken, 1996, ch. 11).

Appendix B: Kohonen's Model of Retinotopic Mapping

INTRODUCTION TO PERCEPTUAL MAPS WITH CARDINAL NEURONS

Biological consideration. The computational model of so-called self-organizing ANN which adapt to perceptual data by evolving brain-maps with preserved input-data relations was established by Teuvo Kohonen in early eighties (Kohonen, 1982, 1995). For vision, it might have some biological significance for those levels where the "Mexican-hat" - like receptive fields were found: in retinal and LGN cells. Swindale (1996) presented in detail how this model could fit globally the experimental data on the visual-cortex topology, especially development of ocular-dominance columns and columns selective to orientation of stimuli, and also on global retinotopic mapping.

DOI: 10.4018/978-1-61520-785-5.ch013

The model should be used only as a "metaphor", although it performs very well in computer simulations. Namely, it might give us some rough intuition about global perceptual computation if applied to particular levels of a long multi-level computation along the retino-geniculo-striate visual pathway. However, it cannot be applied directly to any level of the visual pathway, and also not as an implementation model for the global retinotopic mapping onto V1 as a whole, because it neglects biological details in many stages (e.g., LGN), it ignores experimentally-supported V1-profiles of receptive fields (Gabor filters), and also phase-information has not yet been incorporated. This situation was not satisfactorily improved also after many followers of Kohonen (overviews in: Swindale, 1996; Kohonen, 1995) have extensively tried to fit particular biological details.

The main strength of the model is computational flexibility which approximates cortical map formation, i.e. e ectively reproduces it in simulations. Computational power is accompanied by mainly unjustified modularity (as with many popular and applicative computational models). The main weakness is lack of biological plausibility on the level of individual neurons and especially sub-neuronal processes. Therefore, there are significant differences between Kohonen-based mapping models and the holonomic theory, although some other models like the "infomax" perceptual-net model (Linsker, 1988) may suggest compromising relations.

Introduction to the model. Let us now present the basic ideas of the *self-organizing topology-preserving ANN* (Ritter, Martinetz & Schulten, 1992; Peretto, 1992). Encoding is constructed by *reducing the difference (error) between the external state and the network's internal representation of that state or similar previous states*. Both, internal and external states, are presented as activity patterns and are mathematically described by vectors. The individual components of each pattern-vector corresponds to an individual element of the pattern. The synaptic connections, where patterns are stored, change according to the rate of this disagreement between the prototype (internal representation) $\vec{w}$ and ordinary pattern (pattern of environmental stimuli detected by sensory cells) $\vec{x}$. Prototype can be prescribed by a "teacher", which could be provided by a higher-order cortical center (this is supervised learning), or can be established by a self-organizing procedure (unsupervised learning). In the second case, which is biologically more relevant, the prototype has been constituted by the previous experience and is encoded by the "winning neuron". This is the neuron which wins the competition between neurons and more-or-less individually takes the task of coding a specific pattern. The winning neuron is also called *cardinal neuron*, or *order-parameter cell* (because it orders the whole layer of neurons after it has "taken all the power") (Haken, 1991). To be more specific, as long as such a neuron competes with other dominant neurons we will refer to it as *potentially-cardinal neuron*. At that stage also its competitors are potentially-cardinal neurons. Only one of these potentially-cardinal neurons can win the com-

Figure 1. Kohonen's Self-Organizing Mapping network with the first (input) layer of sensory neurons and the second (target, output) layer of cardinal neurons which constitute maps. Potentially-cardinal neurons inhibit each other.

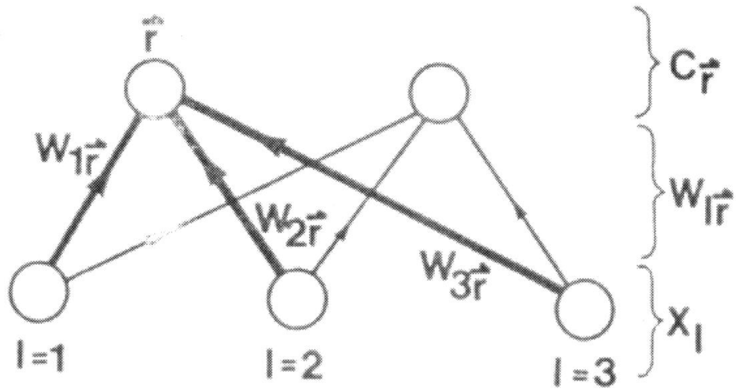

petition and become the most activated one. In that case it is called actually-cardinal neuron. Such a winning neuron, located at $\vec{r}$, represents its corresponding prototype-pattern $\vec{w}_{\vec{r}}$. All new-incoming patterns $\vec{x}$ are compared to this "ideal or typical" memory representation (called also *target-pattern* $\vec{w}$) and are classified according to this most similar prototype $\vec{w}$. Of course, a neuron can win the competition only if it has support of the actual state in the environment.

Connectivity. Kohonen's network producing the so-called *Self-Organizing Maps* is a combination of a fully-interconnected network, if we look the target layer only, and a feed-forward network if we look vertically through the hierarchy of layers (Figure 4: more layers could also be added) (Peretto, 1992). Within the target-layer (output-layer) connections are non-plastic and *inhibitory*, whereas inter-layer connections are plastic. In the model of Self- Organizing Maps, it is usually taken that neurons have activity-values between 1 (excited) and 0 (quiescent), and that synapses have positive and negative real values including 0.

Receptive field is a population of cells in an earlier/lower level of the visual pathway (let us call them sensory cells) which are connected to a specific potentially-cardinal neuron (in the later/higher level). This cardinal neuron is specialized to be most sensitive to a specific input-pattern only. A cardinal neuron with its corresponding receptive field classifies and stores a specific input-pattern. A receptive field of a potentially-cardinal neuron is a domain of (its) sensory cells, e.g. on our model-retina, whereas a *map* is a domain of all potentially- cardinal neurons in our model-cortex.

Filtering. Each receptive field of a neuron has a form of a tree-like net (many sensory cells connected to that neuron) which was called perceptron in early ANN models. Such perceptron effectively performs a role of an *adaptive filter*: it selects those pattern which is the most similar to the right target-pattern. Each target-pattern is represented by a potentially-cardinal neuron. The cardinal neuron whose target-pattern w is the closest to the actual input-pattern x becomes the most activated one. The "filter" is mathematically illustrated by a "weighting profile", usually similar to a Mexican hat. Therefore we often simply talk about "Mexican hat" receptive fields, although the "Mexican hat" is just a filtering profile (like a Gabor wavelet is in V1) corresponding to the receptive field of a particular neuron (in retina or LGN).

TOPOLOGICALLY-CORRECT FEATURE MAPPING

Similar inputs coded in near-by neurons. Neurophysiologists have discovered specific brain regions which encode and process specific kinds of perceptual information.[1] For this localized encoding, topologically-correct maps are essential. In this case, mapping means a transformation of an input-pattern $\vec{x}$ to a corresponding internal representation or memory of that input. Topology-preserving mapping means that *adjacent points of an input-pattern are connected to adjacent neurons in the target-layer*. So, topological relations are conserved while the input-pattern $\vec{x}$ is projected to the target-pattern (output-pattern) $\vec{w}$.

Self-organization means that the mapping is a result of interactions between neurons and "spontaneous search of collective agreement" (Kohonen, 1984; AuxLit 9). The trigger of this process is a stimulus from environment. This perturbation is enough for forcing the system into search for a new equilibrium. Stability is found by forming a state where the actual input-pattern $\vec{x}$ and the target-pattern $\vec{w}$ are as close as possible, i.e. $\| \vec{w} - \vec{x} \|$ is minimal. Actually, input-pattern $\vec{x}$ selects those target-pattern $\vec{w}$ which is the most similar to it. This is done by selective activation of those sensory cells x_l which "by chance" have at that moment such an activity which matches the stimulus.

Self-organizing mapping is an optimization process where the similarity of input signals *is projected to proximity of excited neurons*. So, similar input patterns will be encoded into topologically close neurons. "Topologically correct" mapping means that neurons which are physically located close to each other will react similarly to similar inputs, but neurons located far apart of each other will react differently to similar inputs.

Multiple optimization. In neural networks, dynamics is in general governed by minimizing a co-called *cost function* which is analogous to energy-minimization in physics: the cost-function has to be minimized if the system has to reach equilibrium.[2] In Kohonen's model, the cost function is proportional to the square of difference between the input-pattern $\vec{x}$ and the target-pattern $\vec{w}$. The cost-function of

Self-Organizing Maps is analogous to the elasticity energy which is proportional to the dislocation from the equilibrium point (here $\vec{w}$). The network can "save energy" if neurons with similar tasks communicate over very short connection paths. Therefore the mapping where *similarity relationships among the input-patterns* are transformed into *spatial relationships among the responding neurons* is an "energy"-optimization process (Ritter, Martinetz & Schulten, 1992).

At the same time, the self-organized topology-conserving mapping means *dimensionality reduction of the representation space.* A high-dimensional space of input-patterns is reduced usually to two-dimensional maps which maintain neighborhood relationships. Input-pattern has a dimension equal to the number *n* of constituting elements - excited sensory cells *l*. Cardinal neurons are determined by x and y coordinates belonging to the cortical map. This mapping means also data compression. Network learns to discern the most important features of input-patterns. Only these main characteristics of the input-information are stored.

The self-organized map is a result of continuous and non-linear approximative mapping which is defined implicitly by a self-organized process of searching stability through maximal possible agreement of neurons. This process is called *learning*. It is driven by learning examples from the network's evolutionary history.

Columns. In the neocortex there are, vertically cutting its layers, the cortical columns and micro-columns. They can also be understood as a pro duct of self-organized topology-conserving mapping. They are thus *specialized to analyze some particular features* of the incoming visual patterns (orientations, periodicity, color shades, and approximately also edges, velocities and directions of movements, etc.), or to organize motor actions (e.g., innervations of a common muscle). In association areas, columns are densely connected to collectively execute complex tasks like recognition of faces, speech comprehension and planning of arm trajectories. Columns provide informational basis for higher brain functions modelled by attractor neural networks.

ADAPTIVE LEARNING BY ERROR MINIMIZATION

Using topology-preserving mapping, a large set of p input-patterns $\vec{x}^k$ ($k = 1,..., p$) gets encoded into a smaller set of p' prototypes $\vec{w}_{\vec{r}}$. Here $\vec{r}$ is used as an index which refers to the location of a (potentially-cardinal) neuron, which receives signals from sensory cells that it is connected with. The kth input-pattern is described by a vector $\vec{x}^k = (x_1^k, x_2^k, ..., x_n^k)$. Its components represent individual pixels or elements of the input-pattern which were projected into activities of individual sensory cells on the retina (vision), cochlea (audition), etc. n is the number of sensory cells.

The prototype is represented by a vector $\vec{w}_{\vec{r}} = (w_{\vec{r}1}r, w_{\vec{r}2}, ..., w_{\vec{r}n})$. Its components $w_{\vec{r}l}$ actually represent synaptic connections between the potentially-cardinal neuron at location $\vec{r}$ and the sensory cells with index *l*. If some prototype $\vec{w}_{\vec{r}}$ matches the input-pattern $\vec{x}^k$ more than other prototypes do, then the corresponding potentially-cardinal neuron at location $\vec{r}$ wins the competition against other potentially-cardinal neurons $\vec{r}\,'$ and becomes actually- cardinal neuron - the winner. In this case we say that the network has recognized the input pattern $\vec{x}^k$ so that it classified this input as an example of the prototype $\vec{w}_{\vec{r}}$. By meeting every new input-pattern, the prototype changes somewhat: the new updated prototype is a result of a com-

Figure 2. Kohonen network presented by a block of synaptic strengths or synaptic weights. Each sensory cell x_l is connected to each neuron in the 7 x 7 output-map. 6-dimensional input x (here presented one only: p = 1) is projected into 2-dimensional map of 7 x 7 neurons (as many as there are squares on the top). Each neuron is presented by a vertical column which corresponds to the vector of synapses between this neuron and all sensory cells x_l (l = 1,..., 6). A "level" of synaptic-weight-values which are given by the 3rd input variable x_3 is shaded. Dark-shaded circles correspond to the 3rd synapse on the neuron (1,1), and the 3rd synapse on the neuron (6,5). In this network-block 49 uncorrelated 6-dimensional input-patterns could simultaneously be stored. If inputs are correlated, memory compression starts. (Courtesy: J. Zupan (1994): Acta Chim. Slov. 41, 327-)

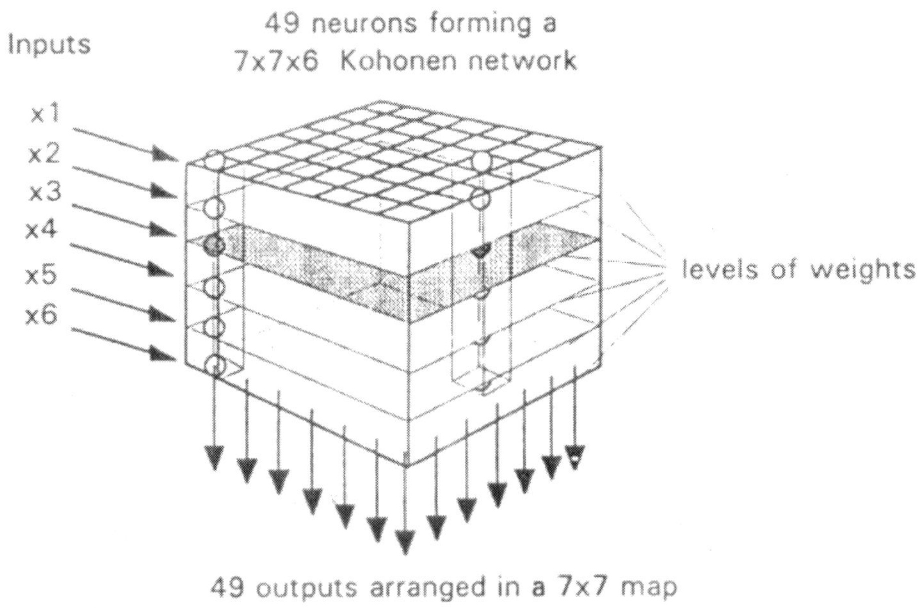

promise between the old prototype and new input. The prototype is encoded in the synaptic connections $w_{\vec{r}l}$ and/or in the cardinal neuron at location $\vec{r}$. We cannot localize the encoding since the synaptic strengths $w_{\vec{r}l}$ and the resulting activity of the cardinal neuron $\vec{r}$ both contribute to the mapping as a *dynamic process*. Statics does not mean anything to the network; only dynamics has meaning in the sense of realizing specific input-output transformations. (Mathematical description, which is static, is therefore not enough for understanding neural network processing. One needs a lot of intuitive, flexible "mental modelling" also.)

"Algorithm." All equations, written from now on, will be valid for all *sequentially occurring adaptation steps following each arrival of an input-pattern* $\vec{x}^k$, therefore the index *k* will always be omitted. The activity of some potentially-cardinal neuron in the second layer (target-layer or map) is given by (Ritter et al., 1992):

$$c_{\vec{r}} = f(\sum_{l=1}^{n} w_{\vec{r}l} x_l + \sum_{\vec{r}\,'} J_{\vec{r}\vec{r}\,'} c_{\vec{r}\,'} - T_{\vec{r}}) \tag{13.1}$$

f is the non-linear (tangens-hyperbolicus-like) activity function of the neuron $\vec{r}$ called *sigmoid function*. T is the threshold of the neuron $\vec{r}$. n is the number of sensory cells x_l in the first layer which are connected with each potentially-cardinal neuron $\vec{r}\,'$ in the second layer (map). The neuron's activity $c_{\vec{r}}$ is updated by a "sigmoid-shaped" sum of three contributions:

1. A sum of all inter-layer synapse-weighted signals x_l from sensory cells *l*;
2. A sum of all intra-layer synapse-weighted signals $c_{\vec{r}\,'}$ from other neurons $\vec{r}\,'$ in the map;
3. The threshold *T* of the receiving neuron $\vec{r}$.

$w_{\vec{r}l}$ is the strength of an inter-layer sensor-neuron synaptic connection. $J_{\vec{r}\vec{r}\,'}$ is the strength of an intra-layer neuron-neuron synaptic connection.

Let us assume that the neuron $\vec{r}\,'$ is the winner. Then usually, $J_{\vec{r}\vec{r}\,'} > 0$ (excitatory synapses) for neighboring neurons, and $J_{\vec{r}\vec{r}\,'} < 0$ (inhibitory synapses) for neurons more far away. So, around each neuron there is first a concentric circle of positive synapses (support of similarly-oriented neurons), and then a concentric circle of negative synapses (*lateral inhibition*, found in biological perceptual systems). Those neuron who has the most strong support of the nearest neurons (its inside domain), and is at the same time able to inhibit neurons in the neighboring domains (outside its domain), will become the winner. The concentric order of synaptic strengths is illustrated by the "Mexican hat" function (Figure 6). The response of such a self- organizing- mapping network to a stimulus *is spatially-localized to an oligarchy of neurons* around the "winner" neuron $\vec{r}\,'$. Because it is the strongest at the cardinal neuron $\vec{r}\,'$, Kohonen (1982) and successors have for simplicity often made approximation by considering only the response of the "winning" neuron as if other neurons $\vec{r}$ would not respond at all ("winner takes all"):

$$\sum_{l=1}^{n} w_{\vec{r}l} x_l = \max_{\vec{r}} \sum_{l=1}^{n} w_{\vec{r}l} x_l \tag{13.2}$$

If for each neuron $\| \vec{w}_{\vec{r}} \|$ (that is $\sqrt{\sum_l w_{\vec{r}l}^2}$ - total synaptic strength per neuron) is constant, and if $\| \vec{x} \| = 1$ (all input-patterns have the same "intensity"), then the approximation can be written also as follows:

$$\| \vec{w}_{\vec{r}'} - \vec{x} \| = \min_{\vec{r}} \| \vec{w}_{\vec{r}'} - \vec{x} \| \tag{13.3}$$

In the model of Self-Organizing Maps, it is usual to represent the target-pattern by a vector of strengths of synaptic connections ($\vec{w}_{\vec{r}}$) from each sensory cell l (l = 1,..., n) to a given neuron $\vec{r}$. By this definition input-pattern $\vec{x}$ and target-pattern

Figure 3. Mexican-hat function: distribution of excitatory and inhibitory synapses to the neurons encircling the center-point neuron

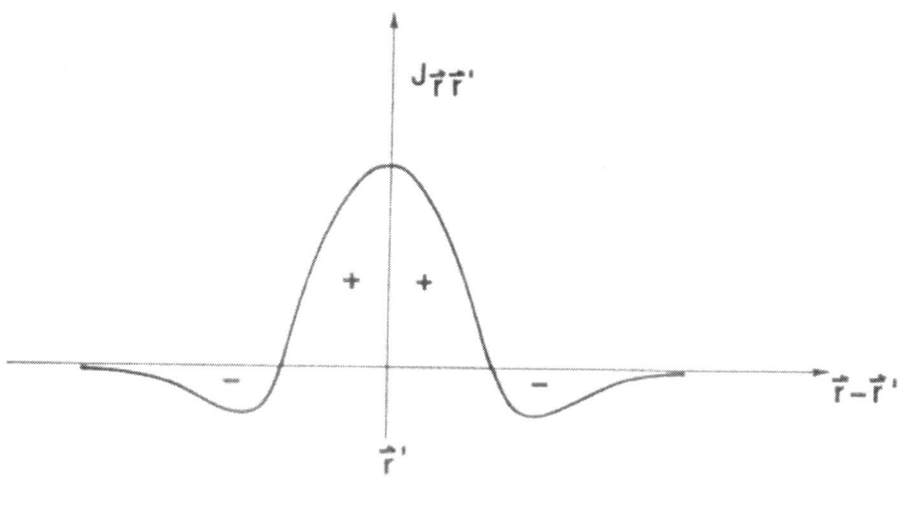

$\vec{w}_{\vec{r}}$ have the same dimension (n) and can be subtracted. Now we see what is advantage of such kind of mathematical description. Activities of neurons $c_{\vec{r}}$ are not important in *this* model. Only their *relations* to configurations of sensory-cells (equivalently, the receptive-field "weighting function") are important. These relations are encoded in configurations of synaptic connections ($\vec{w}_{\vec{r}}$) to a given neuron ($\vec{r}$). It is enough if a neuron is denoted by its location $\vec{r}$ only.

The version (13.2) chooses the most excited neuron to be the winner. The version (13.3) chooses those cardinal neuron which represents the target-pattern closest to the input-pattern $\vec{x}$. Both ways bring the same effect.

The strengths of synaptic connections are changed according to

$$\Delta w_{\vec{r}l} = \varepsilon h_{\vec{r}\vec{r}'}(x_1 - w_{\vec{r}l}) \; or \; \Delta w_{\vec{r}} = \varepsilon h_{\vec{r}\vec{r}'}(x_1 - w_{\vec{r}}) \tag{13.4}$$

where

$$h_{\vec{r}\vec{r}'} = \exp\left(\frac{-(\vec{r} - \vec{r}\,')^2}{2\sigma^2}\right) \tag{13.5}$$

σ is the radius of the Gaussian excitation function $h_{\vec{r}\vec{r}'}$. σ is fixed in the first approximation, but in the second approximation it can be slowly decreased during learning. In equation (13.5) $\vec{r}\,'$ marks the excitation maximum gained by the actually-cardinal neuron. The "shape" of $h_{\vec{r}\vec{r}'}$ is fixed with respect to variation of $\vec{r}$ (locations of the neurons in surroundings). The position of Gaussian function $h_{\vec{r}\vec{r}'}$ is chosen by an incoming stimulus which enables some specific potentially-cardinal neuron to become the actually-cardinal neuron $\vec{r}\,'$. The response of nearby neurons $\vec{r}$ is given by $h_{\vec{r}\vec{r}'}$. Introduction of this function is necessary in order to achieve localized neuronal excitation.

Discussion. The prescription (13.4) originates from neurophysiological and neuropsycho-logical models (Kohonen, 1984). Although details of its biological implementation may be questionable, especially for the visual system, it seems globally and functionally acceptable and has some correspondence to the *Test*-stage of Pribram's TOTE servomechanism (sec. 2.4). The so-called learning rule (7) says that the synapses $w_{\vec{r}l}$ should be changed more if the difference between the individual input-stimuli x_l and target-activities $w_{\vec{r}l}$ is bigger, or that the rates of synaptic transmission x_l should be changed less if the difference is smaller. So, larger disagreement requires more adaptation. This is not all: the spatial distance between neurons is also important. Its role is encoded in the shape of $h_{\vec{r}\vec{r}'}$, which controls the size of the neighborhood region. With that, the number of neurons participating in the localized response, after a single adaptation step (13.4), is determined as well as the rate of bilateral influence between them. The target-patterns $\vec{w}_{\vec{r}}$ associated with the neighboring potentially-cardinal neurons $\vec{r}$ are coupled more strongly than target-patterns $\vec{w}_{\vec{r}}$ corresponding to more distant potentially-cardinal neurons $\vec{r}$.

Learning is the process where the adjustments (13.2) or (13.3), respectively, and the adjustment (7) are following each other in an *iteration*-procedure (Peretto, 1992; Ritter et al., 1992). Equations (13.2), (13.3) and (13.4)[3] are equally valid for all adaptation steps marked by k if we keep in mind that $\vec{x}^k$ is new in each step. In the beginning of each (k^{th}) cycle of iteration, the new input-pattern $\vec{x}^k$ is presented to the network. Then the winning neuron is chosen by equation (13.2) or equation (13.3), and synapses are changed according to the rate of similarity between the input-pattern and the target-pattern. In a computer simulation it can be prescribed whether only one learning cycle or many cycles are following each input. Self-organizing adaptation causes a shift of the input-vector $\vec{x}^k$ and the target-vector $\vec{w}_{\vec{r}}$ closer to one another. The adaptation process is finished when $\vec{w}_{\vec{r}} = \vec{x}^k$. We say that in this moment the input-pattern was *recognized* and simultaneously it gets *stored* in the new vector of synapses $\vec{w}_{\vec{r}}$. In such a way we get a new target-vector which is a new memory representation. If the network is already filled with old memories, we at least get a better memory representation, because it considers also the last input-data. [4]

VECTOR QUANTIZATION

In most images, adjacent pixels are significantly correlated. Also if we have to store many similar input-patterns (e.g., many faces) it is useful to store all correlated pixels close to each other, to store all equal data only once, and to ignore all redundant data. Self-organizing mapping helps to solve this task. It realizes data storage by nearby-coding of similar information and at the same time it executes adaptive data compression. In self-organized maps, data are stored in a compact form, after the most important information has been extracted, so that the recalled data include as little error as possible comparing to the original information. Usually we introduce a measure of recall-error which should be minimized in order to achieve maximal data compression.

A suitable measure is the sum of all differences between the input-patterns $\vec{x}^k$ and the target- or prototype-patterns $\vec{w}_{\vec{r}}$. By introducing a probability density of input-patterns P ($\vec{x}^k$) we can write our measure as an expectation value E of the square error: [5]

$$E(\vec{w}) = \sum_{k=1}^{P} || \vec{x}^k - \vec{w}_{\vec{r}} ||^2 \ P(\vec{x}^k) \tag{13.6}$$

We want to minimize this expression in order to compress data. This procedure is called *optimal vector quantization.* The best way to find the minimum of *E* is *gradient descent*, i.e. movement in the vector-space along a line of maximal reduction of E according to change of prototypes $\vec{w}_{\vec{r}}$:

$$\vec{w}_{\vec{r}}(t+1) = \vec{w}_{\vec{r}}(t) - \frac{\varepsilon}{2}\frac{\partial E}{\partial \vec{w}_{\vec{r}}} = \vec{w}_{\vec{r}}(t) + \varepsilon \sum_{k=1}^{P} (\vec{x}^k - \vec{w}_{\vec{r}}(t)) P(\vec{x}^k) \tag{13.7}$$

This equation, obtained by inserting (13.6), is used for every prototype $\vec{w}_{\vec{r}}$.

We can simplify equation (13.7) if the probability distribution P ($\vec{x}^k$) is not known explicitly as it is often the case:

$$\vec{w}_{\vec{r}'}(t+1) = \vec{w}_{\vec{r}'} \,|\, (t) + \varepsilon(\vec{x}^k - \vec{w}_{\vec{r}'}) \tag{13.8}$$

Here we have focused our attention only to the adjustment of the prototype $\vec{w}_{\vec{r}}$ which matches mostly with the current input-pattern $\vec{x}^k$. Input-patterns are presented successively during learning and in each step (13.8) is used.

Formula (13.8) is equivalent to equation (13.4). In order to get this generalization we would only have to add the neighborhood function $h_{\vec{r}\vec{r}'}$. In the most general case, which includes the neighborhood function and probability distribution, we can write the error function as a functional

$$E' = \sum_{\vec{r}\vec{r}'} h_{\vec{r}\vec{r}'} \sum_{k=1}^{P} || \vec{x}^k - \vec{w}_{\vec{r}} ||^2 \ P(\vec{x}^k) \tag{13.9}$$

Dimension of vectors $\vec{x}^k$ and $\vec{w}_{\vec{r}}$ is *n*. *k* is the ordinary number of an input-pattern which is to be learned or classified, respectively.

It is a special characteristic of Kohonen maps that it is, say, three-times more dense map (three times more neurons) formed where there is three-times higher input- probability-distribution P ($\vec{x}^k$). So, we get a better map-resolution corresponding to those receptive fields where there are more stimuli.

ENDNOTES

1 Such maps, roughly two-dimensional "sheets", are mostly located in the primary layer of the neocortex. Examples are retinotopic map (in the visual cortex) of the eye's retina, somatotopic map (in the somatosensory cortex) of the skin surface, tonotopic map (in the auditory cortex) of the ear's cochlea-spiral, etc. In the map of auditory cortex, pitch (frequency) is encoded in one direction, amplitude in the orthogonal direction. In superior colliculus in midbrain, direction and amplitude of rapid eye movements (saccades) are spatially encoded. Triggering mechanism for flexion and extension movements of muscles are encoded in the motor cortex (ring map). (Ritter, Martinetz & Schulten, 1992)

2 The holonomic theory shares such a cost-function minimization (known in physics as the fundamental law of "least action"). But, in contrast to Self-Organizing Maps, the holonomic theory advocates entropy minimization which is equivalent to information maximization

3 The iteration process (13.3)/(13.4) using function (13.5) is actually an alternative and complementary description of the signal-exchange determined by equation (13.1).

4 The learning rate ε can also decrease gradually with the number (k) of learning or adaptation steps. Without proof we shall mention that two important conditions are necessary for ε in order to achieve convergent adaptation process (Kohonen, 1984): $\lim_{t\to\infty} \varepsilon(t) = 0$ and $\lim_{t\to\infty} \int_0^t \varepsilon(t')dt' = \infty$. Learning- time t flows in parallel with increasing index k which belongs to sequentially arriving learning examples $\vec{x}^k$. The second condition demands that decreasing of ε towards zero in time t (as prescribed by the first condition) must not be too fast.

5 Details see in: E. Domany, J.L. van Hemmen & K. Schulten (Eds.) (1991): Models of Neural Networks (Series "Physics of NNs"); Berlin: Springer.

Appendix C: Iimportant Mathematical Definitions

Convolution. The convolution $f(x,\ y) = g(x,\ y) * h(x,\ y)$ of two 2-dimensional functions $g(x,\ y)$ and $h(x,\ y)$ is defined as:

$$f(x,y) = \int_{-\infty}^{\infty}\int_{-\infty}^{\infty} g(u,v)h(x-u,y-v)dudv$$

Correlation. The cross-correlation function $c(x,\ y) = g(x,\ y) \star h(x,\ y)$ of two 2-dimensional functions $g(x,\ y)$ and $h(x,\ y)$ is defined as:

DOI: 10.4018/978-1-61520-785-5.ch014

$$c(x,y)=\int_{-\infty}^{\infty}\int_{-\infty}^{\infty} g*(u,v)h(x+u,y+v)dudv$$

Corresponding relations. Relation of correlation *c(x, y)* and convolution *:

$c(x, y) = g^*(x, y) *h(-x,-y)$.

If F denotes Fourier transform, and G and H denote Fourier-transformations of g and h, then:

$\int_{-\infty}^{\infty}\int_{-\infty}^{\infty} |\, g(x,y)\,|^2\, dxdy = \int_{-\infty}^{\infty}\int_{-\infty}^{\infty} |\, G(k_x,k_y)\,|^2\, dk_x dk_y$ (Plancherel or Rayleigh theorem),

$F[g(x,y) \star g(x,y,)] = |\, G(k_x,k_y)\,|^2$ (Wiener-Khinchin theorem),
$F[g(x,y) *h(x,y,)] = G(k_x,k_y)H(k_x,k_y)$ (See also MacLennan, 1990.)

Appendix D: Abbreviations and Neurological Adjectives

ANN = Artificial Neural Network
Aux(il). = auxiliary (ch., sec., literature)
ch., chs. = chapter(s)
eq., eqs. = equation(s) / **fig., figs.** = figure(s)
fMRI = functional Magnetic Resonance Imaging
HNeT = Holographic Neural Technology (recently, **HQNeT** = Holographic Quantum Neural Technology)
ICA = Independent Component Analysis
infomax = INFOrmation MAXimization (i.e., a model / algorithm / network that maximally preserves information)
ITC = Inferior Temporal Cortex
J. = (The) Journal (of)(the)… (used in bibliography)
LGN = Lateral Geniculate Nucleus (a part of thalamus)
LTP = Long-Term Potentiation; LTD = Long-Term Depression
MT(C) = Medial Temporal (Cortex)
NMR = Nuclear Magnetic Resonance
PCA = Principal Component Analysis
PDP = Parallel Distributed Processing (many processes take place simultaneously in many areas of a network)

DOI: 10.4018/978-1-61520-785-5.ch015

PET = Positron Emission Tomography
sec., secs. = section(s)

NEUROLOGICAL DESCRIPTIONS OF LOCATIONS IN THE BRAIN

Mainly based on Latin. Adapted from: R.M. Youngson: *Collins Dictionary of Medicine*. Harper Collins Publ., Glasgow, 1992.

basal = pertaining to, situated at or forming a base / **caudal** = pertaining to the tail end (of the brain, body,…)
contralateral = pertaining to the opposite side (of the body)
dorsal = relating to the back
frontal = pertaing to the forehead
inferior = situated below / **superior** = above / **anterior** = at or towards the front of the body (**ante** = before, in front of)
ipsilateral = pertaining to the same side (of the body)
lateral = of, at or towards the side (**contra-lateral** = opposite side; **ipsi-lateral** = same side)
medial = middle / **nasal** = pertaining to, near to, toward the nose
occipital = rear lobe (of the main brain)
parietal = pertaining to the wall or outer surface (of the brain)
peri = round about, surrounding / **pre** = before, preceding to or prior to / **sub** = under, less / **hyper** = above, beyond, over
temporal = pertaining to the temples
ventral = pertaining to the front; venter (Lat.) = belly

Compilation of References

Abbot, L.F. & Nelson, S.B. (2000). Synaptic plasticity: taming the beast. *Nature Neurosci. (Suppl.), 3*, 1178-1183.

Aharonov, Y. & Vaidman, L. (1994): Measurability of the Schrödinger wave. In H. Sakaki & H. Noge (Eds.), *Nanostructures and Quantum Effects* (pp. 28-34 and the next paper on p. 35). Berlin: Springer.

Ahn, J., Weinacht, T.C. & Bucksbaum, P.H. (2000). Information storage and retrieval through quantum phase. *Science, 287*, 463-465.

Aine, C.J., Supek, S. & George, J.S. (1995). Temporal dynamics of visual-evoked neuromagnetic sources: Effects of stimulus parameters and selective attention. *Internat. J. Neuroscience, 80*, 79-104.

Aine, C.J., Supek, S., George, J.S., Ranken, D., Lewine, J., Samders, J., Best, E., Tiee, W., Flynn, E.R., & Wood, C.C. (1996). Retinotopic organization of human visual cortex: Departures from the classical model. *Cerebral Cortex, 6*, 354-361.

Aleksenko, S.V. & Kirvelis, D. et al. (Eds.) (1987): *Zriteljnjie sistemji [= Visual Systems]*. Vilnius: MVSSO LitSSR. (Also: D. Kirvelis (1975): Kvazigolograficheskie principipji funkcionaljnoi organizacii zriteljnogo analizatora formji izobrazhenii [= Quasiholographic principles…]; dissertation, Kaunas.)

Alfinito, E. & Vitello, G. (2000). The dissipative quantum model of brain: how does memory localize in correlated neuronal domains. *Information Sciences, 128*, 217-229.

Alicki, R. (1997). Quantum ergodic theory and communication channels. *Open Systems & Information Dynamics, 4*, 53-69.

Amari, S. & Maginu, K. (1988). Statistical neurodynamics of associative memory. *Neural Networks, 1*, 63-73.

Amit, D. (1989): *Modeling Brain Functions (The world of attractor neural nets).* Cambridge: Cambridge University Press.

Anandan, J. (1992). The geometric phase. *Nature, 360*, 307-313.

Arbib, M.A. (Ed.) (1995). *The Handbook of Brain Theory and Neural Networks.* Cambridge, MA: MIT Press.

Artun, Ö.B., Shouval, H.Z., & Cooper, L.N. (1998). The effect of dynamic synapses on spatiotemporal receptive fields in visual cortex. *Proceedings of the National Academy of Sciences of USA, 95*, 11999-12003.

Aspect, A., Dalibard, P., & Rogier, G. (1982). Experimental test of Bell's inequalities using time-varying analyzers. *Physical Review Letters, 49*, 1804-1807 (and also: A. Aspect, P. Grangier & G. Rogier: *Phys. Rev. Lett., 47* (1981) 460- &., *49* (1982) 91-.)

ASSC (1998). Neural correlates of consciousness. *ASSC conference abstracts.* Bremen: Hanse-Wissenschaftskolleg (from WWW).

Baars, B.J. (1997). *In the Theater of Consciousness.* New York: Oxford Univ. Press.

Baddeley, R. (1996): Searching for filters with "interesting" output distributions: an uninteresting direction to explore? *Network: Computation in Neural Systems, 7*, 409-421.

Baird, B. (1990). Bifurcation and category learning in network models of oscillating cortex. *Physica D, 42*, 365-384.

Ballentine, L.E. (1970). The statistical interpretation of quantum mechanics. *Reviews in Modern Physics, 42*, 358-382.

Barahona da Fonseca, J., Barahona da Fonseca, I., Paz Araujo, C., & Simoes da Fonseca, J. (1999). A quantum theoretical approach to information processing in neural networks. In D. Dubois (Ed.), *AIP Conference Proceedings, 517, Computing Anticipatory Systems – CASYS'99* in Liege (pp. 330-344). Melville, NY: American

Institute of Physics. (Also: Cognition: characteristic waveform correlates. CASYS 2000, Liege: full-length preprint & abstract book.)

Barlow, H. (1996). Intraneuronal information processing, directional selectivity and memory for spatio-temporal sequences. *Network: Computation in Neural Systems,* **7**, 251-259.

Bartlett, M.S. & Sejnowski, T.J. (1997). Viewpoint invariant face recognition using independent component analysis and attractor networks. *Advances in Neural Information Processing Systems, 9*, 817-823.

Bartlett, M.S. & Sejnowski, T.J. (1998): Learning viewpoint-invariant face representation from visual experience in an attractor network. *Network: Comput. in Neural Sys., 9*, 399-417.

Bear, M.F. (1996). A synaptic basis for memory storage in the cerebral cortex. *Proceedings of the National Academy of Sciences of USA, 93*, 13453-13459.

Bell, A.J. & Sejnowski, T.J. (1995). An information-maximization approach to blind separation and blind convolution. *Neural Computation,* **7**, 1129-1159.

Bell, A.J. & Sejnowski, T.J. (1996). Learning the higher-order structure of a natural sound. *Network: Computation in Neural Systems,* **7**, 261-266.

Bell, A.J. & Sejnowski, T.J. (1997). The "independent components" of natural scenes are edge filters. *Vision Research* **37**, 3327-3338.

Berg, R.H., Hvilsted, S. & Ramanujam, P.S. (1996). Peptide oligomers for holographic data storage. *Nature Letters, 383*, 505-508.

Berger, D., Pribram, K., Wild, H., & Bridges, C. (1990). An analysis of neural spike-train distributions: determinants of the response of visual cortex neurons to changes in orientation and spatial frequency. *Experimental Brain Research, 80*, 129-134.

Berger, D.H. & Pribram, K.H. (1992). The relationship between the Gabor elementary function and a stochastic model of the inter-spike interval distribution in the responses of visual cortex neurons. *Biological Cybernetics, 67*, 191-194.

Berman, G.P., Doolen, G.D., Mainieri, R. & Tsifrinovich, V.I. (1998).*Introduction to Quantum Computers.* Singapore: World Scientific.

Berne, R.M. & Levy, M.N. (Eds.) (1993). *Physiology* (3rd ed.). St. Louis: Mosby Year Book.

Bianconi, G. & Barabasi, A.-L. (2001). Bose-Einstein condensation in complex networks. *Physical Review Letters, 86*, 5632-5635.

Bickle, J., Bernstein, M., Heatley, M., Worley, C., & Stiehl, S. (1999): A functional hypothesis for LGN-V1-TRN connectivities suggested by computer simulation. *J. Computational Neuroscience, 6*, 251-261.

Bickle, J., Worley, C., & Bernstein, M. (2000). Vector subtraction implemented neurally: A neurocomputational model of some sequential cognitive and conscious processes. *Consciousness & Cognition.*

Bjorken, J.D. & Drell, S.D. (1964/1965). *Relativistic Quantum Mechanics (I) / Relativistic Quantum Fields (II).* New York: McGraw-Hill (espec. ch. 6: Propagator Theory).

Blais, B.S., Intrator, N., Shouval, H. & Cooper, L.N. (1997). Receptive field formation in natural scene environment: comparison of single cell learning rules. (Preprint, submitted to *Neural Computation.*)

Blanchard, Ph. & Jadczyk, A. (Eds.) (1999). *Quantum Future – Proceed. Xth Max Born Symp.* In Przesieka, PL, 1997. Berlin: Springer.

Bob, P. & Faber, J. (1999). Quantum information in brain neural nets and EEG. *Neural Network World, 9*, 365-372.

Bogacz, R., Brown, M.W., & Giraud-Carrier, C. (2001). Emergence of movement sensitive neurons' properties by learning a sparse code for natural moving images. *Advances in Neural Information Processing Systems, 13.*

Bohm, D. (1954). *Quantum Theory.* London: Constable & Co.

Bohm, D. (1980). *Wholeness and Implicate Order.* London: Routledge & Paul Kegan.

Bohm, D. & Hiley, B. (1993). *The Undivided Universe (An ontological interpretation of quantum theory).* London: Routledge.

Bonnell, G. & Papini, G. (1997). Quantum neural network. *Internat. J. Theoretical Physics, 36*, 2855-2875.

Bray, D. (1995). Protein molecules as computational elements in living cells. *Nature, 376*, 307-312.

Burnod, Y. (1990). *An Adaptive Neural Network: the Cerebral Cortex.* London: Prentice Hall.

Buser, P. & Imbert, M. (1992). *Vision.* Cambridge, MA: MIT Press.

Caelli, T.M. (1976). The prediction of interaction between visual forms by products of Lie operators. *Mathematical Biosciences, 30*, 191-.

Cahill, R.T. (2001). Smart nanostructures and synthetic quantum systems. In L.B. Kish (Ed.), *Proceedings of SPIE Conference no. 4590: BioMEMS and Smart Nanostructures* (pp. 319-328). [And: *ArXiv:quant-ph/0111026*]

Caputo, B. & Dorko, G. (2002). How to combine color and shape information for 3D object recognition: Kernels do the trick. *Advances in Neural Information Processing Systems*, preprint.

Cerf, N.J. & Adami, C. (1997): Quantum information theory of entanglement and measurement. 4th Workshop on Physics and Computation. *ArXiv:quant-ph/9605039 v2.*

Changeux, J.-P. (1983). *L'homme neuronal.* Paris: Fayard.

Chapline, G. (1999). Is theoretical physics the same thing as mathematics? *Physics Reports, 315*, 95-105.

Chapline, G., & Dubois J.L. (2009). Topological quantum image analysis. arXiv:0904.4306v1

Churchland, P.S. & Sejnowski, T.J. (1992). *The Computational Brain.* Cambridge, MA: MIT Press.

Clement, B.E.P., Coveney, P.V., Jessel, M., & Marcer, P.J. (1999). The brain as a Huygens machine. *Informatica, 23*, 389-398.

Comon, P. (1994). Independent component analysis: A new concept? *Signal Processing, 36*, 287-314.

Cory, D.G., Fahmy, A.F. & Havel, T.F. (1997). Ensamble quantum computing by NMR spectroscopy. *Proceedings of the National Academy of Sciences of USA, 94*, 1634-1639.

Cramer, J.G. (1986). The transactional interpretation of quantum mechanics. *Reviews of Modern Physics, 58*, 647-687.

Crick, F. (1984): Function of the thalamic reticular complex: The searchlight hypothesis. *Proceedings of the National Academy of Sciences of USA, 81*, 4586-4590.

Damask, A.C. & Swenberg, C.E. (1984). *Medical Physics.* Orlando: Academic Press. *Vol. III: Synapse, Neuron, Brain*; ch.4: Chemical and electrical properties of synapses; ch. 5: Neuronal integration and Rall theory. *Vol. I: Physiological Physics*; ch. 3: The nerve impulse.

DeAngelis, G.C. (2000). Seeing in 3 dimensions: Neurophysiology of stereopsis. *Trends in Cognitive Sciences, 4*, 80-90.

Daugman, J.G. (1985). Uncertainty relation for resolution in space, spatial frequency, and orientation optimized by 2-D visual cortical filters. *J. Optical Society of America A, 2*, 1160-.

Daugman, J.G. (1988): Complete discrete 2-D Gabor transforms by neural networks for image analysis and compression. *IEEE Transactions on Acoustics, Speech, and Signal Processing 36*, 1169-.

Davies, M. & Humphreys, G.W. (Eds.) (1993). *Consciousness.* Oxford: Blackwell.

Deco, G. & Schürmann, B. (2000). A hierarchical neural system with attentional top-down enhancement of the spatial resolution for object recognition. *Vision Research, 40*, 2845-2859.

Denisyuk, Yu.N. (1984). *Fundamentals of Holography.* Moscow: Mir.

Denschlag, J., Simsarian, J., Feder, D., Clark, C., Collins, L., Cubizolles, J., Deng, L., Hagley, E., Helmerson, K., Reinhardt, W., Rolston, S., Schneider, B. & Phillips, W. (2000). Generating solitons by phase engineering of a Bose-Einstein condensate. *Science, 287*, 97-101.

Desimone, R. (1996). Neural mechanisms for visual memory and their role in attention. *Proceedings of the National Academy of Sciences of USA, 93*, 13494-13499.

De Valois, R.L. & Jacobs, G.H. (1968). Primate color vision. *Science, 162*, 533-540.

De Valois, R.L. & De Valois, K.K. (1990). *Spatial Vision.* New York: Oxford Univ. Press.

De Yoe, E.A. & Van Essen, D.C. (1988). Concurrent processing streams in monkey visual cortex. *Trends in Neurosciences, 11*, 219-226.

Diaz, J.-L. (1997). A patterned process approach to brain, consciousness, and behavior. *Philosophical Psychology, 10*, 179-195.

Dimitrovski, D., Pop-Jordanov, J., Pop-Jordanova, N. & Solov'ev, E. (1998). Quantum transitions between collective rotational states of water molecules in brain cells as a possible mechanism of memory imprinting. *Prilozi MANU = Contributions, Sec. Math. Tech. Sci., MANU, 19*(1-2), 35-58.

Dodwell, P.C. (1983). The Lie transformation group for visual perception. *Perception & Psychophysics, 34,*, 1-16.

Domijan, D. (2001). Visual feature integration with amplitude maximization. *Reviews of Psychology 8*(1-2), 31-40.

Dubois, D.M. (Ed.) (2000a). Proceedings of CASYS'99. *Internat. J. Computing Anticipatory Systems 5, 6, 7.* Liege: CHAOS. Especially, in vol. **7**: Proceedings of the Symposium "Quantum Neural Information Processing: New Technology? New Biology?" (espec. papers by Marcer, Sutherland, Farre, Mitchell, Dubois; also Citko, Luksza & Sienko).

Dubois, D.M. (Ed.) (2000b). *AIP Conference Proceedings* (Vol. *517*: *Computing Anticipatory Systems – CASYS'99*). Melville, NY: American Institute of Physics (espec. Sienko, Hoekstra & Rouw, Santoli, Pribram, Dubois, Araujo).

Ebdon, M. (1993). Is the cerebral neocortex an uniform cognitive architecture? *Mind & Language, 8*, 368-398.

Ezhov A.A, & Berman G.P. (2003). *Introduction to Quantum Neural Technologies.* Princeton, NJ: Rinton Press.

Ezhov, A.A. (2000). Spurious memory, single-class and quantum neural networks. In *Proceedings of the Int. Conf. on Computational Intelligence and Neuroscience 2000* (part of *Proceed. JIC on Information Sci. 2000*, Atlantic City, NJ, pp. 635-638).

Ezhov, A.A. (2001): Pattern recognition with quantum neural networks. In S. Singh, et al (Eds.), *Proceedings of ICAPR '01* (pp. 60-71). Rio de Janeiro.

Ezhov, A.A., Nifanova, A.V. & Ventura, D. (2000). Quantum associative memory with distributed queries. *Information Sciences, 128*, 271-293.

Ezhov, A.A. & Ventura, D. (2000). Quantum neural networks. In N. Kasabov (Ed.), *Future Directions for Intelligent Systems and Information Sciences* (pp. 213-235). Heidelberg: Physica-Verlag (Springer).

Fatmi, H.A. & Resconi, G. (1988). A new computing principle. *Nuovo Cimento*, Note Brevi, *101 B*, 239-242.

Fedorec, A.M. & Marcer, P. (Eds.) (1996). *Living Computers* (symposium proceedings). Dartford: Greenwich Univ. Press.

Fidelman, U. (2002). Kant, the two stages of visual search, quantum mechanics, and antimatter. *Internat. J. Computing Anticipatory Systems, 13*, 274-289.

Flanagan, O. (1992). *Consciousness Reconsidered.* Cambridge, MA: MIT Press.

Freeman, W. (1991). The physiology of perception. *Scientific American, 264,* 78-85.

Freitas da Rocha, A., Pereira Jr., A., & Bezerra Coutinho, F.A. (2001). NMDA channel and consciousness: From signal coincidence detection to quantum computing. *Progress in Neurobiology, 64*, 555-573.

Frith, C., Perry, R. & Lumer, E. (1999). The neural correlates of conscious experience: an experimental framework. *Trends in Cognitive Sciences, 3*, 105-114.

Fröhlich, H. (1968). Long-range coherence and energy storage in biological systems. *Internat. J. Quantum Chemistry, 2*, 641-649.

Fuster, J.M. & Jervey, J.P. (1982). Neuronal firing in the inferotemporal cortex of the monkey in a visual memory task. *J. Neuroscience, 2*, 361-375.

Gardner, D. (Ed.) (1993). *The Neurobiology of Neural Networks.* Cambridge, MA: MIT Press.

Gennaro, R.J. (1995). *Consciousness and Self-Consciousness.* Amsterdam/Philadelphia: John Benjamins.

Georgopoulos, A.P., Lurito, J.T., Petrides, M., Schwartz, A.B. & Massey, J.T. (1989). Mental rotation of the neuronal population vector. *Science, 243*, 234-236.

Gershenfeld, N.A. & Chuang, I.L. (1997): Bulk spin-resonance quantum computation. *Science, 275*, 350-356.

Geszti, T. (1990). *Physical Models of Neural Networks.* Singapore: World Scientific.

Gochin, P.M. (1996): The representation of shape in the temporal lobe. *Behavioral Brain Research, 76*, 99-116.

Goldman-Rakić, P.S. (1996). Memory: Recording experience in cells and circuits. *Proceedings of the National Academy of Sciences of USA, 93*, 13435-13437 (organizer's introduction to the symposium with the same title; whole proceedings in the same vol., pp. 13435-13551.).

Goswami, A. (1990). Consciousness in quantum physics and the mind-body problem. *J. Mind & Behavior, 11*(1990), 75-96.

Gould, L.I. (1989). Nonlocal conserved quantities, balance laws and equations of motion. *Internat. J. Theoretical Physics, 28*, 335-364.

Gould, L.I. (1995). Quantum dynamics and neural dynamics: Analogies between the formalisms of Bohm and Pribram. In J. King & K. Pribram (Eds.), *Scale in Conscious Experience: Is the Brain Too Important To Be Left to Specialists to Study?* (pp. 337-348). Mahwah, NJ: Lawrence Erlbaum Assoc.

Gray, C.M., König, P., Engel, A.K., & Singer, W. (1989). Oscillatory responses in cat visual cortex exhibit inter-columnar synchronization which reflects global stimulus properties. *Nature, 338*, 334-337.

Gray, C.M., König, P., Engel, A.K. & Singer, W. (1990). Synchronization of oscillatory responses in visual cortex: A plausible mechanism for scene segmentation. In H. Haken & M. Stadler (Eds.), *Synergetics of Cognition* (p. 82-). Berlin: Springer.

Gray, M.S., Movellan, J.R. & T.J. Sejnowski (1997): Dynamic features for visual speech-reading: A systematic comparison. *Advances in Neural Information Processing Systems* **9**, 751-757.

Griniasty M., Tsodyks M., Amit D., (1993). Conversion of temporal correlation between stimuli to spatial correlations between attractors. *Neural Computation* 5, 1-17.

Grössing, G. (1988): How does a quantum system perceive its environment? In A. van der Merwe et al. (Eds.), *Microphysical Reality and Quantum Formalism* (vol. 2, pp. 225-238). Dordrecht: Kluwer.

Gupta, M.M. & Knopf, G.K. (Eds.) (1994). *Neuro-Vision Systems.* New York: IEEE Press.

Haken, H. (1991). *Synergetic Computers and Cognition (A Top-Down Approach to Neural Nets).* Berlin: Springer.

Haken, H. (1996). *Principles of Brain Functioning.* Berlin, Springer.

Hameroff, S.R. (1994). Quantum coherence in microtubules: A neural basis for emergent consciousness? *J. Consciousness Studies, 1*, 91-118.

Hameroff, S.R. (1998). "Funda-Mentality": Is the conscious mind subtly linked to a basic level of the Universe? *Trends in Cognitive Science, 2*, 119-127.

Hameroff, S.R., Kaszniak, A.W. & Scott, A.C. (1996). *Towards a Science of Consciousness: Tucson I.* Cambridge, MA: MIT Press. (Especially: K. Pribram: The varieties of conscious experience: Biological roots and social usages (ch. 10) & C. Koch: Towards a neuronal substrate of visual consciousness).

Hameroff, S.R. & Penrose, R. (1995). Orchestrated reduction of quantum coherence in brain microtubules: A model for consciousness. In J. King, & K. Pribram (Eds.), *Scale in Conscious Experience: Is the Brain Too Important To Be Left to Specialists to Study?* (pp. 243-275). Mahwah, NJ: Lawrence Erlbaum Assoc.

Hameroff, S.R., & Penrose, R. (1996). Conscious events as orchestrated space-time selections. *J. Consciousness Studies, 3*, 36-53.

Hariharan, P. (1996). *Optical Holography.* Cambridge: Cambridge Univ. Press.

Harpur, G.F. & Prager, R.W. (1996). Development of low entropy coding in a recurrent network. *Network: Computation in Neural Systems* **7**, 277-284.

Haykin, S. (1994). *Neural Networks (A Comprehensive Foundation).* New York: MacMillan.

Hellwig, B. (2000). A quantitative analysis of the local connectivity between pyramidal neurons in layers 2/3 of the rat visual cortex. *Biological Cybernetics, 82*, 111-121.

Hiley, B.J. & Peat, F.D. (Eds.) (1987). *Quantum Implications (Essays in Honour of David Bohm).* London: Routledge.

Hoffman, W.C. (1966). The Lie algebra of visual perception. *J. Mathematical Psychology, 3*, 65-98.

Hoffman, W.C. (1968): The neuron as a Lie group germ and a Lie product. *Quarterly J. Applied Mathematics, 25*, 423-.

Hoffman, W.C. (1970): Higher visual perception. *Mathematical Biosciences, 6*, 437-.

Hoffman, W.C. (1997): Mind and the geometry of systems. In S.O. Nuallain et al. (Eds.), *Two Sciences of Mind* (pp. 459-483). Amsterdam: John Benjamins.

Hopfield, J.J. (1982). Neural networks and physical systems with emergent collective computational abilities. *Proceedings of the National Academy of Sciences USA, 79*, 2554-2558.

Hopfield, J.J. (1984). Neurons with graded response have collective computational properties like those of two-state neurons. *Proceedings of the National Academy of Sciences of USA, 81*, 3088-3092.

Hopfield, J.J. (1995). Pattern recognition computation using action potential timing for stimulus representation. *Nature, 376*, 33-36.

Huang D.S, Heutte, L., & Loog, M. (2007). Quantum probability distribution network, ICIC 2007 (LNCS 4681, pp. 25–33).

Hucho, F. (1986). *Neurochemistry.* Weinheim: VCH.

Hyvärinen, A. (1999). Fast and robust fixed-point algorithms for independent component analysis. *IEEE Transactions on Neural Networks, 10*, 626-634.

Hyvärinen, A. & P. Hoyer (2000): Emergence of topography and complex cell properties from natural images using extensions of ICA. *Advances in Neural Information Processing Systems, 12*, 827-833 (& NIPS'99).

Hyvärinen, A. & Oja, E. (1997). A fast fixed-point algorithm for independent comp. anal. *Neural Computation, 9*, 1483-1492.

Ingber, L. (1998). Statistical mechanics of neocortical interactions: Training and testing canonical momenta indicators of EEG. *Mathematical & Comput. Modelling, 27*(3), 33-64.

Intrator, N. & Gould, J.I. (1992). Three-dimensional object recognition using an unsupervised BCM network: The usefulness of distinguishing features. (Preprint, to appear in *Neural Computation.*)

Jasper, H., Descarries, L., Castellucci, V. & Rossignol, S. (1998). *Advances in Neurology* (Vol. 77: *Consciousness: At the Frontiers of Neuroscience*). Philadelphia & New York: Lippincott-Raven (espec. ch. 17).

Jibu, M., Pribram, K.H. & Yasue, K. (1996). From conscious experience to memory storage and retrieval: The role of quantum brain dynamics and boson condensation of evanescent photons. *Internat. J. Modern Physics, 10*, 1735-1754.

Jibu, M. & Yasue, K. (1995). *Quantum Brain Dynamics and Consciousness.* Amsterdam/ Philadelphia: John Benjamins.

Jibu, M. & Yasue, K. (1997). Quantum field theory of evanescent photons in brain as quantum theory of consciousness. *Informatica, 21*, 471-490.

Jones, J.A., Vedral, V., Ekert, A. & Castagnoli, G. (2000): Geometric quantum computation using nuclear magnetic resonance. *Nature Letters, 403*, 869-871.

Kainen, P.C. (1992). Orthogonal dimension and tolerance (Tech. report IM-061592). Industrial Math., Washington, DC.

Kainen, P.C. & Kurková, V. (1993). Quasiorthogonal dimension of Euclidean spaces. *Applied Mathematics Letters, 6*(3), 7-10.

Kak, S.C. (1995). On quantum neural computing. *Information Sciences, 83*, 143-160.

Kaminski, W.A. & Linkevich, A.D. (1997/1998). Correlation functions of nonlinear dynamical systems and methods of quantum field theory. *Annales Universitatis Mariae Curie-Sklodowska* (Lublin, PL), Sectio AAA, LII/LIII, 3, 25-35.

Kandel, E.R., Schwartz, J.H. & Jessel, T.M. (1991). *Principles of Neural Science* (3rd ed.). London: Prentice Hall Internat.

King, J.S., Xie, M., Zheng, B. & Pribram, K.H. (2000). Maps of surface distributions of electrical activity in spectrally derived receptive fields of the rat's somatosensory cortex. *Brain and Mind, 1*, 327-349.

Kirvelis, D. (2000b). Visual analyzer as anticipatory system (Functional organization). In D. Dubois (Ed.), *AIP Conference Proceedings* (Vol. 517: *Comput. Anticip. Sys. – CASYS'99* in Liege, pp. 277-288). Melville, NY: Amer. Inst. Phys.

Koch, C. (1997). Computation and the single neuron. *Nature, 385*, 207-210.

Koch, C. & Segev, I. (2000). The role of single neurons in information processing. *Nature Neurosci. (Suppl.), 3*, 1171-1177.

Kohonen, T. (1982). Self-organized formation of topologically correct feature maps. *Biological Cybernetics, 43*, 59-.

Kohonen, T. (1984). *Self-Organization and Associative Memory.* Berlin: Springer.

Kohonen, T. (1995). *Self-Organizing Maps.* Berlin: Springer.

Kononenko, I. (1997). *Machine Learning.* Ljubljana: FRI (in Slovene). (Or see English books by, e.g.: Natarajan; Michie; Weiss & Kulikowski.)

Kosslyn, S.M. (1988). Aspects of a cognitive neuroscience of mental imagery. *Science, 240*, 1621-1626.

Kosslyn, S.M. (1994). *Image and Brain.* Cambridge, MA: MIT Press.

Kosslyn, S.M. & Andersen, R.A. (Eds.) (1992). *Frontiers in Cognitive Neuroscience.* Cambridge, MA: MIT Press.

Körner, E., Gewaltig, M.-O., Körner, U.,Richter, A. & Rodemann, T. (1999). A model of computation in neocortical architecture. *Neural Networks, 12*, 989-1005.

Kovacs, I. (1996). Gestalten of today: early processing of visual contours and surfaces. *Behavioral Brain Research, 82*, 1-11.

Krüger V., & Sommer G. (2002). Gabor wavelet networks for efficient head pose estimation. *Image & Vision* Computing, 20, 665-672.

Kurková, V. & Kainen, P.C. (1996): A geometric method to obtain error-correcting classification by neural networks with fewer hidden units. In *Proceedings of Int. Conf. on Neural Networks '96* (pp. 127-132). Washington, DC: IEEE.

Kwiat, P., Berglund, A., Altepeter, J., & White, A. (2000). Experimental verification of decoherence-free subspaces. *Science, 290*, 498-501.

Lades, M., Vorbrüggen, J., Buhmann, J., Lange, J., v.d. Malsburg Ch., Würtz, R., & Konen, W. (1993). Distortion invariant object recognition in the dynamic link architecture. IEEE Transactions on Computers, 42, 300-311.

Landauer, R. (1991, May). Information is physical. *Physics Today*, 23-29.

Lee, S.-H. & Blake, R. (1999a). Visual form created solely from temporal structure. *Science, 284*, 1165-1168.

Lee, S.-H. & Blake, R. (1999b). Detection of temporal structure depends on spatial structure. *Vision Research, 39*, 3033-3048.

Lee, T.S. (1996). Image representation using 2-dimensional Gabor wavelets. *IEEE Transactions on Pattern Analysis and Machine Intelligence, 18*(10), 1-13.

Lehky, S.R. & Sejnowski, T.J. (1988). Network model of shape-from-shading: Neural function arises from both receptive and projective fields. *Nature, 333*, 452-454.

Lewicki, M.S. & Olshausen, B.A. (1999). A probabilistic framework for the adaptation and comparison of image codes. *J. Optical Society of America A: Optics, Image Science, and Vision, 16*, 1587-1601.

Lewin, R. (1980). Is your brain really necessary? *Science, 210*, 1232-1234.

Licata, I. (2007). Effective Physical processes and active information in Quantum Computing. *Quantum Biosystems 2007, 1,* 51-65.

Liljenström, H. (2002). Regulating cortical neurodynamic for past, present and future. In D. Dubois (Ed.), *AIP Conference Proceedings* (Vol. 627: *Computing Anticipatory Systems – CASYS'01* in Liege, pp. 375-386). Melville, NY: American Institute of Physics (In this vol. also: Ribeiro; Ehresmann & Vanbremeersch; Ömer; Wallace & Narayanan; Binz et al.)

Linkevich, A.D. (2001). Neural fields: An optimal approach to infinite-dimensional systems for information processing. *Informatica, 25*, 235-246.

Linsker, R. (1988, March). Self-organization in a perceptual network. *Computer*, 105-117.

Livingstone, M. & Hubel, D. (1988). Segregation of form, color, movement, and depth: Anatomy, physiology, and perception. *Science, 240*, 740-749.

Lockwood, M. (1989). *Mind, Brain and the Quantum*. Oxford: Blackwell.

Logothetis, N.K. (1999, November). Vision: A window on consciousness. *Scientific American*, 45-51.

Luccio, R. (1993). Gestalt problems in cognitive psychology: Field theory, invariance, and auto-organisation. In V. Roberto (Ed.), *Advances in Computational Perception*. Heidelberg: Springer.

Luria, A.R. (1973). *Fundamentals of Neuropsychology*. Moscow: Izdateljstvo Moskovskogo universiteta (in Russian).

MacLennan, B.J. (1990). *Field computation* (Parts I-IV) (Tech. report CS-90-100). Univ. of Tennessee, Knoxville (also at: http://www.cs.utk.edu/~mclennan).

MacLennan, B.J. (1991). *Gabor representation of spatiotemporal visual images* (Tech. report CS-91-144). Univ. of Tennessee, Knoxville (also at: http://www.cs.utk.edu/~mclennan/neurodynamics.html).

MacLennan, B.J. (1994). Continuous computation and the emergence of discrete. In K. Pribram (Ed.), *Origins: Brain & Self-Organization* (pp. 121-152). Hillsdale, NJ: LEA.

MacLennan, B.J. (1999). Field computation in natural and artificial intelligence. *Information Sciences, 119*, 73-89.

Majewski, A.W. (1999). Separable and entangled states of composite quantum systems – rigorous description. *Open Systems and Information Dynamics, 6*, 79-86.

Maletić-Savatić, M., Malinow, R. & K. Svoboda (1999). Rapid dendritic morphogenesis in CA1 hippocampal dendrites induced by synaptic activity. *Science, 283*, 1923-1927.

Manger, R. (1997). Holographic neural networks and data compression. *Informatica, 21*, 665-673.

Mannion, C.L.T. & Taylor, J.G. (1992). Information processing by oscillating neurons. In J.G. Taylor & C.L.T. Mannion (Eds.), *Coupled Oscillating Neurons* (pp. 98-111). London: Springer.

Marcel, A.J. & Bisiach, E. (Eds.) (1988). *Consciousness in Contemporary Science*. Oxford: Clarendon Press.

Marcer, P. (Ed.) (1999). Proceedings of the Symposium 9: Quantum Neural Information Processing: New Technology, New Biology? In D. Dubois (Ed.), *Abstract book of CASYS'99 (Computing Anticipatory Systems)*. Liege: CHAOS asbl.

Marcer, P. & Schempp, W. (1997). A model of neuron working by quantum holography. *Informatica 21*, 517-532.

Marcer, P. & Schempp, W. (1998). The brain as a conscious system. *Internat. J. General Systems, 27*, 231-248.

Marcer, P. & Schempp, W. (2000). Why space has three dimensions: A quantum mechanical explanation. In D. Dubois (Ed.), *AIP Conference Proceedings* (Vol. 517: *Computing Anticipatory Systems – CASYS'99* in Liege, pp. 430-436). Melville, NY: American Institute of Physics.

Marcer, P. & Rowlands, P. (2007). How Intelligence Evolved? *Quantum Interactions. Papers from the AAAI Spring Symposium* (Technical report SS-07-8).

Marder, E., Abbott, L.F., Turrigiano, G.T., Liu, Z. & Golowasch, J. (1996). Memory from the dynamics of intrinsic membrane currents. *Proceedings of the National Academy of Sciences of USA, 93*, 13481-13486.

Matsuno, K. (2002): Quantum coherence as a marker of synchronous time and its implication on consciousness acting in brain. *Internat. J. Computing Anticipatory Systems, 13*, 360-375.

McClelland, J.L., Rumelhart, D.E. & PDP Res. Group (1986). *Parallel-Distributed Processing — Explorations into the Microstructure of Cognition* (Vols. 1 & 2). Cambridge, MA: MIT Press.

McIntosh, A.R., Rajah, M.N. & Lobaugh, N.J. (1999). Interactions of prefrontal cortex in relation to awareness in sensory learning. *Science, 284*, 1531-1533.

McKeown, M.J., Makeig, S., Brown, G.G., Jung, T.-P., Kindermann, S.S., Bell, A.J., & Sejnowski, T.J. (1998). Analysis of fMRI data by blind separation into independent components. *Human Brain Mapping, 6*, 160-188.

Messiah, A. (1965). *Quantum Mechanics I., II.* Amsterdam: North-Holland.

Mesulam, M.-M. (1981): A cortical network for directed attention and unilateral neglect. *Annals of Neurology, 10*, 309-325.

Mezard, M., Parisi, G. & Virasoro, M.A. (1987). *Spin Glass Theory and Beyond.* Singapore: World Scientific.

Miller, K.D., Keller, J.B. & Stryker, M.B. (1989). Ocular dominance column development: Analysis and simulation. *Science, 245*, 605-615.

Mishkin, M., Ungerleider, L.G. & Macko, K.A. (1983). Object vision and spatial vision: Two cortical pathways. *Trends in Neurosciences, 6*, 414-417.

Miyashita, Y. & Chang, H.S. (1988): Neuronal correlate of pictorial short-term memory in the primate temporal cortex. *Nature, 331*, 68-70.

Montero, V.M. (2000). Attentional activation of the visual thalamic reticular nucleus depends on 'top-down' inputs from the primary visual cortex via corticogeniculate pathways. *Brain Research 864*, 95-104.

Moran, J. & R. Desimone (1985): Selective attention gates visual processing in the extrastriate cortex. *Science, 229*, 782-784.

Mullen, K.T., Beaudot, W.H.A. & McIlhagga, W.H. (2000). Contour integration in color vision: a common process for the blue-yellow, red-green and luminance mechanisms? *Vision Research, 40* , 639-655.

Nadal, J.-P. & Parga, N. (1997). Redundancy reduction and independent component analysis: Conditions on cumulants and adaptive approaches. *Neural Computation, 9*, 1421-1456.

Nanopoulos, D.V. (1995). Theory of brain function, quantum mechanics and superstrings. *Tech. reports: ACT-08/95; CERN-TH/95-128; CTP-TAMU-22/95*; http://xxx.lanl.gov/abs/hep-ph/9505374

Navarro, R., Tabernero, A., & Cristobal, G. (1996). Image representation with Gabor wavelets and its applications. In P.W. Hawkes (ed.): Advances in Imaging and Electron Physics, 97. New York: Academic Press.

Newman, J. (1997). Toward a general theory of the neural correlates of consciousness. *J. Consciousness Studies, 4*, 47-66 (part I), 100-121 (part II).

Nishimori, H. & Nonomura, Y. (1996). Quantum effects in neural networks. *J. Physical Society of Japan, 65*, 3780-3796.

Nobili, R. (1985). Schrödinger wave holography and brain cortex. *Physical Review A, 32*, 3618-3626.

Oakley, D.A. (Ed.) (1985). *Brain and Mind.* London / New York: Methuen.

Obermayer, K., Blasdel, G. & Schulten, K. (1992). Statistical-mechanical analysis of self-organization and pattern formation during development of visual maps. *Physical Review A, 45*, 7568-7589.

Ohya, M. (1989). Some aspects of quantum information theory and their applications to irreversible processes. *Reports on Mathematical Physics, 27*, 19-46.

Okajima, K. (1996): A model visual cortex incorporating intrinsic horizontal neuronal connections. *Neural Networks, 9*, 211-222.

Olshausen, B.A. & Field, D.J. (1996a). Natural image statistics and efficient coding. *Network: Computation in Neural Systems,* **7**, 333-339.

Olshausen, B.A. & Field, D.J. (1996b). Emergence of simple-cell receptive field properties by learning a sparse code for natural images. *Nature Letters, 381*, 607-609.

Olshausen, B.A. & Field, D.J. (1997): Sparse coding with an overcomplete basis set: A strategy employed by V1? *Vision Research, 37*, 3311-3325.

Olshausen, B.A. & Millman, K.J. (2000). Learning sparse codes with a mixture-of-Gaussian prior. *Advances in Neural Information Processing Systems, 12*, preprint.

Palmer, S.E. (1999). *Vision Science (Photons to Phenomenology).* Cambridge, MA: MIT Press.

Peretto, P. (1992). *An Introduction to the Modeling of Neural Networks*. Cambridge: Cambridge Univ. Press.

Perret, D.I. & Oram, M.W. (1998). Visual recognition based on temporal cortex cells: Viewer- centered processing of pattern configuration. *Zeitschr. für Naturforschung, 53c*, 518-541.

Peruš, M. (1995a). *All in One, One in All (Brain and Mind in Analysis and Synthesis)*. Ljubljana: DZS (in Slovene). [Also: Russian translation by A. Ippa of Part III: *Vse v odnom, odno vo vsem (Matematičeskie modeli associativnjih neironnjih setei)*. St. Petersburg: KARO, 2000.]

Peruš, M. (1995b). Synergetic approach to cognition-modeling with neural networks. In K. Sachs-Hombach (Ed.), *Bilder im Geiste* (pp. 183-194). Amsterdam/Atlanta: Rodopi.

Peruš, M. (1996). Neuro-quantum parallelism in mind-brain and computers. *Informatica, 20*, 173-183.

Peruš, M. (1997a): Mind: neural computing plus quantum consciousness. In M. Gams, M. Paprzycki & X. Wu (Eds.), *Mind Versus Computer* (pp. 156-170). Amsterdam: IOS Press & Ohmsha.

Peruš, M. (1997b). System-processual backgrounds of consciousness. *Informatica, 21*(1997) 491-506.

Peruš, M. (1997c). Neuro-quantum coherence and consciousness. *Noetic J., 1*(1), 108-113.

Peruš, M. (1997d). A comparison of neuropsychological learning models with machine learning models within the neural network theory. *Electrotechnical Review: J. Elect. Enginrg. & Computer Sci., 64*, 136-141 (in Slovene).

Peruš, M. (1998a). Common mathematical foundations of neural and quantum informatics. *Zeitschr. für angewandte Mathematik & Mechanik, 78*, S1, 23-26.

Peruš, M. (1998b). Consciousness: network-dynamic, informational, and phenomenal aspects. *Noetic J., 1*(2), 183-197 (Also electronically published in *Dynamical Psychology* (1997): http://goertzel.org/dynapsyc/1997/Perus.html).

Peruš, M. (1998c). Conscious representations, intentionality, judgements, (self) awareness and qualia. *Informatica, 22*, 95-102.

Peruš, M. (1998d). From neural to quantum Hopfield networks – fundamental information processing. In C. Bavec, M. Gams (Eds.), *Proceedings of the Int. Multi-Conf. on Information Society '98*. (Volume *Int. Conf. on Cognitive Sciences,* Eds.

A. Detela, U. Kordeš, A. Ule, pp. 16-19). Ljubljana: IJS.

Peruš, M. & Ečimovič, P. (1998). Memory and pattern recognition in associative neural networks. *Internat. J. Applied Science & Computations, 4*, 283-310.

Peruš, M. (1999). Discussion on neurocomputers (Eds. A.A. Frolov, A.A. Ezhov). *Neural Network World, 1-2*, 103-171 (see also: Cutting, Chrisley, Ezhov, Martinez, Ventura; *et al.*).

Peruš, M. (2000a). *BioNetworks, Cognition and Consciousness.* Maribor: Satjam (printed book) & Ljubljana: DZS (electronic version: http://www.dzs.si) (in Slovene).

Peruš, M. (2000b). From neural to quantum associative networks: A new quantum "algorithm". In D. Dubois (Ed.), *AIP Conference Proceedings, 517: Computing Anticipatory Systems – CASYS'99* in Liege (pp. 289-295). Melville, NY: American Institute of Physics.

Peruš, M. (2000c). Neural networks as a basis for quantum associative networks. *Neural Network World, 10*(6), 1001-1013.

Peruš, M. & Dey, S.K. (2000). Quantum systems can realize content-addressable associative memory. *Applied Mathematics Letters, 13*(8), 31-36.

Peruš M., (2001a). A synthesis of the Pribram holonomic theory of vision with quantum associative nets after preprocessing using ICA and other computational models. *Internat. J. Computing Anticipatory Sys.,* 10, 352-367.

Peruš M., (2001b). Image processing and becoming conscious of its result. *Informatica,* 25, 575-592.

Peruš, M. (2001c). Multi-level synergetic computation in brain. *Nonlinear Phenomena in Complex Systems,* 4, 157-193.

Peruš, M. (2002). A computational study of reconstruction (from partial data) and anticipation capabilities of an associative neural net with large stored data-bases. *Internat. J. Computing Anticipatory Systems,* 13, 376-391.

Perus, M., Loo, C.K, Bischof, H. (2004a). Associative Memory based Image and Object Recognition by Quantum Holography. *Open Systems and Information Dynamics, 11*(3), 277-289.

Peruš, M., Bischof, H., Caulfield, J., & Loo, C.K. (2004b). Quantum-Implementable Selective Reconstruction of High-Resolution Images. *Applied Optics, 43*(33), 6134-6138.

Peruš, M., Bischof, H., & Loo, C.K. (2005a). Biocomputational model of object recognition: Quantum Hebbian Processing with Gabor Wavelets. *BioSystems*, 82, 116-126.

Perus, M., Loo, C.K, & Bischof, H. (2005b). Simulated quantum-optical object-recognition from high-resolution images. *Optics & Spectroscopy,* 99, 233-238.

Perus, M., Loo, C.K., & Bischof H. (2005c). Object recognition using quantum holography with neural-net pre-processing. *J. Optical Technol.*, *72*(5), 358-363.

Pessa, E., & Vitiello, G. (1999). Quantum dissipation and neural net dynamics. *Bioelectrochemistry & Bioenergetics, 48*, 339-342.

Phattanasri, P., Chiel, H.J., & Beer R.D. (2007). The dynamics of associative learning in evolved model circuits, Adaptive Behavior, December 2007, 15, 377-396.

Poggio, T., Gamble, E.B., & Little, J.J. (1988). Parallel integration of vision modules. *Science, 242*, 436-440.

Poggio, T., Torre, V., Koch, C. (1985). Computational vision and regularization theory. *Nature, 317*, 314-319.

Porrill, J., Frisby, J.P., Adams, W.J. & Buckley, D. (1999). Robust and optimal use of information in stereo vision. *Nature Letters, 397* (1999) 63-66.

Pötzsch, M., Krüger, N. & von der Malsburg, C. (1996). Improving object recognition by transforming Gabor filter responses. *Network: Computation in Neural Systems,* **7**, 341-347.

Pribram, K.H. (1971): *Languages of the Brain (Experimental Paradoxes and Principles in Neuropsychology).* Orig.: Englewood Cliffs, NJ: Prentice-Hall; 5th publ.: New York: Brandon House.

Pribram, K.H., Lassonde, M.C., & Ptito, M. (1981). Classification of receptive field properties in cat visual cortex. *Experimental Brain Research, 43*, 119-130.

Pribram, K.H. & Carlton, E.H. (1986). Holonomic brain theory in imaging and object perception. *Acta Psychologica, 63*, 175-210.

Pribram, K.H. (1991). *Brain and Perception (Holonomy and Structure in Figural Processing).* Hillsdale, NJ: Lawrence Erlbaum Associates.

Pribram, K.H. (Ed.) (1993). *Rethinking Neural Networks: Quantum Fields and Biological Data.* Hillsdale, NJ: Lawrence Erlbaum Associates.

Pribram, K.H. (1995). Brain and perception: From Köhler's fields to Gabor's quanta of information. *Proceedings of the 39th Congress of German Society for Psychology* (pp. 53-69).

Pribram, K.H. (1997a). What is mind that the brain may order it? *Proceedings of Symposia in Applied Mathematics, 52*, 301-329 (Vol. 2: Proceed. of the Norbert Wiener Centenary Congress, 1994, eds. V. Mandrekar & P.R. Masani; Providence: Am. Math. Soc.).

Pribram, K.H. (1997b). The deep and surface structure of memory and conscious learning: Toward a 21st century model. In R.L. Solso (Ed.): *Mind and Brain Sciences in the 21st Century* (pp. 127-156). Cambridge, MA: MIT Press.

Pribram, K.H. (1998a). Brain and the composition of conscious experience. *J. Consciousness Studies 6*(5), 19-42.

Pribram, K.H. (1998b). Status report: Quantum holography and the brain. In *Proceedings of the ECHO conference*, Helsinki. Or see: Quantum holography: Is it relevant to brain function? *Information Sciences, 115*(1999) 97-102.

Pribram, K.H. & Meade, S.D. (1999). Conscious awareness: processing in the synaptodendritic web. *New Ideas in Psychology, 17*, 205-214.

Pribram K.H. (2004). Consciousness reassessed. *Mind and Matter, 2*(1), 7–35

Pribram K.H. (2009). Minding quanta and cosmology. *Zygon, 44*(2), 451-466(16).

Prideaux, J. (1996). Comparison between Karl Pribram's "Holographic Brain Theory" and more conventional models of neural computation (from WWW).

Psaltis, D., Brady, D., Gu, X.-G., & Lin, S. (1990). Holography in artificial neural networks. *Nature, 343*, 325-330.

Pylkkänen, P., & Pylkkö, P. (Eds.) (1995). *New Directions in Cognitive Science*. Helsinki: Finnish AI Soc. (especially articles by Chrisley, Globus, Gould, Hiley, Peruš, Revonsuo).

Rabitz, H., de Vivie-Riedle, R., Motzkus, M. & Kompa, K. (2000). Whither the future of controlling quantum phenomena? *Science, 288*, 824-828.

Rainer, G. & Miller, E.K. (2000). Effects of visual experience on the representation of objects in the prefrontal cortex. *Neuron, 27*, 179-189.

Rakić, Lj., Kostopoulos, G., Raković, D., & Koruga, Dj. (Eds.) (1997). *Brain and Consciousness — Proceedings of the First ECPD Int. Symposium* (vol. I) & *Workshop* (vol. II) *on Scientific Bases of Consciousness '97*. Belgrade: ECPD.

Rao, R.P.N. (1999). An optimal estimation approach to visual perception and learning. *Vision Research, 39*, 1963-1989.

Resconi, G., & van der Wal, A. (2002). Morphogenetic neural networks encode abstract rules by data. Information Sci., 142, 249-273.

Riesenhuber, M., & Poggio, T. (2000). Models of object recognition. *Nature Neurosci. (Suppl.), 3*, 1199-1204.

Rigatos, G.G., & Tzafestas, S.G (2006). Quantum learning for neural associative memories. *Fuzzy Sets and Systems, 157*(13), 1797-1813.

Rigatos, G.G, & Tzafestas, S.G. (2007). *Neurodynamics and attractors in quantum associative memories, 14*(3), 225-242.

Rigatos G.G. (2007). Quantum wave-packets in fuzzy automata and neural associative memories. *IJMPC, 18*(10), 1551-1569.

Ritter, H., Martinetz, T., & Schulten, K. (1992). *Neural Computation and Self-Organizing Maps*. Reading, MA: Addison-Wesley.

Robertson, B. (1966). Equations of motion in nonequilibrium statistical mechanics. *Physical Review 144*, 151-161 (especially pp. 156 left)

Roelfsema, P.R. (1998). Solutions for the binding problem. *Zeitschr. für Naturforschung, 53c*, 691-715.

Savtchenko, L.P. & Korogod, S.M. (1997). Spatial electrical patterns in simulated neuronal dendrites. *European Biophysics J., 36*, 337-348.

Schempp, W. (1993). Bohr's indeterminacy principle in quantum holography, self-adaptive neural network architectures, cortical self-organization, molecular computers, magnetic resonance imaging and solitonic nanotechnology. *Nanobiology, 2*, 109-164.

Schempp, W. (1994). Analog VLSI network models, cortical linking neural network models, and quantum holographic neural technology. In J.S. Byrnes et al. (Eds.), *Wavelets and Their Applications* (pp. 213-260). Amsterdam: Kluwer.

Schempp, W. (1995). Phase coherent wavelets, Fourier transform resonance imaging, and synchronized time-domain neural networks. *Proceedings of the Steklov Institute of Mathematics, 3*, 323-351.

Schempp, W. (1997). Quantum holography and magnetic resonance tomography: an ensemble quantum computing approach to human brain mapping. *Informatica, 21*, 541-562.

Schiller, P.H. & Logothetis, N.K. (1990). The color-opponent and broad-band channels of the primate visual system. *Trends in Neurosciences, 13*, 392-398.

Shepherd, G., Brayton, R., Miller, J., Segev, I., Rinzel, J. & Rall, W. (1985). Signal enhancement in distal cortical dendrites by means of interactions between active dendritic spines. *Proceedings of the National Academy of Sciences of USA, 82*, 2192-2195.

Smith, S.J. (1999). Dissecting dendrite dynamics. *Science, 283*, 1860-1861.

Snider, G., Orlov, A., Amlani, I., Zuo, X., Bernstein, G., Lent, C., Merz, J. & Porod, W. (1999). Quantum-dot cellular automata: Review and recent experiments. *J. Applied Physics, 85*, 4283-4285.

Sompolinsky, H. & Tsodyks, M. (1994). Segmentation by a network of oscillators with stored memories. *Neural Computation, 6*, 642-657.

Spencer, R.G. (2001a). Bipolar spectral associative memories. *IEEE Transactions on Neural Networks, 12*(3), 463-474.

Spencer, R.G. (2001b). Multi-pattern real-valued spectral associative memories. *Proceedings of the Int. Joint Conf. on Neural Nets '01* (Vol. 2, pp. 1180-1195). Washington, DC.

Stapp, H.P. (1993). *Mind, Brain and Quantum Mechanics*. Berlin: Springer.

Stapp, H.P. (2002). Von Neumann's formulation of quantum theory and the role of mind in nature. *Noetic J., 3*, 363-380.

Stillings, N.A., Weisler, S.E., Chase, C.H., Feinstein, M.H., Garfield, J.L., & Rissland, E.L. (1995). Vision. In *Cognitive Science*. Cambridge, MA: MIT Press.

Storkey, A., & Valabregue, R. (1999). The basins of attractions of a new Hopfield learning rule. *Neural Netw.*, 12, 869-876.

Supek, S., Aine, C., Ranken, D., Best, E., Flynn, E. & Wood, C. (1999). Single vs. Paired visual stimulation: superposition of early neuromagnetic responses and retinotopy in extrastriate cortex of humans. *Brain Research, 830*, 43-55.

Sutherland, J.G. (1990). A holographic model of memory, learning and expression. *Internat. J. Neural Systems, 1*, 259-267. (See also: www.andcorporation.com).

Sutherland, J.G. (1994). The holographic cell: A quantum perspective. In V.L. Plantamura, B. Souček & G. Visaggio (Eds.), *Frontier Decision Support Systems*, chapter 3. New York: J. Wiley & Sons (see also chapters 4, 5).

Swindale, N.V. (1996). The development of topography in the visual cortex. *Network: Computation in Neural Systems, 7*, 161-247.

Šlechta, J. (1993). On a quantum-statistical theory of pair interaction between memory traces in the brain. *Informatica, 17*, 109-115.

Tallon-Baudry, C. & Bertrand, O. (1999). Oscillatory gamma activity in humans and its role in object representation. *Trends in Cognitive Sciences, 3*(4), 151-161.

Tang, K.-W. & S.C. Kak (1998): A new corner classification approach to neural network training. *Circuit Systems & Signal Processing, 17*, 459-469.

Tang, Y.-P., E. Shimizu, G. Dube, C. Rampon, G. Kerchner, M. Zhuo, G. Liu & J. Tsien (1999). Genetic enhancement of learning and memory. *Nature Letters, 401*, 63-69.

Tegmark, M. (2000): Importance of quantum decoherence in brain processes. *Physical Review E, 61*, 4194-4206.

Teh, Y.W. & Hinton, G.E. (2001). Rate-coded restricted Bolzmann machines for face recognition. *Advances in Neural Information Processing Systems, 13*, preprint.

Tootell, R.B.H., Hadjikhani, N.K. Mendola, J.D., Marrett, S. & Dale, A.M. (1998). From retinotopy to recognition: fMRI in human visual cortex. *Trends in Cognitive Sciences, 2*, 174-183.

Trstenjak, A. (1996). *Psychology of Colors*. Ljubljana: Inštitut A. Trstenjaka (Trstenjak's collected works no. 5) (in Slovene).

Tsao, T.-R., Shyu, H.-J., Libert, J.M. & Chen, V.C. (1991). A Lie group approach to a neural system for 3-dimensional interpretation of visual motion. *IEEE Transactions on Neural Networks, 2*, 149-155.

Ullman, S. & Soloviev, S. (1999). Computation of pattern invariance in brain-like structures. *Neural Networks, 12*, 1021-1036.

van der Wal, C., et. al. (2000): Quantum superposition of macroscopic persistent-current states. *Science, 290*, 773-777.

van Hateren, J.H. (1992). Real and optimal images in early vision. *Nature, 360*, 68-70.

van Hateren, J.H. & Ruderman, D.L. (1998). Independent component analysis of natural image sequences yields spatiotemporal filters similar to simple cells in primary visual cortex. *Proceedings of the Royal Society of London B* (Biol. Sci.), *265*(1412), 2315-2320.

van Hateren, J.H. & van der Schaaf, A. (1998). Independent component filters of natural images compared with simple cells in primary visual cortex. *Proceedings of the Royal Society of London B, 265*, 359-366.

Vapnik, V.N. (1998). Sec. 11.12 of: *Statistical Learning Theory*. New York: John Wiley & Sons. Espec. subsec. 11.12.1: Reproducing kernel Hilbert spaces.

Ventura, D. (1999). Initializing the amplitude distribution of a quantum state. *Foundations of Physics Letters, 12*, 547-559.

Vidyasagar, T.R. (1999). A neuronal model of attentional spotlight: parietal guiding the temporal. *Brain Research Review, 30*, 66-76.

Vitiello, G. (1992). Coherence and electromagnetic fields in living matter. *Nanobiology, 1*, 221-228.

Vitiello, G. (1996). Living matter physics and the quantum brain model. *Physics Essays, 9*, 548-555.

Vitiello, G. (2001). *My Double Unveiled*. Amsterdam & Philadelphia: John Benjamins.

von der Heydt, R., Peterhans, E. & Baumgartner, G. (1984). Illusory contours and cortical neuron responses. *Science, 224*, 1260-1262.

Wackermann, J., Seiter, Ch., Keibel, H. & Walach, H. (2003). Correlations between brain electrical activities of two spatially separated human subjects. *Neuroscience Letters, 336*, 60-64.

Wainwright, M. J. (1999). Visual adaptation as optimal information transmission. *Vision Research, 39*, 3960-3974.

Wallis, G. & Bülthoff, H. (1999). Learning to recognize objects. *Trends in Cognitive Sciences, 3*, 22-31.

Wang, D.L. (1999). Object selection based on oscillatory correlation. *Neural Networks, 12*, 579-592.

Wang, P.P., et al. (Eds.) (1998). *Proceedings of the 4th Joint Conference on Information Sciences '98*. Research Triangle Park (NC, USA): Assoc. Intellig. Machinery. *Volume II*: Proceedings of the 3rd Internat. Conf. on Computatational Intelligence & Neuroscience (Ed. G. Georgiou); sections on neuro-quantum information processing (pp. 167-224).

Wechsler, H. (1990). *Computational Vision*. Boston: Academic Press.

Weinacht, T.C., Ahn, J. & Bucksbaum, P.H. (1999). Controlling the shape of a quantum wavefunction. *Nature Letters, 397*, 233-235. (See also pp. 207-208 of the same vol.: Quantum control: Sculpting a wavepacket; by W. Schleich).

Weliky, M., & Katz L.C. (1999). Correlational structure of spontaneous neuronal activity in the developing lateral geniculate nucleus in vivo. *Science, 285*, 599-604.

Wheeler, J.A. & Zurek, W.H. (Eds.) (1983). *Quantum Theory and Measurement*. Princeton, NJ: Princeton Univ. Press.

Wolpert, D.H., & MacLennan, B.J. (1993). A computationally universal field computer that is purely linear (Santa Fe Institute Tech. report 93-09-056).

Wundrich, I., v.d. Malsburg, C., & Würtz, R. (2002). Image reconstruction from Gabor magnitudes (LNCS 2525, pp. 117-126).

Wurtz, R.H., Goldberg, M.E. & Robinson, D.L. (1980). Bahavioral modulation of visual responses in the monkey: Stimulus selection for attention and movement. *Progress in Psychobiology and Physiological Psychology, 9*, 43-83.

Yadid-Pecht, O. & Gur, M. (1996). Efficient biologically-based pattern-recognizing networks. *Neural Networks, 9*, 1061-1070.

Young, R.K. (1993). *Wavelet Theory and its Applications*. Boston: Kluwer.

Zak, M. & Williams, C.P. (1998). Quantum neural nets. *Internat. J. Theoretical Physics, 37*, 651-684.

Zeeman, E.C. (1961). The topology of the brain and visual perception. In M.K. Fort, Jr. (Ed.), *Topology of 3-Manifolds* (pp. 240-256). Englewood Cliffs, NJ: Prentice Hall.

Zeki, S.M. (1976). The functional organization of projections from striate to prestriate visual cortex in rhesus monkey. *Cold Spring Harbor Symposia on Quantitative Biology, 40*, 591-600.

Zipser, D. & Andersen, R.A. (1988). A back-propagation programmed network that simulates response properties of a subset of posterior parietal neurons. *Nature, 331*, 679-684.

Zornetzer, S.F., Davis, J.L., & Lau, C. (1990). *An Introduction to Neural and Electronic Networks*. San Diego: Academic Press.

Železnikar, A.P. & Peruš, M. (Eds.) (1998). Consciousness in Science and Philosophy '98 (Math. Dept., Eastern Illinois Univ., Charleston, IL) – abstracts. *Informatica, 22*, 373-403.

Zheng L. (2008). Quantum Associative Memory based on subspace projection, *Intelligent Networks and Intelligent Systems, International Workshop on* (pp. 127-130).

AUXILIARY LITERATURE

*Sets of auxiliary references are cited in the text by "**AuxLit** [the number of set below]".*

1. Cognitive neuroscience: R.J. MacGregor: Neural and Brain Modeling; Academic Press, San Diego, 1987. Z. Pirtošek (Ed.): Proceed. Symp. on Cognitive Neuroscience; UKC, Ljubljana, 1997. P. Rakić, W. Singer (Eds.): Neurobiology of Neocortex; John Wiley & Sons, Chichester, 1988. A.B. Scheibel, A.F. Wechsler (Eds.): Neurobiology of Higher Cognitive Functions; Guilford Press, London, New York, 1990. W. von Seelen, G. Shaw, U.M. Leinhos (Eds.): Organization of Neural Networks (Structures and Models); VCH, Weinheim, 1988.
M.T. Alkire, R.J. Hayer, J.H. Fallon, S.J. Barker: *J. Consc. Stud.* 3 (1996) 448-462. J. Allman, F. Miezin & E. McGuinness: *Perception* 14 (1985) 105-126. H. Eichenbaum, G. Schoenbaum, B. Young, M. Bunsey: *Proc. Natl. Acad. Sci. USA* 93 (1996) 13500-13507. A. Gierer: *Biol. Cybern.* 59 (1988) 13-21. C.G. Gross, C. E Rocha-Miranda, D.B. Bender: *J. Neurophysiol.* 35 (1972) 96-111. R.E. Hampson, J.D. Simeral, S.A. Deadwyler: *Nature* 402 (1999) 610-614. P. Jung, K. Wiesenfeld: *Nature* 385 (1997) 291. H.S. Kudrimoti, C.A. Barnes, B.L. McNaughton: *J. Neurosci.* 19 (1999) 4090-4101. D.N. Levine (1982): in *Analysis of Visual Behavior* (Ed. by D.J. Ingle et al.); MIT Press, Cambridge (MA). J.E. Lisman, J.R. Fallon: *Science* 283 (1999) 339-340. M. Marin-Padilla: *J. Cognit. Neurosci.* 2 (1980) 180. B.C. Motter, V.B. Mountcastle: *J. Neurosci.* 1 (1981) 3-26 M.E. Nelson, J.M. Bower: *Trends Neurosci.* 13 (1990) 403. M.C. Raff et al.: *Science* 262 (1993) 695-700. B.J. Richmond, T. Sato: *J. Neurophysiol.* 58 (1987) 1292-1306. W.E. Skaggs, B.L. McNaughton: *J. Neurosci.* 18 (1998) 8455-8466. J.G. Taylor, F.N. Alavi in: J.G. Taylor (Ed.): *Mathematical Approaches to Neural Networks*; Elsevier, Amsterdam, 1993. R.F. Thompson, J.J. Kim: *Proc. Natl. Acad. Sci. USA* 93 (1996) 13438-13444. T.-R. Tsao, H.-J. Shyu, J.M. Libert, V.C. Chen: *IEEE Transact. Neural Net.* 2 (1991) 149.

2. General biosciences and (neuro)physiology: B. Alberts, D. Bray, J. Lewis, M. Raff, K. Roberts, J.D. Watson: *Molecular Biology of the Cell*; Garland, New York, 1983. E.D.P. De Robertis, E.M.F. De Robertis Jr.: *Cell and Molecular Biology*; Lea & Febiger, Philadelphia, 1987, 8th ed. I. Jerman, A. Štern: *The Gene in Waves*; Znanst. in public. sred., Ljubljana, 1996 (in Slovene). I. Jerman, A. Štern: *Evolution, with Theoretical Biology and Fundamentals of Molecular Biology*; ŠOU, Ljubljana, 1996 (in Slovene). V. Kobe et al.: *Anatomy* (Part 2: *Nervous and Sensory Systems*); MF, Ljubljana, 1988 (in Slovene). A.L. Lehninger: *Principles of Biochemistry*; Worth, New York, 1982. R.F. Schmidt, G.Thews: *Human*

Physiology; Springer, Berlin, 1983. M. Štrucl: *Physiology of the Nervous System*; Medicinski razgledi, Ljubljana, 1989 (in Slovene). R.M. Youngson: *Dictionary of Medicine*; Collins, Glasgow, 1992.

3. Psychology of perception – general: V. Pečjak: *Psychology of Knowledge-Acquisition*; DZS, Ljubljana, 1977 (in Slovene). M. Polič: *Chapters on Perception*; FF, Ljubljana, 1989 (in Slovene). A. Trstenjak: *Presentation of Modern Psychology*, vol. 1; Zal. obzorja, Maribor, 1974 (in Slovene). M. Butina: *Cognition in Visual Art*; Cankar. založba, Ljubljana, 1984 (in Slovene). T. Lamb, J. Bourriau (Ed.): *Colour: Art & Science*; Cambridge Univ. Press, Cambridge; espec. ch. 3, 4, 5.
U. Fidelman: *J. Mind & Behav*. 20 (1999) 43-56. U. Fidelman: *Kybernetes* 29 (2000) 313-331. R. Hess, D. Field: *Trends Cognit. Sci*. 3 (1999) 480-6. K.H. Pribram: *Consc. & Cogn*. 10 (2001) 214-6.

4. Mind and brain – general (mind-body problem): K. Popper, J.C. Eccles: *The Self and its Brain*; Springer, London, 1976. Mind and Brain; *Scientific American*, September 1992 (special issue).

5. Visual perception: S. Ullman: *High-Level Vision*. MIT Press, Cambridge (MA), 2996. S. Zeki: *A Vision of the Brain*; Blackwell, London, 1993.
M. Bernstein, S. Stiehl, J. Bickle in: R. Ellis, N. Newton (Eds.): *The Cauldron of Consciousness*; John Benjamins, New York, 2000. M.A. Brown et al.: Pattern Recogn. 31 (1998) 1479-90. D. Domijan: *Rev. Psych*. 8 (2001) no., 1-2, p. 3-11. M. Egmont-Petersen et al.: Pattern Recogn. 35 (2002) 2279-2301. J.M. Foley, C.-C. Chen: *Vision Res*. 39 (1999) 3855-3872. R. Hamdan et al.: Pattern Recogn. 36 (2003) 1107-18. W.C. Hoffman: *Math. Model*. 1 (1980) 349. J.S. Jordan & W.A. Hershberger: *Percept. & Psychophys*. 55 (1994) 657-666. S.M. Kosslyn, C.F. Chabris, C.J. Marsolek, O. Koenig: *J. Exp. Psych.: Hum. Percept. & Perform*. 18 (1992) 562-577. V. Krüger, G. Sommer: Image & Vision Comput. 20 (2002) 665-672. D. Marr, H.K. Nishihara: *Technol. Rev*. 81 (1978) 2-23. J.D. Victor, M.M. Comte: *Vision Res*. 39 (1999) 3351-3371. N. Walter: *Network (SMN)* 59 (1995) 15-19. W.A.R. Wolf: *Perception Suppl*. 24 (1995) 139. R. Wolf: *Scept. Inquirer* 20 (1996) 23-30. R. Wolf: Depth illusion by delayed 3D perception. *Perception* (1997) preprint. C. Yuan, H. Niemann: Neurocomputing 51 (2003) 249-264.

6. Cognitive neuropsychology: A.W. Ellis, A.W. Young: *Human Cognitive Neuropsychology*; Lawrence Erlbaum Assoc., Hove/London, 1988.
J.D. Cohen, D. Servan-Schreiber: *Psych. Rev*. 99 (1992) 45-77. T. Duka, H.V. Curran, J.M. Rusted, H.J. Weingartner: *Behav. Pharmacol*. 7 (1996) 401-410. N. Geschwind: *Science* 170 (1965) 940-944. D. Hestenes: *Behav. & Brain Sci*. 14 (1991) 31-32. D.Y. Kimberg, M.J. Farah: *J. Exp. Psych.: Gen*. 122 (1993) 411-428. B. Milner, M. Petrides: *Trends Neurosci*. 7 (1984) 403-407. M.I. Posner,

A.W. Inhoff, F.J. Friedrich, A. Cohen: *Psychobiol.* 15 (1987) 107-121. L.R. Squire, S.M. Zola: *Proc. Natl. Acad. Sci. USA* 93 (1996) 13515-13522.

7. Computational neuroscience: M.A. Arbib: *Brains, Machines and Mathematics*; Springer, Berlin, 1987. M.A. Gluck, D.E. Rumelhart (Eds.): *Neuroscience and Connectionist Theory*; Lawrence Erlbaum Assoc., Hillsdale (NJ), 1990.
M.F. Bear: *Nature* 385 (1997) 481-482. P. Dayan, S. Kakade, P.Read Montague: *Nature Neurosci. (Suppl.)* 3 (2000) 1218-1223. D. Durstewitz, J Seamans, T. Sejnowski: *Nature Neurosci. (Suppl.)* 3 (2000) 1184-1191. H.P. Killackey: *J. Cognit. Neurosci.* 2 (1990) 1-. A.G. Knapp, J.A. Anderson: *J. Exp. Psych.: Learn., Mem. & Cogn.* 10 (1984) 616-637. R.C. O'Reilly: *Trends Cognit. Sci.* 2 (1998) 455-462. W. Schultz, P. Dayan, P. Read Montague: *Science* 275 (1997) 1593-1598. S. Supek, C.J. Aine: *IEEE Transact. Biomed. Enginrg.* 40 (1993) 529-540.

8. Neural correlates of conscious experience: G. Globus: *J. Cognit. Neurosci.* 4 (1992) 299. B.L. Lancaster: *J. Consc. Stud.* 4 (1997) 122-142. B. Libet in: *Experimental and Theoretical Studies of Consciousness*; John Wiley & Sons, Chichester, 1993 (pp. 123). D.W. Orme-Johnson, C.T. Haynes: *Neuroscience* 13 (1981) 211.

9. Neural synergetics: H. Haken, M. Stadler (Eds.): *Synergetics of Cognition*; Springer, Berlin, 1989. H. Haken (Ed.): *Neural and Synergetic Computers*; Springer, Berlin, 1988.
W. Banzhaf, H. Haken: *Physica D* 42 (1990) 257. I. Grabec: *Biol. Cybern.* 63 (1990) 403. S.A. Kauffman: *The Origin of Order*; Oxford Univ. Press, New York, 1993; ch. 5. M. Peruš: *Anthropos* 5-6 (1993) 146-177 (in Slovene). M. Peruš: *J. Critics Sci.* 176 (1995) 155-172 (in Slovene).

10. (Artificial) neural network models: J. A. Anderson & E. Rosenfeld: *Neurocomputing – Foundations of Research*; MIT Press, Cambridge (MA), 1988. R. Beale, T. Jackson: *Neural Computing*; Adam Hilger, Bristol, 1990. A. Dobnikar: *Neural Networks*; Didakta, Radovljica (in Slovene). E. Domany, J.L. van Hemmen, K. Schulten (Eds.): *Models of Neural Networks* (Series »Physics of NN«); Springer, Berlin, 1991. R. Hecht-Nielsen: *Neurocomputing*; Addison-Wesley, Reading, 1990. R.G. Palmer, J. Hertz, A. Krogh: *Introduction to the Theory of Neural Computation*; Addison-Wesley, Redwood City, 1991. *Machine Learn.*, vol. 8 (1992). P.D. Wasserman: *Neural Computing (Theory and Practice)*; Van Nostrand Reinhold, New York, 1989. *Advan. Synerget.*, vol. 5 (1995), vol. 7 (1996).
Other PDP monographs: Ashby (Design for Brain); Hillis (Connection Machine); Aleksander; Zupan.
S.V.B. Aiyer, M. Niranjan, F. Fallside: *IEEE Transact. Neural Net.* 1 (1990) 204.

P. Baldi, K. Hornik: *Neural Net.* 2 (1989) 53. S.C. Kak: *Physics Lett. A* 143 (1990) 293. H. Bourlard, Y. Kamp: *Biol. Cybern.* 59 (1988) 291. J.S. Denker: *Physica D* 22 (1986) 216. D.C. Dracopoulos, A.J. Jones: *Neural Comput. & Applic.* 1 (1993) 268. M.B. Gordon, P. Peretto, M. Rodrigues-Girones: *Physica A* 185 (1992) 402. S. Grossberg: *Neural Net.* 1 (1988) 17. T.M. Heskes, B. Kappen: *Phys. Rev. A* 45 (1992) 8885 A.J. Jones: *Neural Comput. & Applic.* 1 (1993) 32. P. Kainen, V. Kurkova, A. Vogt: *Neural Netw. World* 3 (2000) 313-319. T. Kohonen: *IEEE Transac. Computers* C-21 (1972) 353. T. Kohonen: *Appl. Opt.* 26 (Dec. 1987) 23. T. Kohonen: *Neural Net.* 1 (1988) 3. Y. Liu, H. Shouval: *Network: Computat. in Neural Sys.* (1995). M. Peruš: *Programer* 21 (1994) 40 & 22 (1994) 32 (in Slovene). M.D. Plumbley: *Network: Computat. in Neural Sys.* 7 (1996) 301-305. N.J. Radcliffe: *Neural Comput. & Applic.* 1 (1993) 67. D.E. Rumelhart, G.E. Hinton, R.J. Williams: *Nature* 323 (1986) 533. T.D. Sanger: *Neural Net.* 2 (1989) 459. T.J. Sejnowski, P.K. Kienker: *Physica D* 22 (1986) 260. J.S. Shawe-Taylor, D.A. Cohen: *Neural Net.* 3 (1990) 575. S.L. Thaler: *Neural Net.* 8 (1995) 55. I. Ul-Haq, J. Ahmad in *Proceed. 13th Congress Cybernetics*, Namur. R.H. White: *Neural Net.* 5 (1992) 261. K.Y.M. Wong, D. Sherrington: *Physica A* 185 (1992) 466. K.Y.M. Wong, C. Ho: *J. Phys. A* 27 (1994) 5167. M. Wynne-Jones: *Neural Comput. & Applic.* 1 (1993) 17.

11. Oscillatory and temporal neural network models: Y.V. Andreyev, A.S. Dmitriev, L.O. Chua, C.W. Wu: *Int. J. Bifurcation & Chaos* 2 (1992) 483-504. Y.V. Andreyev, Y.L. Belsky, A.S. Dmitriev, D.A. Kuminov: *IEEE Transac. Neural Net.* 7 (1996) 290-299. V. Bondarenko: *Int.J. Comput. Anticipatory Sys.* 12 (2002) 46-61. Y. Hayashi: *Neural Computat.* 6 (1994) 658-667. V.V. Kapelko, A.D. Linkevich: *Phys. Rev. E* 54 (1996) 2802-2806. L. Shastri, V. Ajjanagadde: *Behav. & Brain Sci.* 16 (1993) 417-494.

12. Holography: R.J. Collier, C.B. Burckhardt, L.H. Lin: *Optical Holography*; Academic Press, New York / London, 1971. M. Francon: *Holography*; Academic Press, London / New York, 1974. G. Saxby: *Practical Holography*; Prentice Hall, New York, 1988. L.P. Yaroslavskii, N.S. Merzlyakov: *Methods of Digital Holography*; Consult. Bureau, New York, 1980. E. Menzel, W. Mirande, I. Weingärtner: *Fourier-Optik und Holographie*; Springer, Wien/New York, 1973. D.C. Champeney: *Fourier Transforms and their Physical Applications*; Academic Press, London / New York, 1973.
D. Psaltis, F. Mok: *Sci. Amer.* (Nov. 1995) 52-58.

13. ICA and holographic neural networks: A. Hyvärinen, J. Karhunen & E. Oja: *Independent Component Analysis*; John Wiley & Sons, New York, 2001. T.-W. Lee: *Independent Component Analysis*; Kluwer, Boston, 1998. V.L. Plantamura, B. Souček, G. Visaggio (Eds.): *Frontier Decision Support Concepts*; John Wiley

& Sons, New York, 1994; chapters devoted to Holographic Neural Tech.: 3, 4, 5, 9, 10. Ch. 2 in: B. Souček & IRIS Group (Eds.): *Fuzzy, Holographic and Parallel Intelligence*; John Wiley & Sons, New York, 1992.
J. Segman, W. Schempp: *J. Math. Imaging & Vision* 3 (1993) 79-94. J. Segman, W. Schempp in: J.S. Byrnes et al. (Eds.): *Wavelets and Their Applications*; Kluwer, Amsterdam, 1994 (pp. 347-361).

14. Spin-glass models and analogy with neural networks: W.J. Caspers: *Spin Systems*; World Scientific, Singapore, 1989. D. Chowdhury: *Spin Glasses and Other Frustrated Systems*; World Scientific, Singapore, 1986. V. Dotsenko: *An Introduction to the Theory of Spin Glasses and Neural Networks*; World Scientific, Singapore, 1994. K.H. Fisher, J.A. Hertz: *Spin Glasses*; Cambridge Univ. Press, Cambridge, 1991. D.L. Stein: *Spin Glasses and Biology*; World Scientific, Singapore, 1992.
D.J. Amit, H. Gutfreund, H. Sompolinsky: *Phys. Rev. Lett.* 55 (1985) 428. D.S. Fisher, G.M. Grinstein, A. Khurana: *Phys. Today* (Dec. 1988) 56. M.S. Friedrichs, P.G. Wolynes: *Science* 246 (1989) 371-373. R.A. Goldstein, Z.A. Luthey-Schulten, P.G. Wolynes: *Proc. Natl. Acad. Sci. USA* 89 (1992) 4918-4922. R.A. Goldstein, Z.A. Luthey-Schulten, P.G. Wolynes: *Proc. Natl. Acad. Sci. USA* 89 (1992) 9029-9033. I. Kondor: *Sci. Prog. Oxf.* 71 (1987) 145 D.L. Stein: *Sci. Amer.* (July 1989) 36. J. Šlechta: *Phys. status solidi B* 67 (1975) 595. J. Šlechta: *Phys. status solidi B* 70 (1975) 531.

15. Statistical physics of neural networks: D. Bolle, R. Cools, P. Dupont, J. Huyghebaert: *J. Phys. A* 26 (1993) 549. D. Bolle, G.M. Shim, B. Vinck, V.A. Zagrebnov: *J. Statist. Phys.* 74 (1994) 565 (paper 1); 583 (paper 2). J.W. Clark: *Phys. Reports* 158 (1988) 91. J.W. Clark, J. Rafelski, J.W. Winston: *Phys. Reports* 123 (1985) 215. H. Gutfreund: *Physica A* 163 (1990) 373. A.V.M. Herz, C.M. Marcus: *Phys. Rev. E* 47 (1993) 2155. L. Ingber: *Phys. Rev. A* 8 (1983) 395. I. Kanter: *Phys. Rev. A* 37 (1988) 2739. W. Kinzel: *Europhys. News* 21 (1990) 108. R. Kühn, S. Bös: *J. Phys. A* 26 (1993) 831. H. Nishimori, T. Ozeki: *J. Phys. A* 26 (1993) 859. H. Sompolinsky: *Phys. Today* (Dec. 1988) 70. J.L. van Hemmen, D. Grensing, A. Huber, R. Kühn: *J. Statist. Phys.* 50 (1988) 231. H. Vogt, A. Zippelius: *J. Phys. A* 25 (1992) 2209-2226. K.Y.M. Wong, D. Sherrington: *Phys. Rev.* E 47 (1993) 4465. Yu-q. Ma, Yue-m. Zhang, Yu-g. Ma, C.-de Gong: *Phys. Rev. E* 47 (1993) 3985. A. Zippelius: *Physica A* 194 (1993) 471.

16. Neuro/bio-physical and complex-system backgrounds: E.R. Caianiello (Ed.): *Physics of Cognitive Processes*; World Scientific, Singapore, 1987. H. Haken: *Synergetics, An Introduction*; Springer, Berlin, 1983. H. Haken: *Advanced Synergetics (Instability Hierarchies of Self-Organizing Systems and Devices)*; Springer, Berlin, 1987 D. Raković: *Fundamentals of Biophysics*; Grosknjiga,

Beograd, 1995 (in Serb.). A. Stern: *Matrix Logic and Mind (A probe into the unified theory of mind and matter)*; North-Holland / Elsevier, Amsterdam, 1992. L. Vodovnik: *Neurocybernetics*; ZAFER, Ljubljana, 1991 (in Slovene).
Classical textbooks of quantum physics: Schiff (Quantum Mech.); Loudon (QT of Light); Kaku (Quantum Field Th.); Landau & Lifshitz (Quantum Mech.; Relativist. Quantum Mech.; Statist. Phys.); Feynman & Hibbs (Quantum Mech. and Path Integrals); Fetter & Walecka (QT of Many-Particle Sys.); Robinett (Quantum Mech.); Walls & Milburn, Orszag (Quantum Optics); Morse & Feshbach; Ziman; Dirac; Jauch; Wu; and also Davies (New Physics); Primas & Müller-Herold (Quanten-Chemie); Omnes (Understanding Qu. Mech.; Interpretation of Qu. Mech.); B. Thaller (Visual QM).
J.B. Anderson: *J. Chem. Phys.* 63 (1975) 1499-1503. V. Berkaline, I. Miklashevich: *Int.J. Comput. Anticip. Sys.* 11 (2002) 19-28. D. Derbes: *Am. J. Phys.* 64 (1996) 881-884. J.D. Farmer: *Physica D* 42 (1990) 153-187. R.P. Feynman: *Phys. Rev.* 76 (1949) 749- & 769-. P.J. Lewis: *Brit. J. Phil. Sci.* 48 (1997) 313-228. M.Q. Phan, H. Rabitz: *J. Chem. Phys.* 110 (1999) 34-41. I. Prigogine: *Int. J. Quantum Chem.* 53 (1995) 105-118. W. Schempp: *Results Math.* 28 (1995) 303-344. W. Schempp: *Acta Appl. Math.* (1998) preprint. M.P. Silverman: *Am. J. Phys.* 61 (1993) 514-523. E.J. Squires: *Phys. Lett. A* 148 (1990) 381. E.J. Squires: *Foundat. Phys. Lett.* 5 (1992) 279. E.J. Squires: *Phys. Lett. A* 178 (1993) 22. E.J. Squires: DTP-93/21 (April 1993) & DTP-93/3 (April 1993). H.P. Stapp: *Foundat. Phys.* 21 (1991) 1. H.P. Stapp: *Phys. Rev. D* 3 (1971) 1303.

17. Informational interactions of classical and quantum systems; decoherence: Y. Aharonov, P. Bergmann, J. Lebowitz: *Phys. Rev.* 134 (1964) B 1410-1416. M. Brune et al.: *Phys. Rev. Lett.* 77 (1996) 4887-4890. M. Freyberger et al.: Phys. World (Nov. 1997) 41-45. A.R. Kolovsky: *Phys. Rev. Lett.* 76 (1996) 340-344. C. Monroe, D.M. Meekhof, B.E. King, D.J. Wineland: *Science* 272 (1996) 1131-1136. K. Nugent et al.: Phys. Today (Aug. 2001) 27-32. Z. Onyszkiewicz: *Am. J. Phys.* 52 (1984) 817-. M. Schneider, I. LaPuma: Am. J. Phys. 70(3) (2002) 266-271. P. Törmä at al.: *Phys. Rev. A* 65 (2002) 052110. W.H. Zurek: *Phys. Rev. D.* 26 (1982) 1862-1880. W.H. Zurek: *Phys. World* 10(1) (Jan. 1997) 24-25. W.H. Zurek: *Phys. Today* 44(10) (Oct. 1991) 36-44.

18. Quantum entanglement; EPR; "bootstrap": J.S. Bell: *Speakable and Unspeakable in Quantum Mechanics*; Cambridge Univ. Press, Cambridge, 1987. T. Maudlin: *Quantum Non-Locality and Relativity*; Blackwell, Oxford, 1994 M.P. Silverman: *More Than One Mystery (Explorations in Quantum Interference)*; Springer, New York, 1995.
C. Bennett et al.: *Phys. Rev. Lett.* 70 (1993) 1895-1898. R.A. Bertlmann: *Foundat. Phys.* 20 (1990) 1191-1209. M. Buchanan: *New Scientist* 125 (March 1998) 27-30 G.F. Chew: *Science* 161 (1968) 762-765. G.F. Chew, M. Gell-Mann, A.H.

Rosenfeld: *Sci. Amer.* 210 (1964) 74-93. A. Ekert, P.L. Knight: *Am. J. Phys.* 63 (1995) 415-. D.M. Greenberger, M.A. Horne, A. Zeilinger: *Phys. Today* (Aug. 1993) 22-29. R.B. Griffiths: *Am. J. Phys.* 55 (1987) 11-. E. Hagley et al.: *Phys. Rev. Lett.* 79 (1997) 1-4. R. Penrose: Phil. Tr. Roy. Soc. Lond. (Math., Phys., Eng.) 356(1743) (1998) 1927-39. B. Yurke, D. Stoler: *Phys. Rev. A* 46 (1992) 2229-. A. Zeilinger: Foundat. Phys. 29 (1999) 631-643.

19. Subcellular neurobiology; biological coherence: *Neural Net. World*, vol. 4, no. 3 (1994) (issue devoted to Fröhlich coherent systems).
R.L. Amoroso: *Bioelectrochem.* 40 (1996) 39-42. C.H. Bailey, D. Bartsch, E.R. Kandel: *Proc. Natl. Acad. Sci. USA* 93 (1996) 13445-13452. S.Y. Berkovich: *Nanobiol.* 2 (1993) 99. M.J. Dudziak: *Nanobiol.* 1 (1992) 351-359. I. Jerman: *Acta Analytica* 7 (1991) 55-. I. Jerman: *J. Critics Sci.* 176 (1995) 141-148. D.A. Miller: *Nanobiol.* 1 (1992) 361-371. J. Tuszynski, R. Paul, R. Chatterjee, S. Sreenivasan: *Phys. Rev. A* 30 (1984) 2666-2674. J.A. Tuszynski, H. Bolterauer, M.V. Satarić: *Nanobiol.* 1 (1992) 177-190.

20. Information processing in microtubules: E. Del Giudice, S. Doglia, M. Milani, G. Vitiello in: H. Fröhlich (Ed.): *Biological Coherence and Response to External Stimuli.* R. Grush, P.C. Churchland: *J. Consc. Stud.* 2 (1995) 10-29. E.M. Insinna: *Nanobiol.* 1 (1992) 191-208. R. Penrose, S. Hameroff: *J. Consc. Stud.* 2 (1995) 98-111. A. Samsonovich, A. Scott, S. Hameroff: *Nanobiol.* 1 (1992) 457-468. M.V. Satarić, R.B. Žakula, J.A. Tuszynski: *Nanobiol.* 1 (1992) 445-456.

21. Quantum features of neurophysiological processes: F. Beck, J.C. Eccles: *Proc. Natl. Acad. Sci. USA* 89 (1992) 11357-11361. P.C. Bressloff: *Phys. Rev. A* (1992) 7549-. E.H. Walker: *Noetic J.* 1 (1997) 100-107.

22. Quantum neural information processing: Amlani et al.: *Appl. Phys. Lett.* 72 (1998) 2179-2181. E.C. Behrman, J. Niemel, J.E. Steck, S.R. Skinner: A quantum dot neural network; *Proceed. Workshop on Physics of Computation*, 1996, pp. 22-24 (full-length paper submitted to *IEEE Transac. Neural Net.*). S.C. Kak: *Advances Imaging & Electron Phys.* 94 (1995) 259-313. S.C. Kak in: *Proc. Intl Conf. Info. Sci.*, North Carolina, 1997 (pp. 141-144). S.C. Kak: in K.H. Pribram, J. King (Eds.): *Learning as Self-Organization*; Lawrence Erlbaum A., Mahwah (NJ), 1996; ch. 7 (and also ch. 21). S.C. Kak: Tech. rep. ECE/LSE 92-13; 94-42/1994. T. Menneer, A.Narayanan: Quantum-inspired neural nets; Tech. rep. R329, Univ. of Exeter (Comp. Sci. Dept.), 1995/6 (& Tucson II Conf.). J.E. Steck, S.R. Skinner, A.A. Cruz-Cabrera, M. Yang, E.C. Behrman: Field computation for ANN hardware: Examples in nonlinear optical materials; in *Proceed. IJC Info. Sci. 2000*, Atlantic City (NJ) (there see also: Gould, MacLennan, Menneer & Narayanan, Ventura). V. Tryba, K. Goser: A modified algorithm for self-organizing maps based on Schrödinger equation; in *Proceed. Int. Workshop ANN '91*. D. Ventura, T. Martinez: in *Proc. ICANNGA '97*, Norwich (pp. 482-485). D. Ventura, T. Martinez: in Proc. *JIC NN '98*, Anchorage (pp. 509-513).

23. Artificial quantum neurodynamics: R.L. Dawes: Parametric avalanche stochastic filter; Martingale Res. Corp. (Allen, TX) final tech. report, contract no. N60921-91-C-0071, for NSWC in White Oak Lab., 1993. R.L. Dawes, H.L. Bodine, E.v.K. Hill: Automatic detection and classification…; Martingale Res. Corp. final report of Phase I SBIR Project, contract no. III-9361410, for NSF, 1994.

24. Quantum information theory: N.J. Cerf, C. Adami: *Phys. Rev. Lett.* 79 (1997) 5194-5197. S.C. Kak: *Nuovo Cimento* 33 B (1976) 530-534. S.C. Kak: *Foundat. Phys.* 28 (1998) 1005-1012. M. Heather, B. Rossiter: Int.J. Comput. Anticip. Sys. 13 (2002) 397-326. R. Landauer: *Phys. Today* (May 1991) 23-29. T. Matsuoka, M. Ohya: *Reports Math. Phys.* 36 (1995) 365-379. M. Ohya: *Lett. Nuovo Cimento* 38 (1983) 402-404. M. Ohya: *IEEE Transac. Info. Th.* 29 (1983) 770-774. M. Ohya: *J. Math. Anal.& Appl.* 100 (1984) 222-235. M. Ohya: in *Quantum Probability and Related Topics VI*; World Scientific, Singapore, 1991; pp. 359-369. M. Ohya, T. Matsuoka, K. Inoue: *IEEE Transac. Info. Th.* XX, no. Y (1999). L. Viola: Science 293 (2001) 2059-63. J.A. Wheeler: *Int. J. Theor. Phys.* 21 (1982) 557-578. A. Zeilinger: Phys. World (March 1998) 35-40. W.H. Zurek: *Nature* 341 (1989) 119-124.

25. Quantum computation: C.H. Bennett: *Phys. Today* (Oct. 1995) 24-30. J.I. Cirac, P. Zoller: *Phys. Rev. Lett.* 74 (1995) 4091-4094. I. Chuang et al.: *Proc. Roy. Soc. London A* 454 (1998) 447-467 & *Phys. Rev. Lett.* 80 (1998) 3408-3411. D. Deutsch: *Phys. World* (June 1992) 57-61. D. Deutsch, A. Ekert: *Phys. World* 11(3) (March 1998) 47-52. R.P. Feynman: *Int.J. Theor. Phys.* 21 (1982) 467-488. B. Goertzel: *Noetic J.* 1(2) (1998) 198-206. B. Goertzel: *Dynam. Psych.* (1997) www. S.C. Kak: *Foundat. Phys.* 26 (1996) 127-137. S.C. Kak: *Foundat. Phys.* 26 (1996) 1375-1386. S.C. Kak: *Foundat. Phys.* 29 (1999) 267-279. S. Lloyd: *Science* 273 (1996) 1073-1078. E. Knill, R. Laflamme, W. Zurek: *Science* 279 (1998) 342-345. S. Lloyd: *Sci. Amer.* (Oct. 1995) 44-50. V. Scarani: *Am. J. Phys.* 66 (1998) 956-960. A. Steane: *Rep. Progress Phys.* 61 (1998) 117-173. L. Vandersypen, C. Yannoni, M. Sherwood, I. Chuang: *Phys. Rev Lett.* 83 (1999) 3085-3088.

26. Sub-quantum backgrounds of consciousness: non-locality, implicate order: P.C.W. Davies, J.R. Brown (Eds.): *The Ghost in the Atom*; Cambridge Univ. Press, Cambridge, 1986. B.S. DeWitt, H. Graham (Eds.): *The Many-Worlds Interpretation of Quantum Mechanics*; Princeton Univ. Press, Princeton, 1973. G.I. Shipov: *A Theory of Physical Vacuum* (espec. chs. 2.4, 2.5); Moscow, 1997. D.Z. Albert: *Sci. Amer.* (May 1994) 34. R.L. Amoroso: *Informatica* 19 (1995)

585-590. D. Bohm: *Foundat. Phys.* 17 (1987) 667. D. Bohm, B.J. Hiley: *Foundat. Phys.* 11 (1981) 529. C. Dewdney, P.R. Holland, A. Kyprianidis, J.P. Vigier: *Nature* 336 (1988) 536. D. Deutsch: *Phys. Rev. D* 44 (1991) 3197-3217. F.A.M. Frescura: *Foundat. Phys.* 18 (1988) 777-808. F.A.M. Frescura, B.J. Hiley: *Foundat. Phys.* 10 (1980) 705. F.A.M. Frescura, B.J. Hiley: *Foundat. Phys.* 10 (1980) 7. L.I. Gould: *Arkhimedes* 2 (1993) 144-157. J.S. Hagelin: *Mod. Sci. & Vedic Sci.* 1 (1987) 29-87. B. Hiley: *Annal. Fondation L. de Broglie* 5 (1980) 75. B. Hiley, N. Monk: *Mod. Phys. Lett. A* 8 (1993) 3625; *Foundat Phys. Lett.* 11 (1998) 371-7. B. Hiley, M. Peruš: *J. Critics Sci.* 174 (1995) 33-49 (in Slovene) & *Dynam. Psych.* (1997) www (Engl. orig.). B. Hiley: in Saunders, Brown (Eds.): *The Philosophy of the Vacuum*; Oxford Univ. Press, 1991. B.D. Josephson, F. Pallikari-Viras: *Foundat. Phys.* 21 (1991) 197. P. Kwiat, H. Weinfurter, A. Zeilinger: *Sci. Amer.* (Nov. 1996) 52-58. I. Pitowsky: *Foundat. Phys.* 21 (1991) 343. G. Taubes: *Science* 285 (1999) 512-517.

27. Quantum backgrounds of consciousness: M. Kafatos, R. Nadeau: *The Conscious Universe*; Springer, New York, 1990. H. Moritz: *Science, Mind and the Universe*; Wichmann, Heidelberg, 1995. R. Penrose: *The Emperor's New Mind (Concerning Computers, Minds, and Laws of Physics)*; Oxford Univ. Press, London, 1989. R. Penrose: *Shadows of the Mind (A Search for the Missing Science of Consciousness)*; Oxford Univ. Press, Oxford, 1994. J. Polkinghorne: *The Quantum World*; Princeton Univ. Press, Princeton (NJ), 1984. H.P. Stapp: *Mind, Matter and Quantum Mechanics*; Springer, Berlin, 1993. A. Stern: *The Quantum Brain (Theory and Implications)*; North Holland / Elsevier, Amsterdam, 1994.
G. Blommestijn: in *Proceed. Conf. Consc.*- Tucson II, 1996 (abstract & full-length preprint). J.S. Hagelin: *Mod. Sci. & Vedic Sci.* 3 (1989) 3-72. R.S. Ingarden: *Open Sys. & Info. Dyn.* 7 (2000) 139-156. K. Ludwig: *Psyche* 2(16) (1995) www. P. Marcer: *Proceed. 14th Int. Congress Cybernetics*, Namur, 1995 (pp. 429-434 & 435-440). P. Marcer: *World Futures: J. Gen. Evol.* 44 (1995) 149-159. M. Pavšič: ch. 12 of *The Landscape of Theoretical Physics: A Global View*; Kluwer, Dordrecht. M. Peruš: *Sci. Tribune* (1997) www. D. Raković: *Informatica* 21 (1997) 507-517 E.J. Squires: *Synthese* 97 (1993) 109-123. A. Scott: *J. Consc. Stud.* 3 (1996) 484-491. H.P. Stapp: *Foundat. Phys.* 21 (1991) 1451-1477. H.P. Stapp: *Psyche* 2(5) (1995) www. V. Weiss: *Biol. Cybern.* 68 (1992) 165-172. V. Weiss: *Cahiers de Psych. Cogn.* 14 (1995) 387-408.

28. Informational and cybernetic aspects of cognition and consciousness: F.J. Varela: *Principles of Biological Autonomy*; North Holland, New York, 1979. M. Gams, C. Bavec (Eds.): *Proceed. Int. Multi-Conf. Info. Soc. '98 & '99*; Ljubljana, IJS. D. Dubois (Ed.): *CASYS'99 Book of (Extended) Abstracts*; CHAOS asbl, Liege (espec. Symp. 9 & Pribram; Sienko; Kirvelis; Dubois; Hoekstra & Rouw; Araujo; Tsirigotis).

A.J. Bell: *Phil. Transac. Roy. Soc. London B* 354 (1999) 2013-2020. D. Bojadžiev: *Informatica* 19 (1995) 627-634. G.A. Christos: *Noetic J.* 1 (1998) 149-161. G. Farre: *Informatica* 21 (1997) 533-540. D. Gernert: *BioSys.* 54 (2000) 165-180. C. Green, G. Gillet: *Philos. Psych.* 8 (1995) 333-340. M.W. Ho: *J. Consc. Stud.* 3 (1996) 231-244. J. Horne: *Informatica* 21 (1997) 675-682. J. Horne: *Informatica* 24 (2000) 275-279. S.C. Kak: *J. Intell. Sys.* 6 (1996) 133-144. M. Peruš: *Anthropos* 1-2 (1996) 84-89 (in Slovene) M. Peruš: *World Futures: J. Gen. Evol.* 51 (1997) 95-110. J.D. Valentine: *Psychoenerget.* 4 (1982) 257-274. A.P. Železnikar: *Cybernetica* 40 (1997) 261-296.

29. Cognitive aspects: E. & W. Baumgartner *et al.* (Eds.): *Phenomenology and Cognitive Science*; Röll, Würzburg, 1997. D.C. Dennett: *Content and Consciousness*; Routledge & Kegan Paul, London, 1969. A.R. Luria: *Fundamentals of Neuro-linguistics*; Izdat. Moskovskogo univ., Moscow, 1975 (in Russian). W.G. Lycan (Ed.): *Mind and Cognition (A Reader)*; Blackwell, Oxford / Cambridge (MA), 1992. H.L. Pick, P. van den Broek, D.C. Knill (Eds.): *Cognition*; APA, Washington (DC), 1992 (espec. ch. by E.J. Gibson). M. Potrč: *Thought, Language and Object*; DZS, Ljubljana, 1988 (in Slovene). A. Ule: *Logos of Knowledge-Acquisition*; Znanst. in publ. središče, Ljubljana, 2001 (in Slovene).
B. Mangan: *Psyche* 7 (2001) no. 18, www. A. Prince, P. Smolensky: *Science* 275 (1997) 1604-1610. J. Seppänen in: *Language and Synergetics*; Univ. of Oulu, Oulu, 1992. P. Smolensky: *Behav. & Brain Sci.* 11 (1988) 1.

30. Consciousness – implementation; complex-system background processing: R. Cahill, M. Klinger, K. Kitto: *The Physicist* 37 (2000) 191-195. A.D. Linkevich: *Informatica* 21 (1997) 435-464. M. Peruš: *J. Critics Sci.* 174 (1995) 11-22 (in Slovene). M. Peruš: *Horizons Psych.* 4 (1996) 73-84 (in Slovene).

31. Consciousness – general: *Scientific American* spec. ed. 31. Aug. 2002: *The Hidden Mind.* T. Nagel (Ed.): *Experimental and Theoretical Studies of Consciousness*; John Wiley & Sons, Chichester, 1993; especially: M. Kinsbourne (pp. 45), B. Libet (pp. 123). Lj. Rakić, D. Raković, Đ. Koruga, A. Marjanović (Ed.): *Consciousness*; ECPD, Beograd, 1996 (in Serbian). D. Raković, Đ. Koruga (Eds.): *Consciousness*; ECPD, Belgrade, 1996. J.R. Smythies, J. Beloff (Eds.): *The Case for Dualism*; Univ. Press of Virginia, Charlottesville, 1989. *Towards a Science of Consciousness -«Tucson II«*; Consciousness Research Abstracts, JCS, 1996.
J.F. Kihlstrom: *Cognit. & Consc.* 2 (1993) 334. J.R. Searle: *Cognit. & Consc.* 2 (1993) 310. *Trends Cog. Sci.* 4 (2000) 372-382. C. Tart: *J. Consc. Stud.* 4 (1997) 71-92. A. Ule: *J. Critics Sci.* 176 (1995) 117-130. A. Ule: *Informatica* 21 (1997) 683-688. A.P. Železnikar: *Informatica* 20 (1996) 475-484.

32. Consciousness – phenomenal aspects: F. Brentano: *Psychology from an Empirical Standpoint*; Routledge & Kegan Paul, London, 1973 (German orig. 1874). D.W. Hamlyn: *Perception and Sensation*; Routledge, London, 1961. N. Nelkin: *Consciousness and the Origins of Thought*; Cambridge Univ. Press, Cambridge, 1996.
S. Blackmore: *J. Mental Imagery* 11 (1987) 53. A.J. Deikman: *J. Consc. Stud.* 3 (1996) 350-356. M. Draganescu: *Noetic J.* 1 (1997) 28-33. H. Hendriks-Jansen: *Informatica* 21 (1997) 389-404. J.J. Kupperman: *Philos. Psych.* 8 (1995) 341-351. M. Peruš: *Poligrafi* 7-8 (1997) 115-124 (in Slovene).

33. Qualia problem: W.P. Banks: *Consc. & Cognit.* 5 (1996) 368-380. D.J. Chalmers: *Sci. Amer.* (Dec. 1995) 62-68. T.L. Hubbard: *Consc. & Cognit.* 5 (1996) 327-358. P. Hut, R.N. Shepard: *J. Consc. Stud.* 3 (1996) 313-329.

About the Authors

Mitja Peruš has developed the ideas presented in this book since the time of his PhD studies of physics and cognitive (neuro)science at the University of Ljubljana, Slovenia, and research work at the National Institute of Chemistry and Institute BION, both in Ljubljana, Royal Holloway and Bedford New College of the University of London, Eastern Illinois University at Charleston, and during post-doc (EU's Marie Curie project) at the Institute for Computer Vision and Graphics of the Graz University of Technology, Austria. He has also been a consultant to the director of the IT company Stroka Ltd., among others. He is affiliated with the Laboratory for Cognitive Modelling of the Faculty for Computer and Information Science in Ljubljana, among others, and lectures also at the Sigmund Freud University in Vienna. This book emerged out of Peruš's PhD dissertation "Conscious Image Processing – A Holonomic Model" under supervision of the world-renowned pioneer brain scientist Professor Karl H. Pribram, Stanford and Georgetown Universities, USA. It is acknowledged in the literature that Peruš pioneered (since 1992-5), together with researchers like Professors Subhash Kak, Walter Schempp and Dan Ventura, the field of quantum neural networks and organized the first conferences on the topics. This had a huge echo in the nineties, leading to many publications even before, and independently to, the quantum computing boom. Collaboration with Assoc.-Prof. Dr. Chu Kiong Loo and his team brought further simulational and applicational development of this model. Dr. Peruš published three books and about 60 papers on topics related to this book; he lectured at about 40 scientific conferences, received two awards for these topics, which are also widely cited. He worked and lectured on the model during visits at several scientific institutions all over the world. He established the Slovene Society for Cognitive Sciences. Peruš's research experience includes neural networks, cognitive, information and vision sciences (object recognition, especially), biophysical foundations of brain and cognitive sciences, statistical physics of spin systems and other complex (bio)systems, foundations of quantum theory, quantum information theory.

Chu Kiong Loo obtained his PhD (University Sains Malaysia), B.Eng (First class Hons in Mechanical Engineering from University Malaya). Formerly he was a design Engineer in various industrial firms in different capacities as well as he is the former chairman of Centre for Robotics and Automation in Multimedia University. He has been involved in the application research of Perus's Quantum Associative Model and Pribram's Holonomic Brain Model in humanoid vision projects. Currently he is the Dean of Faculty of Information Science and Technology, Multimedia University, Malaysia. He has completed many funded projects by ministry of science in Malaysia and currently leading few hi-tech projects funded by Intel Corporation. Loo's research experience includes brain inspired quantum neural network, constructivism inspired neural network, synergetic neural networks and humanoid research.

Index

F

G

H

I

J

K

L

M

N

O

P

Q

Breinigsville, PA USA
14 February 2011

255560BV00005B/4/P

9 781615 207855